D1291431

Fireground Strategies

Fireground Strategies

Anthony Avillo

PennWell®

Copyright ©2002 by
PennWell Corporation
1421 South Sheridan Road
Tulsa, Oklahoma 74112-6600 USA

800.752.9764
+1.918.831.9421
sales@pennwell.com
www.FireEngineeringBooks.com
www.pennwellbooks.com
www.pennwell.com

Marketing Manager: Julie Simmons
National Account Executive Executive: Francie Halcomb

Director: Mary McGee
Production / Operations Manager: Traci Huntsman
Editor: Jared Wicklund
Cover and Book Designer: Joey Zielazinski

Library of Congress Cataloging-in-Publication Data Available on Request

Avillo, Anthony L.
Fireground Strategies
ISBN13 978-0-87814-840-0

Printed in the United States of America

2 3 4 5 6   11 10 09 08 07

DEDICATION

For my beautiful wife, Nikki, and my precious daughters, Stephanie and Lindsay. You have all of my love and thanks.

TABLE OF CONTENTS

Acknowledgements

Writing this book has been one of the greatest challenges of my career. I thank Bill Manning, Diane Feldman, Fire Engineering, and PennWell for their confidence in my abilities.

Currently, I am working as a career Deputy Chief with North Hudson (NJ) Regional Fire & Rescue, located in Northeastern New Jersey. North Hudson Regional Fire & Rescue provides fire service to the communities of North Bergen, Union City, Weehawken, West New York, and Guttenberg. Hudson County is located directly across the Hudson River from Manhattan, New York. This urban jurisdiction includes the New Jersey location of the Lincoln Tunnel, as well as industry, rail, shipping, and highway. Hudson County also happens to be the most densely populated county in the United States. Life hazard and exposure problems associated with old, closely-spaced, combustible structures are the rule in Hudson County, which also includes the cities (and fire departments) of Jersey City, Hoboken, Harrison, Kearney, Bayonne, Secaucus, and East Newark.

In addition to North Hudson Regional Fire & Rescue, I teach at the Bergen County (NJ) Fire Academy (BCFA). I am also a partner in a business that helps firefighters and officers of career departments prepare for promotional exams. We also aid prospective firefighters in preparing for the fire department entrance exam.

I had envisioned a book that would not only educate firefighters on strategies and the associated tactics, but also help promotional candidates achieve success in the examination process. I discussed my ideas along with my vision for the format of the book with a fellow instructor at BCFA, Peter Hodge. Instructor Hodge also happens to be the coordinator for the Fire Department Instructor Conference, a Division of Fire Engineering. He told me that Fire Engineering and Pennwell Publishing has always been interested in these types of user-friendly books, for they are few and far between. He said he would get back to

Welcome to Hudson County, New Jersey, the mostly densely populated county in the United States. Almost 60,000 people per square mile call this area home.

me. About a week later, I got a call from Fire Engineering's publisher, Pennwell, and was asked to submit a proposal. I did so, and, after review, was told to go ahead and write the book. I thank Peter for his help and guidance.

My career in the fire service has its roots in my childhood. I lived up the block from a firehouse in Union City, New Jersey, where, from my porch, I would watch the apparatus respond. I knew all the apparatus designations of all the houses in the city. I told everyone I wanted to be a firefighter or a "fireman" as it was called in those days. As I grew older, my family moved to a neighboring town, North Bergen, where my aspirations to become a firefighter diminished as I became involved in other activities, such as sports. Becoming a firefighter was the farthest thing from my mind. I was in college when my interest began again, almost by accident. I was home from college with a friend who was going to the post office to mail in his application for the North Bergen Fire Department. I figured that I would give the test a shot since there was nothing really motivating or challenging me at college, other than playing football. However, there were no more applications left in North Bergen. I called my cousin, who was a Captain in North Bergen. He told me to go over to Weehawken, which is another Hudson County town. He explained to me that it is the same application, and, in addition to applying for North Bergen, I should apply for the Weehawken Fire Department because they sometimes hired from a county entrance list. None of my friends got hired in North Bergen. I got hired in Weehawken.

Prior to being hired, I was working in a health club, designing weightlifting programs for the clientele. I met a guy who was wearing a Weehawken Fire Department t-shirt. He said he was a captain. His name was Ed Flood. I told him I was on the list. He informed me that they were soon to be hiring. We became friends, and when I got hired he took me under his wing and showed me the ropes, both figuratively and literally. Without his guidance, I would never have become the firefighter I am today. I worked under him for six years on the ladder company. He told me that "if you leave the firehouse at the end of the shift and it is not a better place because you were there the last 24 hours, then you did not do your job." I believed it then and I believe it now. We worked on many projects together and my career was nurtured by the guidance he provided. This same guidance steered me in getting promoted to captain and later to battalion chief. Captain Flood has since become chief of the North Hudson Regional Fire & Rescue. He has always set a high standard for himself and his officers. Even today, I constantly strive to meet the standards he has set for me as an officer. I am both a better man and a firefighter for having been involved with him. I thank him for his guidance and for allowing me to "snatch the pebble from his hand."

In January of 1999, my small department of Weehawken was involved in a major consolidation of services in North Hudson, which created the North Hudson Regional Fire & Rescue. This consolidation, the largest of its kind in New Jersey, merged the services of the fire departments of North Bergen, Union City, Weehawken, West New York, and Guttenberg, making it the third largest department in New Jersey. I was fortunate to be involved in the writing and development of many of the new Standard Operating Procedures (SOP), training division policies, and lesson plans, which aided in the molding of this new department. Being involved in the birth and creative direction process of this new department has been a huge challenge and an enhancement to my career. At the time this was written, I was assigned to the very firehouse that I used to gaze at from my porch as a child.

I would like to acknowledge my first company officer, Captain Tim Finnegan for his guidance during the incipient phase of my career, and most of all, for his undying patience with my almost comical attempt to learn how to drive a manual stickshift.

Acknowledgment also to Captain Mike Hern, whose advice and input into my work has helped me to develop into a better teacher and officer.

I would be remiss not to acknowledge the members of the former Weehawken Fire Department who set the bar for how a small fire department should operate, and to my new brothers in the North Hudson Regional Fire & Rescue. We have the unique opportunity to make a model department from scratch and to set the standard for excellence and professionalism as the fire service heads into the new millennium. I hope to serve you well and support your efforts.

I would also like to thank the following people who have helped influence and guide my career and life: Gerald Huelbig, Art Certisimo, Frank Nagurka, Frank Pizzuta, Andrew Scott, Bob Teta, Vin Ascolese, and Jay Lally.

I thank Bill Hamilton of the Fire Department of New York (FDNY) for his guidance and information from the brothers on the other side of the river.

I thank the men I work with at the Fire Department Instructor's Conference, namely Mike Nasta, Joe Berchtold, Jim Weiss, Joe Alvarez, Tim Hetzel, and Scott Sherman. Thanks for your contribution to my skill and knowledge portfolio.

I would like to thank the instructor staff at the Bergen County Fire Academy for guidance and support without which this book could not have been written.

In addition, I would also like to acknowledge and thank the myriad of fire-fighters that have been my students both at the Bergen County Fire Academy and through my teachings at Study Group, Inc. I hope I was able to pass on the torch with the same passion and excitement that has always been the spirit of the fire service for me.

I must acknowledge Bob Moran, Jack Murphy, and John Lewis for their input and advice into this project.

I also owe a great deal to Ron Jeffers for his great photos, access to his massive archives, and his selfless contributions to the fire service of North Jersey. I would also like to thank Bob Scollan for his photos and the New Jersey Metro Fire Photographers Association (NJMFPA) for their support and contributions.

Thanks also to Christine and Angelika at Fire Engineering for their help and for letting me use the copying machines and to Jared Wicklund, my editor, for making sense out of all this stuff.

Finally, I would like to thank my family and friends for their unfailing support during this project. I would especially like to thank my brothers, my mother, the strongest woman I have ever met, and my father, a builder of youth and a true leader of men.

Deputy Chief Anthony Avillo commands operations at a multiple alarm fire in a vacant warehouse.
(Ron Jeffers, NJMFPA)

Anthony Avillo
August, 2002

Introduction

To the reader who has been "in the books", many of the strategies and tactics used in this book will be familiar. There are several reasons for this. First, fires are more similar than most people think. While there will be some differences in each particular fire, the field of play, the buildings we wage our battles in, allow the application of standard operations to begin the suppression process. This application will act as both a solid footing and as a springboard in adapting to the various contingencies presented by each incident. Second, while there are "many ways to skin the cat", firefighting is firefighting. Depending on the given conditions, basics like apparatus positioning, line placement, support activities, and command procedures are universal in their application. There is very little new material being developed by fire service authors. There are, however, many different slants on tried and true strategies and tactics. Some strategies and tactics have been modified over the last 25 years due to the many factors and pressures placed on today's fire service including:

1. Legislation placed on the fire service at the national and state levels. The two-in, two-out mandate and Haz Mat incident handling are two examples.
2. Changes in building construction. The use of lightweight construction methods has prompted the fire service to review and modify our strategy and tactics in these structures.
3. The use of plastics in building contents (which, in the past were predominately made of wood and other natural materials) cause them to burn faster, hotter, more intensely, and produce darker and more toxic smoke. These conditions also promote earlier flashover.

4. Technological advances in communications, firefighting protective clothing, and equipment. While these advances have, for the most part, made our job better and safer, over-reliance on these new systems and equipment without proper training and hands-on experience has caused needless firefighter fatalities and injuries.

Any new innovation relating to the job of fighting fires should prompt a review of the basics, for this is where all strategy and the accompanying tactics are built. The key to the basics is training. With all of the highly specialized aspects of the job, we tend to get away from the basic information that got us where we are today, the bread and butter of the job.

Over the last 25 years, there have been many fire service "fads" that have caused a reaction (sometimes over-reaction) in the fire service. There has been Haz Mat hysteria, confined-space psychosis, and trench rescue mania. All have found and will continue to find their place in the modern fire service, and rightfully so, because we must be prepared for such incidents, but the down and dirty business of actually going in and putting the fire out aggressively is still the most common response and the best way to get the job done. This is where the heart of firefighter training should focus.

Recently, I was teaching a class for deputy chief candidates. In New Jersey, as in other places, the chief officer positions are usually determined by an oral exam, where the candidate is tested on dimensions such as leadership, judgment, oral communication, and most of all, knowledge. I was dis-

The fire in this lightweight wood truss condominium complex under construction completely destroyed the structure as well as several other area buildings. Fires spread rapidly and pose a conflagration threat in crowded urban areas.

cussing the format of the exam they would be taking and I placed an over-head question to the class. It was a very straightforward question that any level officer should know, "What are the five types of building construction?" I was astonished at the fact that not one person in that class could name all five correctly. I asked them how any one of them, in all honesty and with a clear conscience, could stand in front of a fire building with a dozen or so men inside and not know how a fire is going to spread in that building, how that building is going to react to the insult of fire, and, worse yet, how the building is going to fall apart. I found out that for their captain's test, all they were required to do was memorize some SOPs of a fictitious city and regurgitate them back in the written portion of the test. It was a pass/fail exam. The candidates who passed were allowed to take a two-part oral exam from which the rankings were tabulated. They were never required to open a book, as there was no book list. They weren't required to study for the position, so most of them did not. They seemed to fail to realize that modern firefighting is not a game for those who "dabble" in the field. It should never be a hobby. To be successful and remain alive in today's fire service all personnel must become information-rich to deal with the demands of the profession. This is only possible through dedication to the study of firematics.

Cycle of Competence

The cycle of competence is repeated many times throughout one's life, starting from when we are children. Learning in the fire service should also be based on this cycle. There are four stages to the Cycle of Competence:

1. Unconscious incompetence
2. Conscious incompetence
3. Conscious competence
4. Unconscious competence

As we learn and master a new skill, we pass through the four stages of the cycle. For example, during Firefighter I training, knot skills is an area that many recruit firefighters find to be difficult. Say, for instance, the firefighter is expected to learn the bowline-on-a-bight knot in that day's scheduled training. Prior to the class, the student will more than likely not even be aware that this knot exists and will not have even heard of it. He is, therefore, *unconsciously incompetent*, in that he is not aware he does not even know about the knot.

Hands-on training as well as staying "in the books" are vital to safety and success on the fireground. A healthy dose of both throughout one's career cannot be overemphasized.

At the beginning of the training session, the instructor may introduce the knot and may even show how it is tied in "normal" time. At this point, when the knot is first introduced, the student will realize that he doesn't have a clue as to how to tie this knot or what it is used for. At this point, he becomes *consciously incompetent*.

As the skill of tying the knot is acquired during the training session, the firefighter strives to master the new knot. This may take days or even weeks, and in this time, the firefighter has to consciously work out the intricacies of tying the knot in his mind as he ties it. In other words, he has to think about it. The firefighter has now reached the third stage, that of being *consciously competent*. He can tie the knot, but has to think about it to do it. At this stage, constant reinforcement and effective coaching by the company or training officer is critical so that bad habits, which may be difficult to correct later, are not formed. The goal of the firefighter in regard to this knot, as well as in all fire service skills, is to make the tying of this knot routine or second nature. He can then consider himself to be *unconsciously competent* at tying this knot. He doesn't even have to think about it because he knows the skill so well. All firefighters and officers should endeavor to make this the goal of all training.

It has been said and played out many times over, that in a stressful situation, people revert back to habit. *Unconscious competence* is that point where the skill becomes habit. If a firefighter can reach that level, then under stressful, and sometimes desperate, conditions on the fireground, the firefighter is more likely to operate as he was trained, with effectiveness and efficiency to successfully meet the

A working fire is no place for the uninformed. Many times, the only thing that stands between life and loss is a solid incident commandar with a firm grasp on strategies and tactics.

objective of the situation. The Cycle of Competence can be applied to almost anything, both on and off the fireground. The key is the desire of the student to learn through self-motivation and the guidance of competent instructors/officers.

This text and accompanying workbook will address many fire situations. While it is not a book on procedures regarding how to accomplish specific tactics, it invokes and explains tactics that are relevant to the situation that the scenario presents. It is written from the point of view of the fire officer who must identify a strategy, develop an action plan, and implement tactics to meet the objectives of that plan. The *Fireground Strategies Scenarios Workbook* will allow the strategist to apply information learned in this textbook to make decisions about such activities as line placement, ventilation considerations, and resource distribution, among other things. It will also allow the tactician to choose proper tactics in a given situation, thus enhancing the decision-making process on the fireground. It is the intent of this text, through diligent study and lesson reinforcement, to motivate, challenge, and strengthen both the fireground strategist/tactician and/or the promotional candidate.

How to Use This Book

As I mentioned in the introduction, this text is to be used as both a guide for the fireground strategist/tactician and the promotional candidate in preparing for a promotional exam. The first part of the text contains general information regarding areas such as size-up, building construction, heat transfer, and strategic operational modes. The remainder covers occupancy-specific firefighting problems as well as a chapter addressing operational safety. This information will be applied in the accompanying work book.

The workbook contains scenarios answered in both short answer and multiple-choice formats. More than one testing format is presented to allow the student to be more versatile in test taking. Each answer is explained in depth to help the reader understand the reason for the strategy or tactic presented. Where applicable, supplemental information is added to further get the point across. It is suggested that the student not only check why the best answer is correct, but also understand why the other answers were not correct. In this way, no vital information will be missed. To properly absorb and reinforce the information, I urge the student to read this book, one chapter at a time, and then reinforce each chapter by answering and checking the scenarios presented in the workbook.

The only way to absorb the information is through diligent study and commitment. Remember that the only thing in life achieved without effort is failure.

There are no shortcuts. I have found that if you are going to be tested in a certain way, such as a multiple-choice scenario or a question and answer format, then this is how you must prepare. It can be compared to trying to master a certain skill such as catching a football. You don't master catching a football by taking hand-offs. You must try catching it over and over again until the skill is mastered. The same is true of tests. If you are taking a multiple-choice test, then you have to practice taking these types of tests. To practice what you are good at will not properly prepare you. You must strive to eliminate any and all of your weaknesses. This takes a certain amount of mental toughness and what I like to call "testicular fortitude". Toughness is one's ability to work outside his own comfort zone. This is where the challenge lies, outside the comfort zone, for it is here that forward strides are made and real accomplishments and success are realized.

Personalizing the book

I further recommend to the owner of this book to make it your own. What I mean by this is to personalize it. This means highlighting sections, sketching in apparatus, ladders, and attack lines in the diagrams, putting notes in the margins, developing acronyms to aid in remembering some of the information, and anything else that will aid you in learning. Each time you read the book try to get something else out of it. Knowing this book, as well as other books you read, has the ability to enhance your career inside and out.

Fire Service leadership and command is a skill and an art best left to the experts. There is no room for guessers and shoot-from-the-hip non-actors. This skill/art must be learned through intense commitment and diligent study.

Becoming a student of the job is a journey, not a destination. There is much to learn and to apply. Preparing for a promotional exam takes perseverance and commitment. The preparation is akin to running a marathon, while the exam itself is like a 100-yard dash. Those who are dedicated and are in it for the long haul will do the best. For my promotional exams, I read some of the books as many as 15 times. My books now look like roadmaps and scribble, but I understand them. Using textbooks in this manner helps maximize your chances for success both on the fireground as you relate what you have learned to real life situations and in the promotional arena.

Another benefit of this personalization is that if anyone steals your book, it will be impossible for the low-life to understand the incomprehensible-except-to-the-owner hieroglyphics entered into it to guide you through the information in the book.

Learning By the Case Study Method

There is possibly no other profession where case study is more valuable than in the fire service. In the case study method, details of past incidents are reviewed. Using this method, opportunities to learn arise from every emergency. All incidents should be a learning experience to the participants. Case studies allow those who were not at the actual incident to both experience it and learn from it. The benefits of learning from the successes and mistakes of others cannot be underestimated. It is a reality that we just do not get to go to as many fires as in the past. It therefore becomes mandatory that we learn from the experience of others. This text uses case studies to better explain the learning points.

Command of the fire scene requires both knowledge and confidence. Above all, command requires a leadership ability honed through study, experience, and sometimes treachery. It is not a game for the meek. *(Ron Jeffers, NJMFPA)*

Challenge-Based Learning

The work book takes advantage of a method called "challenge-based learning". The text is a source of information regarding various areas of study. The work book functions as a workable study guide. In the write-in areas and the multiple-choice scenario tests, the student is challenged to excel at the presented scenarios. After reading the prevalent information in each chapter of the text, the scenarios in the work book are presented to bring the information to life where the student is challenged to manipulate the previously read information and to apply it to the scenario. The explanations and scoring that follow provide the student with a way to gauge his strengths and learn from his weaknesses. I always found this application of material a most rewarding way to learn the information. I would always keep a notebook with my scores from the various study guides, so that when I read the particular book again, I could check my retention from the last time I read it. The motivation for me was to get better each time I read it. The goal was to thoroughly understand the information and material.

Guide for Answering Questions In This Book

When determining answers to questions in the workbook, as well as determining strategy and tactics on the fireground, there is a rule of thumb that basically applies to all fireground objectives and evolutions regardless of whether or not a firefighter is entering the building for attack, exiting the building for evacuation or removal of victims, applying hosestreams, ventilating structures, placing apparatus and ladders, protecting exposures, and opening the building up for overhaul purposes. For convenience reasons, it is called the "Point of Entry Rule of Thumb". The Point of Entry Rule of Thumb states that objectives should be accomplished utilizing *"the safest, most effective, path of least resistance:"*

1. To accomplish the desired objective.
2. To place the line between the fire and the victims and/or the vertical arteries.
3. To maintain a viable path of escape/egress/retreat.
4. Ideally attack from the unburned side.
5. To access and remove victims.
6. To effectively remove the products of combustion.
7. To effectively perform salvage operations.

It is interesting to note that the behavior of the fire will most often follow this rule of thumb. This allows the strategist to evaluate and not only head off fire spread, but also take steps to localize it, based on his knowledge of these paths of least resistance. This comes in handy when determining fire attack, ventilation operations, and overhaul strategies. For example, the path of least resistance for fire travel in an old tenement is likely to be the open stairwell. With this knowledge in hand, the strategist can vent the building over the stairs to localize this likely path of spread and place lines to head it off. He can also predict, based on the building construction, where the fire is most likely to initially travel.

In direct contrast to fire, which will primarily spread vertically before it spreads horizontally, water takes the opposite path of least resistance. This is valuable in setting a salvage strategy. Water takes the path of least resistance vertically in the opposite direction that fire will, which is downward. It will only move horizontally when downward travel is blocked.

The key word here is "safest". Although the most effective path of least resistance may be the most tempting way to stretch a line, it may not be the safest. Consider a fire in a building where the power lines have been burned and are now laying on the ground at the front of the building. Although the front door is the most effective path of least resistance to stretch the line, the presence of the power line clearly makes it unsafe. The safety of firefighters is always, without exception, the highest priority. In this case, stretching the line via a rear or side door is safer, and therefore the best tactic.

Another example is to consider that the usual primary means of access to the building is through the front door. If that front door is a heavily fortified steel door with a steel gate over it, but there is a wooden side door to the occupancy, then the attack should be directed toward there. By the time that the steel door is forced, it may be all that is left of the building. It was not the path of least resistance, nor was it an easy way in. These are extreme examples, but using this guideline in your decision-making process on the fireground and on a promotional exam will usually direct you toward the proper way of accom-

Venting as directly over the main body of fire as possible utilizes the most accessible path of least resistance to channel the products of combustion up and out of the building. Knowledge of building construction is critical in determining where to apply the "Point of Entry" (or in this case, "Exit") Rule of Thumb.

plishing the objective. Safety is the overriding concern of all fireground operations and must always be foremost in the mind of all personnel.

As mentioned above, fire and the accompanying products of combustion will, incidentally, also take the path of least resistance. In some buildings such as private and multiple dwellings, this path will be the open stairwell. In build-

THE POINT OF ENTRY RULE OF THUMB:
"THE SAFEST, MOST EFFECTIVE PATH OF LEAST RESISTANCE"

1. To accomplish the objective
2. To place the line between the fire and the victims and/or the vertical arteries
3. To maintain a viable path of escape/egress/retreat
4. Ideally attack from the unburned side
5. To access and remove victims
6. To effectively remove the products of combustion
7. To effectively perform salvage operations

ings of fire-resistive construction, this may be the Heating, Ventilating, and Air Conditioning (HVAC) system. Still, in other buildings, if the fire cannot spread vertically, it will take the path of least resistance in a horizontal direction such as in a common cockloft in row houses or garden apartments. The knowledge possessed by the incident commander regarding these paths in each type of the building construction class will have a major impact on his ability to cut off, channel, and control the fire or whether the fire will control him by being allowed to spread via these paths unimpeded.

Guide for Answering the Short Answer Questions

When answering short answer or oral questions, there are six questions that should guide your answer and, subsequently, your actions on the fireground. These are "who?", "what?", "when?", "where?", "how?", and "why?". The Point of Entry Rule of Thumb applies here and should guide you in the manner in which you address your tactics. For instance, at a top floor fire in an attached 5-story Class 3 multiple dwelling, the question may be about how you would order the building to be ventilated. Following the format, it could be correctly answered covering all bases:

Regarding victim removal, the more the rescue operation deviates from the normal means of egress (the path of least resistance), the more manpower, time, and danger is involved. *(Ron Jeffers, NJMFPA)*

WHO?	Ladder 1
WHEN?	After properly positioning the apparatus at the front of the building
WHERE?	Access the roof via the interior stairs of the Exposure B and
WHAT?	Provide both horizontal ventilation of the fire floor and vertical ventilation of the roof

If fire reaches the cockloft area of this garden apartment, failure to cut the roof over the area of most fire involvement will allow the fire to take the horizontal path of least resistance and burn the roof off the entire complex.

HOW?   By opening the bulkhead door and other natural openings at the roof level, horizontally ventilating the top floor windows from the roof, and cutting a large vent hole in the roof as directly over the fire as is safely possible

WHY?   These actions will clear the vertical arteries of smoke and heat, allow the attack team to access the seat of the fire, and provide a means for the flame, smoke, and superheated gases to escape from the building

This same format can be used for any question asked, including questions regarding:

1. Hose placement for confinement and extinguishment of the fire
2. Primary search and rescue operations
3. Exposure protection
4. Command establishment, transfer, and termination
5. Ventilation operations
6. Overhaul operations
7. Salvage operations

CHAPTER ONE

Size-Up, Pre-Fire Planning, & Incident Communications

Size-up, also known as situation evaluation, is something that we do every day, both on and off the fireground. In the fire service, size-up of potential fire problems in the response area should begin the first day a new recruit gets on the job and should continue until he retires. Thorough knowledge of this area of response is imperative to enable a firefighter to operate in a safe and efficient manner. Obtaining all of this information takes time and should be part of a continuing process throughout a firefighter's career. Every time a firefighter leaves the firehouse, for whatever reason, he or she should be gathering information on the area of response. This includes building types, changing occupancies, street conditions such as repair work, overhead obstructions, and the presence and location of auxiliary appliances, to name just a few. Many of these features are temporary in nature and/or will change over time. Fewer surprises occur on the fireground if personnel and the department keep abreast of any changes to the profile of their area.

Size-up of the fire scene is also part of this routine of gathering facts through observation. Size-up can be described as a continuous evaluation of all facts that have an impact on operations. The accuracy of size-up is a key factor in determination of a proper strategy and often the difference between success and failure on the fireground.

Fireground Strategies

It is the responsibility of the first officer on the scene to size-up the situation and begin the fireground management process. Pre-fire planning and area familiarization are a vital part of the size-up process. *(Ron Jeffers, NJMFPA)*

It has been said that size-up begins with the receipt of the alarm and continues until the emergency is under control. While this is true, size-up can be analyzed as being a part of a bigger picture that goes much deeper.

Size-up begins with pre-fire planning. In fact, pre-fire planning can be thought of as the size-up before the fire. If you fail to plan, you plan to fail. Pre-fire planning, like fireground size-up, is an information-gathering process about the building, its contents, and our ability to fight a fire in it. Information is power. How much power that the incident commander can apply to the fire situation has a direct impact on his ability to operate. An effective pre-planning process should ascertain much of the information used in the initial fireground size-up beforehand. The superior advantage of pre-fire planning is that, unlike the dynamic situation that is a working fire, it can be done in a calculating, thorough, and deliberate manner. Generally, it is a plan that is drawn up under the best conditions and used under the worst.

COAL WAS WEALTH

One of the best information-gathering processes is the **COAL WAS WEALTH** size-up process consisting of:

Construction
Occupancy
Apparatus and Manpower
Life Hazard

Water Supply
Auxiliary Appliances
Street Conditions

Weather
Exposures
Area and Height
Location and Extent
Time
Hazardous Materials

Applying this acronym to the pre-fire planning process allows us to answer many questions about a building or area prior to an incident.

Construction

Knowledge of buildings and building features enables the incident commander to enact a strategy and an action plan that will locate, cut off, and extinguish the fire. It's more important to know the building that is on fire than it is to know the fire that is in the building. Having information about how the building is put together allows us to predict how the building will come apart under the destruction of a fire. A building on fire is essentially a building under demolition. Fires spread in relatively predictable patterns in the different types of building construction, usually in the paths of least resistance. It is the goal of the fire strategist to intervene by properly utilizing the suppression forces to put a stop to this demolition process.

Pre-fire planning trips can supply valuable information about a building. The best way to ascertain how a building is constructed is to visit the site during the construction phase. Visit often because buildings change quickly during this phase and vital building components will be covered up quickly, thus eliminat-

Determining how the building will fall apart under the insult of fire is one of the most critical aspects of both the prefire planning and the size-up process. This wood overhang collapsed as a result of fire exposure. Careful and continuous evaluation of fire conditions is critical to firefighter safety. *(Joe Berchtold, Teaneck Fire)*

This fire-resistive low-rise building is being covered with a brick veneer curtain wall. When finished, it will look like an H-shaped apartment building of ordinary construction. Even false lintels are in place above the windows.

ing a chance to detect structural deficiencies from a firefighting point of view, potential mantraps, and other problem areas that may hinder firefighting operations. From a construction standpoint, surveys of older buildings are a little tougher as the buildings may have been renovated many times. Veneer and other wall coverings, such as siding, may conceal the true construction type. It is sometimes advantageous to check the basements or attics because you will most likely find original building components in these areas. For instance, a wood-frame building covered with a brick or stone veneer will show the true wood-frame construction either in the basement or in the attic. Checking the sides and the rear of the building will also reveal the true construction type. Many buildings have masonry or brick veneer on the front of the building, but a quick look down an alley on the side of the building will often reveal old shingles covering wood-frame walls. By the same token, the opposite may be true. The front of the building may be covered with some type of decorative material, such as vinyl siding or stucco, while the sides and rear of the building show that the building is actually of ordinary construction. For economic purposes, many times the owner will only renovate the front and maybe the rear of the building. Always take a second and inspect the sides of the building. Check for multiple drop ceilings. This information is invaluable because it is often hard to ascertain

The masonry front of this building is brick veneer and does not show the true construction of the building. A look at the side wall (Side B) reveals shingles, indicating wood-frame construction.

A look at the rear of this corner building reveals the original brick on Side C. Shown here on Side B, as well as at the front and Side D, brick veneer and siding cover the bottom and top.

on the fireground, being, more than likely, obscured by smoke. This tactic also allows you to check the roof's construction from below if the joists are exposed. Don't forget those rain roofs or newly renovated roofs that were once constructed of solid joists, but are now metal or lightweight wood-truss construction. The fire scene is no place to be surprised by this.

Check for the potential for forcible entry problems and/or availability of alternate means of egress. This ensures that once you've overcome these forcible entry problems and you are inside, you know as many ways out of the building as possible.

There are many variables among buildings and there are exceptions to every rule. Take the time to survey buildings each time you enter one for an alarm. Getting to know one building is like getting to know one fish in the sea. It is better to know a little about a lot of buildings than to know everything about just one building. Always be versatile and flexible in your approach. Learn something valuable every time you enter a building.

Occupancy

A building's occupancy can clue firefighters in to such factors as fire load, presence of hazardous materials, and most of all, expected life hazard. In fact, the occupancy type will, to great extent, determine the life hazard. To a lesser extent, it gives clues to other hazards that may impact the chosen strategy of the incident. Knowing what and who are inside a building is one of the critical keys to the life hazard problem. This information can tip you off to the most critical times for rescue, forcible entry, and possible delayed alarms. Other information that may be gained is the presence of handicapped occupants, hazardous materials, and/or processes. Keep abreast of changing occupancies and the ability of the building to accommodate the changes. A building that housed a pillow-storage warehouse may not be able to handle the load of the MachoMan Barbell and Porcelain Emporium without some kind of structural reinforcement. If there is none, a collapse may be in their future, even without a fire.

The occupancy of this building should signify a heavy content load. Although this building was probably built for this occupancy, it is old and its dried-out wood floors should not be expected to withstand the ravages of a serious fire for a prolonged period of time.

Apparatus and manpower

The success or failure of an operation depends, to a great degree, on how well you manage apparatus and manpower. Inadequate manpower at the outset of operations is the cause for more firefighter injuries and property loss due to firefighters overexerting themselves. Incident commanders who do not have enough men on the scene to carry out the tactical objectives to match the chosen strategy must either request additional alarms immediately or consider changing the strategy. When the incident commander is out of men at the command post or in staging, and tasks have yet to be completed, it is already too late to call for additional alarms.

In a career department, the response and accompanying compliment of manpower can be ascertained beforehand with a good deal of accuracy, but the response in a volunteer, part-paid, or on-call department is another story. The manpower may change with the time of day or year. However, certain buildings, if properly preplanned, can show deficiencies in the apparatus and manpower compliment. Plans for automatic aid should be worked out to handle incidents in these buildings beforehand. For example, if a municipality has only one or two buildings that call for the use of an aerial device, but don't have one, it is a

Request additional manpower and apparatus as soon as the need becomes apparent, and even before. A building that will create a major challenge for the fire forces or present an unusual need may be preplanned to include a specific mutual aid response upon confirmation of a working fire. Once the fire breaks out, it may be difficult to locate and effectively position later-arriving, but essential pieces of apparatus. *(Joe Berchtold, Teaneck Fire Dept.)*

good idea to have an agreement with a neighboring community who has an aerial. The most efficient way to cut down on aerial reflex time, in this instance, is to have the mutual-aid aerial device dispatched on the first alarm response. This will make apparatus positioning as well as support operations easier. The same is true for areas on waterfronts that do not have fire boat service. Due to the long response times, upon the confirmation of a working fire in a water-front property, a fireboat should be dispatched.

Manpower (the lack of it) has always been an issue in the fire service. As of this writing, NFPA 1710 (Standard for the Organization and Deployment of Fire Service Suppression Operations, Emergency Medical Operations, and Special Operations to the Public by Career Fire Departments) was passed. The standard suggests the following requirements:

1. A minimum of four firefighters to staff engine and ladder companies.
2. Jurisdictions with tactical or high-hazard occupancy areas should staff their apparatus with five or six firefighters.
3. A minimum response time of four minutes for the first-arriving company.
4. Rapid Intervention Teams (FAST or RIC) are to have first-responder capability.
5. Staffing of chief's aides.
6. The public must be informed about the department's response capabilities and the consequences when deployment criteria are not met.

NFPA 1720 is the volunteer department's equivalent of NFPA 1710 and states that "the department having jurisdiction determines if the standard is applicable to their department".

Suggested requirements of NFPA 1720 include the following:

1. Minimum staffing requirements to ensure a sufficient number of members are available to operate safely and effectively.
2. Once the necessary resources are in place, the department should have the capability to institute an initial attack within two minutes.
3. At least four members should be assembled before an interior fire suppression operation may be initiated.
4. Upon arrival, if a life-threatening situation exists, and immediate intervention may prevent injury or fatality, initial attack operations may be conducted in accordance with NFPA 1500.
5. The department should have the capacity to implement a Rapid Intervention Crew (FAST or RIC) during operations that would subject firefighters to immediate danger of injury.
6. A firefighter occupational safety and health program should be provided in accordance with NFPA 1500.

The adoption of these standards is a definite step in the right direction. It should be interesting to see how they affect the fire service. Standards are nice. Laws are better. Remember that NFPA standards are not mandatory but suggested as requirements of the service. They have, however, held up in court as the ideals by which fire departments operate. Departments operating contrary to the standards have been found liable for their actions.

Life hazard

The protection of life is our #1 priority. We must determine, based on the location and extent of the fire, how to best utilize apparatus and manpower to accomplish this crucial benchmark.

Many times, the life hazard is not determined until scene arrival, with arrival cues aiding in strategy determination. Steps, however, can be taken beforehand during pre-fire planning, to simplify the location of these occupants (potential victims) in certain occupancies. For instance, in a day care center or nursing home, the majority of the population may be located in a specific area of the building at certain times of the day, such as a lunchroom during the noon hour or sleeping areas at night. Having this knowledge, along with the best routes for access to and egress (evacuation) from these areas, will simplify the search and removal operation.

In addition, determining the life hazard before an incident allows us to begin to safeguard the occupants by taking steps to prevent incidents rather than to just respond to them and hope we can make a quick impact to mitigate a problem. The best way to prevent incidents in these occupancies, and many others like them, is through an effective and

Be alert for signs of potential life hazard. The open garage door as well as the car in the driveway suggests that someone is home. Also, toys strewn about points to the potential presence of children in the structure.

aggressive fire educational program. More lives are saved by occupants with information than by any responding fire department.

Water supply

Without sufficient water, the game is over before it begins. It is essential that this be determined in the pre-fire planning phase. Fire-flow formulas should be used to determine the needed water for the building and exposures. Next, you must determine if the needed flow is available from the water supplies in the area. You cannot possibly begin to address any aspect of water-delivery methods and tactics until this is determined.

Selecting the proper hydrant by virtue of prior knowledge of the area can be the difference in stopping a fire or complete destruction of the building, or worse, an entire block. This prior knowledge includes areas of sufficient or poor water supply. Hydrants directly in front of the fire building are usually not the best choice because access to this hydrant will not allow the ladder company to position themselves in the most advantageous place. If the building begins to show signs of collapse, it's easier to move the ladder company than to shut down the lines, break them from the engine discharge, disconnect from the hydrant, and move the engine company. The moral: Leave the front of the building for the ladder company. It will make everyone's life easier.

Sometimes hydrants in close proximity to each other may be significantly different in terms of water available and residual pressures due to the size of the mains that serve them. Also, know where the dead-end hydrants are in your

This restaurant fire at the end of a pier on the Hudson River required two separate water supplies from the main road over 1000' away as well as a supply from draft. Pre-fire plans called for staging apparatus on the main road while two relays were established using both mutual aid and local companies. *(Jeff Richards, NHRFR)*

To supplement the water supply at the fire pictured above, a drafting operation was necessary. It is essential to have drafting sites preplanned so these positions can be reserved in case the need arises. *(Bob Scollan, NJMFPA)*

area. These can surprise you by being in the middle of a block. Possibly, one side of the block is newer than the other and the hydrants on the newer side are fed from a different main that does not connect to the last hydrant on the old grid. This will be impossible to tell from the street. The only way to ascertain this information is from up-to-date hydrant maps. Hydrant maps should be provided to companies and should be included in the pre-fire information available to the incident commander. If the first company does not pick the best hydrant, at least the incident commander can have subsequent engine companies secure water supply from better hydrants in case conditions deteriorate.

Another aspect of water supply to consider is the thread type of the hose couplings. There may be times when hydrants and fire department connections will not be compatible with the thread used by the fire department. This may be the case when responding on mutual aid to a neighboring city. If threads are not the same, arrangements should have been made beforehand to secure proper adapters to ensure compatibility. This is the case in North Hudson when responding into Hoboken and Jersey City. North Hudson uses National Standard Thread (NST) on hose couplings, hydrants, and fire department connections. Jersey City, Hoboken, and New York City use

To mate up mismatched couplings, adapters must be used. Preplan the thread type used by neighboring departments to avoid severe problems at the fire scene.

New York Corporation (NYC) threads. The problem with New York Corporation threads also occurs for North Hudson when responding into the Lincoln Tunnel. The solution is the use of adapters. When operating in the above-mentioned jurisdictions, we use adapters to mate with their fittings. When the incident is in North Hudson, the departments from Jersey City and Hoboken use adapters to work with our water supply features.

If this is the case in your jurisdiction, ensure the proper adapters are available and be cognizant of the thread type on hydrants at the city borders. A few feet either way may mean different threads. Know this before incidents occur in order to avoid such problems.

These hydrants are on the Union City–Jersey City border. Although they look identical, the hydrant on the left is National Standard Thread while the hydrant on the right is New York Corporation thread. Not knowing this beforehand can cause water supply problems.

Auxiliary appliances

The availability of auxiliary appliances is one of the most critical factors to consider when determining strategy and tactics. The presence and service status of these systems will either allow fire forces to quickly attack a fire or cause an excessive delay, which endangers firefighters, occupants, and the existence of the building. An out-of-service system will cause the need for additional men and apparatus to accomplish the objectives required to bring the fire under control.

In regard to these systems, fire service personnel must know prior to the incident:

1. The type of system (sprinkler, standpipe, combination)
2. The location(s) of the fire department connections and which areas of the building they serve
3. The location(s) of fire pumps and how to manually override them if they are not functioning
4. The location(s) of the hydrants that will be utilized to supply the system
5. Which hydrants are connected to the same main that supplies the auxiliary system (should not be used to also supply attack lines)
6. The presence and location of any supplemental water supply systems and how to access and operate them (gravity tanks, pressure tanks, etc.)
7. Where the interior standpipe connections are located
8. What class of standpipe system the building has installed
9. Whether there are any pressure-reducing devices on the standpipe
10. How to shut the system down, drain it, and restore it to service (if possible)

Remember that auxiliary systems also include specialized communications equipment such as fire command stations and standpipe telephones. Personnel should be thoroughly familiar with their use.

Be aware of the presence, location, and serviceability of all auxiliary appliances. Note that this fire department connection is facing the wrong way. To supply this connection, it would be necessary to first break out the window glass.

Signs posted on the fire department connection make life easier. When there are no signs and multiple connections, only thorough preplanning will ensure the water gets to the right place.

A fire command station is just as much an auxiliary system as a standpipe in high-rise buildings. This is essentially where the incident commander sets up shop. Communications are essential to safety and operational coordination between the command post and the fire floors.

Street conditions

Knowing the street locations, addresses, and the best way to get to them is the job of all personnel, especially the officer. Certain routes are better at certain times of the day, while others, such as a school zone or crowded business district, are best avoided at certain times of the day or year. In North Hudson, there is an avenue that extends through three municipalities. On a Saturday, and especially during certain holiday seasons, it is almost impossible to maneuver due to the many double and even triple-parked cars. Unless an alarm is received for that particular street, companies avoid it because being delayed by traffic is inevitable. The same is true for other well-traveled streets during rush hour. Personnel should know these areas and routes and how to get around them.

Other areas to consider are areas under construction, railroad crossings, and low-clearance areas. If an engine company "just makes it" under a low-clearance, it does not mean it will make it back safely. For example, if the engine company responds to a car or brush fire under one of these low clearances, they will usually use tank water to extinguish the fire. Five hundred gallons of water weighs over 4000 lbs., and when it is gone, it takes a large load off the springs of the apparatus, causing it to rise. If the company does not refill the tank before they attempt to make it under the low-clearance on their return, they may be in for a surprise that will liberate their light bar from the roof of the apparatus. It might be better to avoid these areas entirely, if possible. This applies especially to aerial devices that also have pumping capabilities and water tanks such as telesqurts and quints.

This Telesquirt, which just makes it under this bridge, may not be so lucky when the booster tank is empty, raising the chassis. Also, the highest part of this sloping bridge is on the right hand side, forcing the apparatus to stay to the right. It will not fit under the left side of this bridge, even with a full booster tank. Always be aware of the limitations of your apparatus.

Steep hills may also cause the response to be altered. In icy weather, steep hills cause treacherous driving conditions, causing response routes and positioning to be altered. A fire occurring on a steep hill may cause upper floor access problems at the front and rear of the building. Unless you are using an aerial device, laddering the front or rear of the building may be impossible. It will not be possible to set ground ladders on a grade. It may only be possible to ladder the sides of the building.

Another condition that may cause aerial positioning to be altered is the presence of overhead power lines. Generally, the first-arriving ladder company positions at the front of the building. When responding to a corner building, if it is not possible to ladder the building or the roof from the front, it may be necessary to ladder it from the side on the adjacent street. This is best addressed prior to the incident. Always position apparatus to take the best advantage of the equipment available. This requires flexibility and efficient size-up on the part of the incident commander and officers.

Weather

Weather is almost impossible to preplan for. This is something that must be sized-up constantly during the course of the day, even when not in the response mode. Companies must listen to weather reports and speculate on how it might affect operations. Some areas are

Ground ladder placement at the front of this building is impossible due to the steep grade. In addition, the presence of the power lines may also make aerial access difficult.

Placement of the aerial to the roof of this unattached, corner building may not be possible from Side A. It might be best to position the ladder truck on Side D. If two ladder companies are responding, the first ladder truck positions on side A while the second aerial unit ladders the roof from Side D.

more affected by the extremes of weather or have prevalent weather patterns that can have an impact on an incident and should be taken into account even before an incident. For example, some streets always appear to be windier than others due to the wind-tunnel effect produced by large buildings that create canyons of the streets in the area. Other areas that always seem to be affected by weather, in a more intense manner, are waterfront areas, higher elevations, and extremely hilly areas. Wind in these areas can whip and spread a fire in a frighteningly fast manner. Be prepared for these in advance. Certain areas are also conducive to flooding when heavy rains occur. Decide how these and other unfriendly weather conditions could impact your response, apparatus positioning, and operations.

One weather-related factor that must be evaluated is the presence of snow and ice. Snow loads on a roof can either be uniformly distributed or piled in a

drift up against a parapet wall, forming a dangerous, concentrated load. Firefighters operating on the roof should report this condition as soon as it is recognized. In cold weather, water, which would normally leave the building as run-off, may freeze and place a heavy load on an already fire-weakened structure, one that it may not withstand. The incident commander should always be aware of the effect ice accumulation may be having on the building.

Ice accumulation on the exterior of a building creates an eccentric load on the walls that the building may not be able to withstand. Always be cognizant of the final disposition of fire streams. When the ice thawed on this building, it collapsed. *(Ron Jeffers, NJMFPA)*

Heavy rains and/or high humidity will also have a detrimental effect on the buoyancy of smoke, possibly masking the true seat of the fire and causing ventilation operations to be more difficult. It may be necessary to use mechanical or other unorthodox ventilation strategies when these conditions are present.

Keep in mind that weather extremes will necessitate the need for additional personnel. Weather is always a valid reason for additional alarms as the manpower struggles to accomplish tasks that, due to the weather, require more time and extra hands to accomplish.

The extremes of weather will also have a profound impact on both the firefighters and the equipment. The incident commander and safety officers must be trained to recognize the symptoms of weather-related injuries, such as heat exhaustion and heat stroke in the summer or frostbite and hypothermia in the winter. Remember that hot weather is not absolutely necessary to cause heat exhaustion due to the amount of protective clothing worn by firefighters.

Cold-related injuries are avoidable. In extreme conditions, ensure that a proper rehab post is established. *(Ron Jeffers, NJMFPA)*

A rehab post with supplies appropriate for the season should be on the response at every major fire. All apparatus should carry water coolers so that personnel can constantly re-hydrate themselves during an incident, no matter how minor. In addition, the incident commander must see to it that shelter is provided in the extremes of weather. A warm shelter in the winter, a place where firefighters can change into dry clothes, etc., are

Rehab of personnel in hot weather is equally important in regard to firefighter safety. Cool-air misters, fluids, and a shady place to rest are all required at the rehab post. In any weather, incident command must ensure medical monitoring is available. *(North Bergen Fire)*

necessities. Fire-fighters should be urged during all seasons to carry extra clothes and especially gloves on the apparatus. While the firefight is continuing, you tend not to feel the cold, but once the fire is knocked down and it is time to pick up equipment or when faced with a long, drawn-out defensive battle, if you are wet, it can be a nightmare. In the summer months, provide shade and fans to provide relief. Large fan-driven misters are excellent for cooling down firefighters during the battle. The incident commander who does not provide for his firefighters is not meeting the most important fireground priority, that of providing for firefighter safety.

Apparatus and equipment are also affected by weather extremes. Overheating apparatus in the summer and frozen hose couplings and equipment in the winter are just some of the problems associated with weather.

Some weather-related problems are also beyond our control. Frozen hydrants have caused many fire operations to be lost, while ice accumulation and broken pipes inside poorly maintained sprinkler and standpipe systems cause additional problems. These problems, in turn, affect the firefighters who must overcome the problems caused by the malfunctioning systems. In the summer, power outages caused by excessive electricity use during heat waves have made the fireground less safe as a result of darkness and have stranded elevators, forcing firefighters to walk up countless flights during high-rise fires. Weather problems and firefighting have never been a complimentary combination. Take steps to minimize any negative effects caused by weather in a proactive manner by having a plan in place.

Exposures

Exposures are our #2 priority after life hazard. Exposures represent where the fire is going and sometimes create a greater problem than the parent fire. Failure to realize this fact (caused more than likely by incident commander "tunnel-vision" and "candle-moth" syndrome) has led to many conflagrations and needless deaths, injuries, and property losses.

Many of the facts you will be required to know for the target building will carry over to the exposures. This includes the construction, occupancy, presence of auxiliary appliances and/or hazardous materials, and areas of access

and egress, to name just a few. Your preplanning activities may reveal that, in the event of a serious fire in the target building, it may be more practical to focus your efforts on protecting the exposure. Some exposures may change with time of day, such as rail cars or trucks parked alongside or inside the structure. These should be included in your plans to protect the exposed building.

Pre-piped deluge guns mounted on the roof, which can be operated unmanned, protect this hospital. The exposure at the rear is a large one-story storage facility used to store hospital supplies. Rear access is limited.

It is important to know the difference between the most severe exposure and the exposure most severely threatened. Being able to properly evaluate this factor will guide the incident commander to the correct protection strategies. It may be that, on the initial response, manpower may only be sufficient to protect one exposure or accomplish one task. A decision will have to be made based on priorities. That decision may be dependent on your ability to differentiate risks based on a sound analysis of conditions and contingencies. For example, a 2-story, wood-frame building under construction is fully involved in fire. The wind is moderate. On the leeward side are two more identical wood-frame buildings under construction. Behind the fire building is a 1-story, wood-frame construction shed containing ten cases of dynamite. You only have sufficient manpower and water to protect one area. Which one should it be? The dynamite shed is the most severe exposure, even though the wood-frame buildings on the leeward side are most severely threatened. They also represent the path of least resistance for fire-spread due to the wind condition. However, (and this is a HUGE "however") the ignition of the construction shed and the detonation of the dynamite will have a much greater impact on the area and on the safety of both fire personnel and area residents than the complete destruction of the wood-frame buildings under construction. Although this is an extreme case, it proves the point.

Even though these buildings are unattached, the close distance requires extensive exposure operations. These will include evacuation, advancement of lines, and extensive opening up on the top floors of the exposure. A large diameter line or deck gun should be directed between the buildings to wash the walls of the exposure and hit extending fire.
(Louis "Gino" Esposito)

Remember that interior exposures in the target building may also impact on your operations, both favorably and unfavorably. Interior exposures are fire threat hazards the building offers to itself. A firewall, for example, is an excellent place to make a stand against a heavy fire condition in one part of the building. On the other hand, propane tanks or other hazardous materials stored inside a building that are being exposed by an interior fire must be brought to the attention of the incident commander for immediate consideration. They will have to be protected in place or moved. Another interior exposure that must be protected is the upper floor of most residential dwellings. The open stairs will most certainly expose these upper floors, as the fire will take the path of least resistance. Stretching a line to this area will protect both occupants on the upper floor and firefighters engaged in primary search.

Area and height

Area should be dealt with in two manners. First, there is the area of the building. Large-area buildings present the problem of long hose stretches and the accompanying friction loss. Using a large diameter line for the attack on large fires, or as a supply to a gated wye where two 1¾" lines could be stretched, may be the answer to the friction loss problem. Knowing in advance the depth and area of the building, as well as setbacks, differences in elevations on different sides of the building, and locations of stairways in relation to the attack entrance can help determine which line and how much of it to stretch in. It can also be an indicator that a search rope may have to be utilized for the primary search due to large, open areas, or worse yet, a maze of cubicles in a large area.

Knowing the response area can help firefighters determine what type of building they are responding to. There may be a need for extra security in some areas. In such cases, it may be wise that when responding to areas that are known to be unfriendly, even to firefighters, to stage the apparatus in a safe position until the police arrive to escort fire personnel into the area.

Other areas may produce "dead spots" in radio communication to the dispatch center. In these areas, accommodations should be made in advance to communicate vital information from the incident commander to the dispatch center. Cellular telephones are the best answer to this problem. If no cell phone is available, a public telephone can be utilized, or possibly a radio relay from the "dead" area to an area in close proximity to the command post, which can communicate with the incident commander and still relay pertinent incident information to the dispatch center. This may have to be utilized in tunnels or other areas below grade. Hard-wire telephone systems may also be a

Setbacks make a four-story building set on a hill actually about seven stories from the street. The presence of power lines strung in front of it makes aerial operations impossible from the street level.

The one-story building on the upper roadway (on left) is actually a five-story building. There are entrances on two separate streets. What might look like an upper floor fire from one street will look like a basement fire on another.

This training exercise in the rail tunnels under the Hudson River required the use of both cell phones and sound-powered phones. Mock incidents are where the effectiveness of these operations can be evaluated.
(Courtesy of Pete Guinchin)

solution, but they tend to take a long time to set up. None of these alternatives will work if they are not straightened out before the incident, and above all, practiced to ensure they are effective.

Height of the building will be a problem in areas of high rises, but fortunately, buildings of this type have many features that allow us to operate without the aid of an aerial, such as fire resistive construction, compartmentalization of areas, enclosed stairways, and smoke-proof towers. Fire service elevators, if used properly, are another aid in such construction. As mentioned earlier, departments without the service of an aerial device can make arrangements for automatic aid to acquire the aerial device when needed. Again, working these agreements out in advance works best for all concerned parties. For departments without the high-rise problem, purchasing a quint with a 50' to 75' aerial device may be the answer to the aerial problem without sacrificing the ability to pump and carry water and hose. It makes absolutely no sense for departments to purchase 110' or higher tower ladders when the tallest structure in the jurisdiction is a 3-story building, yet departments do it everyday. These rigs look great at a parade, but in reality, are an irresponsible purchase and a monumental waste of taxpayer money.

Location and extent

This is the big one. No operation can begin and no action plan can be set into motion until this is determined. It may surprise some to learn that location and extent is the most important size-up factor. Location and extent determine the life hazard, which, in turn, determines strategy and tactics. Take for example, a trash receptacle on fire. If that trash receptacle was located at 2nd base of an empty Yankee Stadium, there would be no life hazard, and subsequently, it could be extinguished easily with no fanfare. Take that same receptacle and place it at 4 a.m. at the foot of the stairs of an old

3-story wood-frame occupied tenement with open wooden stairs and it is, as they say, "a whole 'nother ballgame". The life hazard will be severe. There will be the problem of fire extending up the open stairway, the most accessible path of least resistance, blocking the means of egress, which could lead to panic and resulting injuries and death. In this case, the extinguisher had better be left on the apparatus and lines stretched to ensure a quick knockdown and protection of the vertical arteries. The bulkhead door at the roof level would have to be opened to release the heat and smoke. As you can see, the location of the trash receptacle made the whole difference between a "nothing fire" and a potential critical situation.

Location and extent determines the life hazard and how much of the building is threatened. The location and extent of the fire in this cellar will expose the whole building and cause a severe life hazard. The life hazard and fire spread profile would not be as severe if the fire was on the top floor or in the cockloft. *(Ron Jeffers, NJMFPA)*

As another example, consider a fully involved car fire. If that fire is located in the middle of the street, the problems are minimal compared to that car being located in a below-grade parking garage. The problems here would be multiplied exponentially.

Certain areas in some buildings present more problems than other areas. A fire in a factory, where hazardous chemicals are stored, would be much more difficult to handle than a fire in the lunchroom of the same building. This may lead to an incident that is more problematic in regard to access, fire load, life hazard, exposure potential, and anything else you can name. We all know of buildings in our response area where we think, "what if we had a job in this area of the building?" Firefighters should know that major problems can be anticipated.

What problems will be encountered from a fully involved car in this area in comparison with the same car on fire in the middle of the street? Life hazard, extension potential, and additional alarms are some of the factors to consider in regard to this fire's location and extent.

For example, a fire that occurs in the cafeteria kitchen of a hospital at lunchtime would present a severe life hazard, but other than that, it may be nothing out of the ordinary once occupants are evacuated and lines are stretched. There may even be an easy access point via a service entrance where line positioning and equipment access may be to our advantage. Now place the fire in the area of the operating room where an abundance of oxygen is used. Not only would a fire in this area interrupt the entire hospital, but it may also present access problems, the need for special extinguishing agents, and the nightmare of possibly evacuating a great number of non-ambulatory patients. We must always be aware of special places in each building that will give us extra problems if a fire were to strike.

While location and extent cannot be determined until we arrive on the scene, we can, through pre-fire planning inspections, pinpoint areas where a fire is most likely to start, and then predict, by virtue of building features, how it will spread. By determining this, we may be able to have a better idea of the difficulties that may be anticipated should an incident occur in that area.

Time

This is another tough one to tackle because we never know exactly when a fire will strike, but we do know, based upon our knowledge of occupancies in our response area, which times of the day will be more critical in certain buildings. For instance, from Monday to Friday, a school will present a severe life hazard from about 7:00 a.m. until 4:00 p.m., and then again maybe at night if athletic and other activities are scheduled. However, students will be

Lighting up the area at night operations must be a critical safety concern. The incident commander must be aware of the availability of tactical support units such as this light truck well before the fire ever starts. *(Ron Jeffers, NJMFPA)*

awake and all doors must be unlocked by law. This somewhat helps our rescue profile. The life hazard at a school will not be as severe on a Sunday at 6:00 p.m., when the building can be expected to be virtually empty with the exception of a one or two security guards. Likewise, if we are awakened in the middle of the night to a reported fire in a residential multiple-dwelling, where there are "numerous calls", we know we will have a severe life hazard, and just the fact that it is the middle of the night may be the cause for additional alarms being struck.

As firefighter safety must always be the overriding concern of all fire activities, operations at night must prompt the incident commander to request the response of a vehicle that can provide additional lighting to the fire scene. In addition, officers working in areas such as the roof or the rear of the building must ensure the work area is adequately lighted.

Hazardous materials

It is essential that responders know which buildings contain, or are likely to contain, hazardous chemicals. This information comes to us through, among other things, Hazardous Substance Fact Sheets (HSFS), Material Safety Data Sheets (MSDS), Right To Know laws, and our own building inspections. Prior knowledge of an occupancy containing hazardous materials should prompt a hazardous materials response with the initial dispatch. The Haz Mat experience and training level of most first-arriving companies and chief officers is minimal at best. The best plan in such instances would be to:

1. Position upwind and uphill
2. Isolate the area
3. Deny entry to the area
4. Attempt to identify the product if it can be safely done from a distance
5. Let the experts handle the situation

The NFPA 704 labeling on the exterior wall of this building is a definite indicator of hazardous materials present. The bottom of the rightmost diamond cautions against water use in this building. Failure to heed this warning could have a significant impact on firefighter safety.

Bloodborne pathogens are as hazardous as any chemical. Note the pick-up boxes outside this "residence". Sizing up the occupancy before entering may be a clue as to this hazard.

These same tactics should be utilized when responding to a motor vehicle, rail, air, or vessel incident where the presence of hazardous materials is either suspected or confirmed. Knowledge of what specific chemical or compound is within an occupancy will help determine the strategy to be used at an incident. Heavy fire encountered in buildings that contain hazardous materials, such as pesticides, should be allowed to burn in a controlled manner because runoff from hose streams could damage the environment more than the smoke. In this case, you end up with more of a problem than you set out to solve. This may also be the case where the incident presents such exotic hazards as confined space incidents, high-angle rescues, biological and etiological exposures, etc. It is best to slow things down if possible, think firefighter safety, and let the experts handle it.

The incident commander cannot afford to become a victim of the "tunnel vision" syndrome, and let his guard down. He must constantly be evaluating all size-up factors, for the one that he least expects can sneak up on him and cause him much grief.

Size-up and the initial radio reports

Using size-up in your pre-fire planning activities is, without a doubt, an advantage for the incident commander. However, the information gathered from this process prior to an incident does not relieve the incident commander of making a complete size-up, once on the scene. In fact, the effective incident commander will include previous knowledge in with current information in order to create a better picture of what is actually taking place at the scene upon arrival. This on-scene size-up is important for two reasons. First, it gets the incident commander's head into the game. Only with pertinent, current information about the incident, can he begin to create an action plan to handle the situation and mitigate the incident. Second, it gives the incident commander the information he will need to broadcast the Preliminary Size-up Report to dispatch. The Preliminary Size-up Report is used to establish command and start the process of information evaluation. It also gives vital information and, possibly, direction to the incoming companies. It aids these companies in creating their own mental picture of the incident before they arrive on-scene and it allows them to begin their own scene size-up with more information, courtesy of the incident commander. Basically, the Preliminary Size-up Report gets the entire operation moving (hopefully) in the right direction to mitigate the incident.

The Preliminary Size-up Report is actually the first of two critical reports aimed at coordinating the fireground in the initial phases of the operation. The first, the aforementioned Preliminary Size-up Report, given by the initial officer on the scene, usually a company

The first-arriving officer will usually give the Preliminary Size-up Report to dispatch and set the operation into motion. Establishment of incident command by this officer is mandatory.

commander, is actually a regurgitation of the size-up information in the form of a report to dispatch and to the responding companies. The Preliminary Size-up Report must have a definite structure to be of any use to responding personnel. It should paint a picture, in a logical fashion; of what the incident commander can see from the command post. The report need only include the following information.

- Company designation
- Building information (Height / Construction / Occupancy and Occupancy Status)
- Conditions upon arrival
- Action being taken
- Command statement

An example of such a report would be:

"Engine #17 on the scene. Have a 2-story, occupied, wood-frame residential dwelling. Nothing showing upon arrival. Engine #17 will be investigating. Engine #17 is establishing Main Street Command."

Even if a chief officer is first on the scene, this should be the basic format of the Preliminary Size-up Report. This information provides a brief and basic picture of what is taking place upon arrival, but does not unnecessarily bog down the initial incident commander, especially if it is a company officer, of carrying out his assigned duties, in this case an investigation of the building. Of course, at this time, if the situation requires, an additional alarm may be requested.

Once a chief officer arrives on the scene and assumes command, a more detailed report of conditions must be issued to dispatch. This report should be broadcasted within the first ten minutes of arrival. If no chief officer arrives, the initial incident commander must furnish this report, which is called the Initial Progress Report. The Initial Progress Report provides a more comprehensive picture of the fire building and its surroundings, as well as actions taken. Included in this report will be the need for additional alarms and any other resources required, if not already done.

C-BAR

The acronym **C-BAR** can be used to formulate this picture. **C-BAR** stands for Command, Building, Actions, and Resources. The following is some basic information about each portion of both the Preliminary Size-up and Initial Progress reports.

Command

Command must be established at every incident. No matter what the size or circumstances of the incident, this mechanism to control the fireground and protect and account for the participants cannot be ignored. Even at small incidents, command must be established. At single company responses, command of the incident is understood and need not be announced, but at any incident involving more than one company at any given time, command must be established and announced. In the command portion of the Preliminary Size-up Report, the company establishing command must perform the following duties.

1. Identify their designation and that they have arrived on the scene.
2. Announce the establishment of command using the name of the street that the emergency is occurring. The first-arriving officer will usually broadcast these first two statements.
3. At large incidents, the location of the command post should be identified. This will usually be part of the Initial Progress Report given by the chief officer who assumes command. This information allows later-arriving companies to report to the proper place for assignment. At small incidents, the location of the command post is usually understood to be somewhere in proximity to the front of the building, which will be in a conspicuous place and afford a good view of operations without getting in the way.

Building

Information about the building and what is happening to it is the information that will set your strategy and tactics into motion. You cannot begin to take care of the problem until you know what is going on in the building. Often, this part of the Preliminary Size-up Report will have less than complete information due to the fact that not all of the facts are available at the time of arrival. The report must still be made using the best obtainable data at the time of the report. This report regarding the building should include specific information.

C-HOLES. The acronym C-HOLES can help aid the incident commander in making this portion of the report. Not all of C-HOLES will be used in the Preliminary Size-up Report, but can be used as a guide. The Initial Progress Report, however, must include all of the following information.

Construction.

- Fire-resistive
- Limited/non-combustible
- Ordinary
- Heavy timber
- Wood frame

Height. Height is the amount of stories above grade. This may have to be estimated at a high-rise building. A statement on whether the building is attached or unattached should be included at this time. This information is invaluable for responding ladder companies in determining their access routes to the roof. It doesn't always mean that they will be able to use the attached building as access to the fire building, but at least they will get an idea of what their options might be. For example, a building that is attached, but not of the same height as the fire building is usually unusable for roof access. In this case, the aerial will be the best way to the roof.

Occupancy. Identify what the building is actually used for (*i.e.*, residential, commercial, mixed use, etc.). This information, along with the location and extent, may signal engine companies as to what size of hose may have to be stretched to adequately handle a fire in the building. In addition, the occupancy status is also important information that should be relayed to dispatch. Building status can be narrowed down to three status types. These are occupied, unoccupied (which may be hard to justify in a residential occupancy until a primary search is completed), and vacant.

Location & Extent. Identify where fire and/or smoke is showing. This allows companies to get an estimate of the life hazard present and the potential amount of hose that will have to be stretched to reach that floor. A "nothing showing" report should signal to companies that they would be in the investigation mode upon arrival.

Exposures. This information need not be included in the Preliminary Size-up Report. It is, however, a mandatory part of the Initial Progress Report. The incident commander must identify the areas immediately around the building or area of operation. This is crucial information regarding the exterior firespread problem. When taken in regard to wind speed and direction, this data may become more urgent to the commitment of manpower and apparatus to protect it. Phrases such as "attached" or "separated by a 5-foot alley" will cue responding companies as to the potential for a large incident. Sometimes, the best way to protect these exposures is to extinguish the parent body of fire. Other times, due to factors such as location and extent of fire, life hazard caused by occupancy and/or time of day, and apparatus and manpower, exposure protection may be the most critical objective, and therefore receive priority. These are decisions that will face the incident commander in the initial stages of operation.

A Preliminary Size-up Report for a chief officer who is first to arrive at the building in the center is as follows: "Battalion 1 on scene. Have a five-story ordinary construction, mixed residential-commercial occupancy, attached on both sides. Nothing showing upon arrival. Companies will be investigating. Battalion 1 is establishing Park Avenue Command."

Regardless of the system used to identify exposures, whether it is letters or numbers, a clockwise identification sequence must be used and understood by all. Starting at the front of the building and working clockwise, the Incident Management System used by most jurisdictions labels this front side as Side A, the left side of the building is Side B, the rear is Side C, and the right side as Side D. New York City and some others use 1, 2, 3, and 4 instead of A, B, C, and D. This is fine so long as everyone is on the same page.

Old, closely spaced or attached combustible buildings make urban areas the battleground for the Telesquirt. It is a versatile tool for both heavy fire attack and exposure protection. Early and proper positioning of this weapon can make the difference between small loss and conflagration. *(Ron Jeffers, NJMFPA)*

When identifying exposures, the same lettering or numbering system may be used, however, it is important to distinguish between Side A and Exposure A, Side B and Exposure B, and so on. Side refers to the side of the fire building, such as placing men on Side C to flow streams into the rear of the building. They would operate at the rear of the fire building. Exposure refers to a separate building, structure, or vehicle that you want to protect. If these same men were told to use these streams to protect Exposure C, then the streams would be used to coat the walls of the building at the rear with water to protect it from radiant heat and subsequent ignition.

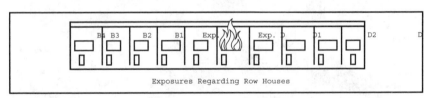

Exposures Regarding Row Houses

Addressing exposures in attached buildings, such as row houses, should also be identified in a systematic manner. Starting to the left of the fire building, the exposures should be lettered and numbered as Exposure B (to the immediate left), then Exposure B1, B2, B3, and so on. The same treatment should be used for the exposures on the D (right) side. Questions often arise regarding these attached exposures. The first is that if Exposure B becomes involved, does it become part of the fire building, and does the building that was initially B1 now become Exposure B? The answer is no. Leave the exposures as they were originally designated in regard to the original fire building. It will save everyone on the fireground a lot of confusion.

Another question often asked is "what about buildings that are catty-corner to the fire building, such as both to the rear and to the left of the fire

Fires that spread beyond the building of origin and/or require extensive operations at the rear and sides require that a systematic approach be taken to identify areas of operation. A method of identifying these areas must be adopted, established and made known to all participants. *(Louis "Gino" Esposito)*

building?" The easiest way to identify these exposures is to use two letters (or numbers if you use that system). Therefore, this building to the rear and the left would be Exposure BC. Incident commanders should ensure that their reports and orders are clear, concise, and direct so that the order is understood to be the same tactic the incident commander intended to accomplish.

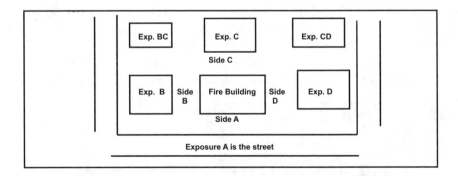

Special circumstances. This portion of the initial radio report is reserved for those circumstances that are crucial in regard to the focus of the firefighting effort. For example, say you are given information from a tenant who has already exited the building that there is someone trapped on the 3rd floor. This information should be part of the initial radio report for it allows rescue teams to focus their attention to a particular area even before they have gotten off the apparatus. Other information that should be included in this portion of the report are the factors that impact firefighter safety. An example of this would be the presence of a truss roof or floor, knowledge of a previous fire where the roof may be open or the floor/stairs may be in question, or even the presence of pit bulls in the basement. This last bit of information would be critical to any firefighter entering the basement. Not having this information at the outset could put him in a position of being on the wrong end of the jaws of a firefighter-hating beast. Ouch!! It is up to the incident commander to decide which information to include. This comes with experience, prior knowledge of the building, or just plain intuition or luck.

Action

In order for incoming companies to know which strategic mode they will be operating in, it is essential to know what initial action the incident commander intends to take. This action will be dependent on arrival conditions and information culled from the initial size-up of the incident commander. This action statement, which is an early indicator of the incident action plan, can include, but is not limited to:

Condition	Accompanying Radio Report
"Nothing Showing"	"Companies will be investigating"
"Fire or Smoke Showing"	"Companies will be in an offensive mode, primary search is underway"
"Fully involved"	"Companies will operate in a defensive posture"

"Nothing showing" must prompt an immediate investigation of all areas of the building. Companies must be positioned in the most strategic manner. The ladder truck is at the front of the building while the forward Engine is positioned to supply the attack Engine.

Fire or smoke showing upon arrival demands that companies operate in an offensive mode. Primary search must be conducted, attack lines must be stretched, and related support functions must be accomplished in a coordinated and timely manner.
(Bob Scollan, NJMFPA)

The terminology used depends largely upon local protocols and SOPs, but, regardless of the terminology, the intent is to begin coordinating fire-fighter thinking in the early stages of the incident.

In the Initial Progress Report, the Action portion of the report should give the information of actions already taken and underway at the time of the report. This includes any lines stretched, operating, and the status of any primary search operation, to name just a few.

Buildings that are fully involved contain no savable victims. No primary search will be conducted. The strategy is defensive from the outset. Command must relay this information to the dispatch center and to all companies.
(Newark, NJ Fire Dept)

35

Resources

This is the part of the Initial Radio Report where the incident commander requests additional alarms to handle the incident, based on his size-up and the ability of the initial responding apparatus and manpower to handle the incident in a safe and judicious manner. This request may be made during the Preliminary Size-up Report or in the Initial Progress Report, or both. There is no magic formula to guide the incident commander in requesting additional alarms. Sometimes you must go with a gut feeling based on past incidents. Judgment and experience can be of assistance as guiding factors in whether to call for additional alarms. But certain situations will demand additional alarms whenever they are encountered. These are:

- A severe life hazard due to occupancy and/or time of day.
- A severe exposure problem.
- An advanced fire condition upon arrival.
- A large structure where additional manpower will be required to search and stretch lines.
- Below-grade fires which are difficult to access, and are showing heavy fire and/or smoke.
- High-rise fires
- Any building with an out-of-service auxiliary appliance.
- Any situation where you have more tasks to accomplish than personnel to assign them to.

The main objective is to avoid letting the demands of the incident-action plan outstrip your responding manpower and apparatus levels. If there is a possibility that you might need additional manpower and apparatus, it is always good practice to request an additional alarm.

Many of today's fire departments respond with less than adequate manpower levels. As a result, firefighters on the initial response often overextend themselves to accomplish the many tasks required at a working structural fire, especially in the initial stages of operation. These tasks require a Herculean effort on the part of the undermanned fire force. Firefighters are dedicated professionals who deserve the chance to do their jobs efficiently and safely. Early reinforcement of initial attack and support positions allows the suppression forces the opportunity to accomplish that which they have been trained to do. Overextended firefighters rapidly become injured firefighters.

High-rise fires will require an expanded incident command organization as well as a large commitment of manpower to safely control the operation. Call for help early or, better yet, have more companies respond if a fire is reported in the building. *(Mike Johnston)*

It is the prudent incident commander who recognizes this fact and provides this crucial support by summoning for help early in the firefight.

If you find that your action plan requires specific tasks that still need to be accomplished and you don't have anyone left to do it, then your tactical reserve has run out. At this point, it may be too late to call for additional alarms to reinforce an operation. This is especially true if it is an offensive situation where companies have not been able to get a handle on the fire, but their low-air alarms are activated and they are exiting the building. If sufficient forces are not on the scene to continue the attack, the operation is doomed and the strategy may have to switch to a defensive mode for the rest of the operation, or at least until adequate manpower is on the scene to continue the attack. This is especially critical in high-rise fires where it might take a while to get the required manpower to the area of operation.

Don't forget to account for reflex time in the additional alarm response. Reflex time is that time it takes for a task force to receive the call, gear up, board the apparatus, respond, stage or position as per the incident commander, and begin operation. This time lag could have a huge impact on the action plan and cause an offensive situation to deteriorate into a defensive mode of attack.

I was in charge of a fire at a large, wood-frame and heavy-timber restaurant and banquet hall on the Hudson River waterfront located at the end of a 1000' pier. Over 600 guests occupied the restaurant at the time of the fire. Heavy fire was showing on arrival and a mass evacuation was underway. A second and a third alarm were struck within three minutes of arrival. Because a hose relay was needed to establish a water supply to supplement the initial attack, most of the additional alarm companies had to stage out on the main road, which was over 1000' away. The reflex time here was substantial. By the time these companies made it to the command post, the initial attack companies had already begun exiting the building, their air having run out. Due to the heavy smoke, they were unable to locate the seat of the fire. It was extremely difficult, at this time, to continue an offensive attack. The best that could be done was to launch into a marginal attack because the fire, accelerated by the wind and some poorly placed master streams from non-fire department vessels, had gained great headway. Eventually the fire forces had to be evacuated and a defensive mode of attack was pursued. The reflex time proved to be one of the factors that doomed this structure.

The policy regarding additional alarm response should include a task force of at least two engine companies and one ladder company respond. Requesting

This is the view of the restaurant fire from where mutual aid companies had to stage apparatus and walk to the command post over 1000' away. The reflex time here was a major factor in the destruction of this building. *(Ron Jeffers, NJMFPA)*

a one-company response is counter-productive. Additional alarms are struck due to the need for manpower. If mutual aid is required to fill out the required compliment for a task force response, then it should be dispatched. The manpower compliment of one company is hardly enough to put a dent in the needs of the incident. If it turns out that you don't need them, they can always be released from the incident or staged.

The utility company must be requested at all large fires to kill the power in the street. Here, ice is causing an additional load to be placed on the overhead power lines. *(Tony Castelluccio, Newark Fire Department)*

All additional alarm companies, unless otherwise directed by the incident commander, due to an immediate need, should stage their apparatus in an uncommitted area that does not congest the scene nor block access/egress points into the incident. Officers and crews with a full compliment of equipment and tools should report to the command post for assignment, unless they are ordered to do otherwise. The concept is to always have a tactical reserve on hand to plug into the action plan as required. When the incident commander runs out of tactical reserve, he essentially runs out of plans.

Firefighter safety includes ensuring that operating personnel are well hydrated at all fires and provided with food at marathon-type incidents. The incident commander must take care of the players if he wants to win the game. *(Ron Jeffers, NJMFPA)*

Resources also include those agencies that are essential to a safe operation. Pre-fire planning often aids in this determination. Some agencies are dispatched as part of the routine response. Those include the police, who have the responsibility for crowd and traffic control, and the Emergency Medical Services (EMS). EMS personnel should, as a first priority, monitor firefighters as they exit the building. Other essential agencies may be required depending on the demands of the incident. These may include, but are not limited to:

- Utility companies (gas, electric, water)
- Fire investigators
- Department of public works
- Firefighter canteen units
- Light units
- Mask service units
- Building department or construction officials
- Health Department
- Red Cross or local or county welfare services
- Bureau of Alcohol, Tobacco, and Firearms

The idea is that if you think you may need it, you had better request it. You can always send a resource back.

The FAST Team Response

This is a crucial response in regard to firefighter safety. The FAST Team (Firefighter Assist and Search Team), also known as Rapid Intervention Teams (RIT) or Rapid Intervention Crews (RIC) should be dispatched as soon as the confirmation of a working fire is received. This team should also be dispatched on any special response that is out of the normal response realm. FAST Team response and operation should be part of a department SOPs and addressed in advance. The duties, responsibilities, and assigned equipment of this team should be addressed in the SOP. For this reason, in the *Fireground Strategies Workbook,* I will not address or include the response of a FAST Team when addressing additional alarms or manpower. As it should be part of a set response, it is understood to be standard procedure in all the scenarios presented in the *Fireground Strategies Workbook.* They will be expected to have responded and to be in place.

A **FAST or RIC team** should respond to every work-ing fire. The team should report to the command post, fully equipped and ready to operate. Tools and equipment appropriate for the incident should be staged at the ready for quicker deployment.
(Ron Jeffers, NJMFPA)

By having this FAST Team response, the OSHA two-in, two-out requirement can be satisfied. Technically, when companies arrive at a working fire that requires an offensive mode of attack along with a primary search, they are in the "rescue" mode. When this "rescue" condition exists, the two-in, two-out does not apply. As soon as the "primary search complete" benchmark is reached, the incident enters the "incident mitigation" or "fire control" phase, and the two-in, two-out requirement applies. By that time, if the department has already addressed this issue in an SOP, the dispatched FAST Team will already be in place.

While I understand the concept of the two-in, two-out requirement and the need for firefighter safety, I must say that I have never seen two people outside a structure ever save two people on the inside of the structure. Usually, the two people inside are rescued by people inside before the outside people ever have a chance to get inside. While I believe this safety requirement is a step in the right direction, I also think that the same people who spearheaded this movement should also push just as hard to get minimum apparatus staffing passed as a mandatory requirement, not just a standard. This will make complying with the two-in, two-out law much easier.

Radio Reports From Companies

The incident commander cannot exist on an island at the fire scene. He must depend on reports from all areas of the building, both inside and outside, to help him evaluate the course of action to take, and, more importantly, to evaluate the current action. The Initial Size-up Report only provides information from the vantage point of the incident commander, usually the front and maybe one side of the building. Recon reports will help fill in the information he does not have and create a more complete account of conditions.

The incident commander must insist on radio reports from all areas. A good SOP regarding initial scene assignments will provide proper coverage of most buildings. This SOP must be enforced on all incidents, all the time, such as at routine alarm activations. If it is not done on a routine basis, it will not be done at the incidents where this information becomes critical. Like an Incident Management System, for it to be effective, it must be used all the time.

Things going on in areas the incident commander cannot see are responsible for most problems caused on the fireground. If he doesn't get information on the conditions in these areas as soon as possible, things can go very badly very quickly.

It is critical that the incident commander receive timely reports from all sides of the incident, including the interior. The command officer can only supply support to these areas if aware of the conditions and needs. *(Bob Scollan, NJMFPA)*

Once personnel are in the building, the incident commander cannot accept a lack of information and must insist that proper and timely reports be issued at all incidents. Places where the incident commander cannot see that are critical to his action plan include the following:

Interior:

- The fire floor.
- The floor above the fire.
- The top floor.
- The basement or cellar.

The conditions at the rear of the building can be much worse than those at the front. Command can only reinforce the operation when information from all areas of the building is obtained. SOPs must assign personnel to check and report on conditions in these areas. *(Bob Scollan, NJMFPA)*

Exterior:

- The sides and the rear.
- The roof.
- Any shafts.
- Significant exposures, especially at the rear, the sides, and of course, the interior.

All of these areas must be checked and a report sent to the incident commander on conditions. The incident commander must insist these details are addressed as quickly as possible. Little details turn into big problems if left unattended.

I have been at several fires where the initial companies have been victims of the "candle-moth" syndrome. Like moths in the summer, everyone

Most problems occur in places the incident commander cannot see from the command post. Conditions in this enclosed shaft must be checked either from the roof level or from a window that borders on the shaft. A report must be issued to incident command as soon as possible.

goes to where the fire is. One such fire was in the basement of a large 5-story tenement. Lines were stretched and roof venting of the bulkhead door was accomplished. However, no one went to recon the rear, the sides, or look over the rear from the roof of the building. A check of the rear of the building would have revealed fire traveling up an air shaft, which eventually entered the building on upper floors, and, subsequently, chased the fire forces out of the building. Just one look at the rear of the building may have been able to save the structure by positioning one line to protect the shaft.

An effective SOP will send men to all the critical areas the incident commander cannot see, but needs information on. As I stated before, the incident commander must insist on reports from all areas he cannot see. In fact, in the absence of such reports, he should prompt the people in that area for a report. It is unacceptable to operate without complete information about all sides of the building.

Task-assignment model

Companies arriving on additional alarm assignments or not assigned by SOP should report to the command post for check-in and assignment. Once at the command post, the following sequence, called a Task-Assignment Model, should take place.

- Assignment—the company receives an assignment from the incident commander.
  This assignment should give the company officer an objective to meet. Basically, all the incident commander should have to do is tell the officer what needs to be done, such as "hold the exposure". He should not necessarily tell him how to get it done. This is micromanagement. Most company officers do not need the incident commander to tell them how to do their job; in fact, they will probably be insulted if you do. This is akin to the old saying, "If you want me to do something, tell me what it is, give me the tools to do it, and get out of my way!"
- Stabilization—the company attempts to stabilize the situation as per the given assignment.

Companies receiving assignments must report the status of that assignment to incident command as soon as possible. This is the only way that incident command can reinforce and support company operations. Progress reports to incident command must be issued at regular intervals. *(Ron Jeffers, NJMFPA)*

- Report — the company commander makes a report to the command post.

Reports should be made to the incident commander as soon as conditions are assessed. At working fires, reports should be provided from these positions as soon as the officer assesses the situation and on a regular basis after that. Included is the completion report when the task is finished.

CAR

These reports should follow the acronym CAR and include:

- Conditions—what is happening at your position?
- Action—what you are doing about it?
- Resources required—what additional manpower and/or equipment do you need to get it done?

From this report, the incident commander will be able to determine if the company will be able to stabilize the situation, if it will require more manpower and/or equipment to stabilize the situation, or if the situation has escalated to a point where the current strategy may need to be modified or changed to fit the current condition.

For example, a company may be assigned to stretch a line to the floor above the fire. Obviously, conditions on the floor above will dictate whether this action will be successful. There are only three things that can happen here:

1. Fire extension to the floor above is controlled or non-existent and the company can handle it themselves.
2. Control of the floor above the fire will require additional support, either in the form of water, ventilation, overhaul, or a combination of the three.
3. The company cannot make the floor above the fire due to untenable conditions, which may cause a change or modification in strategy.

In any case, the incident commander will not be aware of the conditions and needs if a report is not furnished as soon as possible and at regular intervals. The incident commander's responsibility is to provide support to the companies working at the fire. The only way this can be effectively done is by regular and accurate reports from companies operating in different areas of the building. In fact, the chief officer, as a matter of routine, must insist these reports are furnished.

Progress reports from the interior are often the only way that the incident commander can gauge the effectiveness of the current strategy. Activities on the fireground can only be supported and reinforced when the person in charge has current and accurate information. *(Bob Scollan, NJMFPA)*

Command Progress Reports

The incident commander should give progress reports to dispatch at regular intervals, usually every 15 minutes, via radio. A progress report should also be given when significant changes occur on the fireground such as collapse or sudden fire extension. A progress report should also be transmitted when additional alarms are requested. Issuing progress reports force the incident commander to look at the building and take an inventory of operations, completed, in progress, and still required. Basically, it serves to keep the incident commander's head in the game.

Only the well-informed incident commander can give an effective progress report. Prompt and accurate information accrued from all areas of the fireground will allow the incident commander to effectively evaluate conditions and plan the strategy accordingly.

CABS

These progress reports should follow the acronym CABS, and include:

- **C**urrent conditions
- **A**ctions taken
- **B**enchmarks completed
- **S**tatus of operations—fire doubtful; probably will hold; conditions unchanged; conditions improving; under control; or extinguished

Using a tactical worksheet will assist the incident commander in issuing accurate and concise progress reports. Failure to utilize this incident command aide at the routine incident will render it next to useless at a major fire. *(Ron Jeffers, NJMFPA)*

Fire doubtful. That indicates that a fluid and still developing situation exists. It will be understood that the situation remains doubtful until changed in subsequent progress reports by the transmission of "probably will hold" or "under control".

Example Report: *"Dispatch from Central Avenue Command, Progress Report #1. Have heavy fire on the 2nd floor, Side C. Three lines are stretched, two operating, primary search is negative. Natural openings on roof are vented at this time. Checking for extension in cockloft as well as in Exposure B. Fire is doubtful at this time. Additional reports to follow."*

Probably will hold. Understood to indicate that in the judgment of the incident commander there are sufficient apparatus, equipment, and personnel on hand to contain the fire or emergency and prevent any further extension or escalation. However, if an unknown, unusual, or unpredictable condition develops, additional help may be required, but the fire or emergency will not develop to critical or uncontrollable proportions.

Conditions unchanged. Indicates that conditions are the same as in the previous report and the fire force has made little or no progress in the extinguishment of the fire. This term is not used in preliminary and early progress reports while the fire situation is still developing and being defined. This report shall be accompanied by a general description of current operations.

Example Report: *"Dispatch from Central Avenue Command. Progress Report #X. Conditions remain unchanged. Companies are still operating in an offensive mode in previously reported locations and the attack is continuing."*

Conditions improving. Fire forces are making headway but final extinguishment has not yet been achieved. Accompanying this report is a description of areas where fire has been contained.

Example Report: *"Dispatch from Central Avenue Command. Progress Report #X. Conditions improving. Visible fire has been knocked down on the 3rd floor. Interior attack continuing on the 4th floor."*

Under control. Indicates that at this time, in the judgment of the incident commander, final extinguishment of the fire or the control of the emergency will be accomplished by apparatus, equipment, and personnel on the scene.

Conclusion

This chapter has attempted to show how the activities of pre-fire planning, size-up, and incident scene communications intermingle to produce the best approach to the fire incident in regard to data available. Each supports the other and will help create the big picture that will assist the incident commander in filling the informational gaps at the fire scene.

Questions For Discussion

1. How can the pre-fire planning process be linked to the size-up process?
2. Discuss how each of the 13 size-up points can be of value to pre-incident information.
3. What factors should firefighters consider when preplanning a structure with an auxiliary appliance?
4. Who is responsible for giving the Preliminary Size-up Report?
5. What are the elements of the Preliminary Size-up Report?
6. What are the elements of the "building" portion of the Preliminary Size-up Report?
7. What are the elements of the Initial Progress Report?
8. Discuss and give examples of the concept of most severe exposure vs. the exposure most severely threatened.
9. Name some of the conditions that would likely demand additional resources.
10. Name some of the critical areas on the fireground where a report of conditions to incident command must be issued as soon as possible. Give reasons why.
11. What are the elements of the Task-Assignment Model and what should be included in the report to incident command regarding the task assigned?
12. What should be included in the Command Progress Report?

CHAPTER TWO
Heat Transfer: Strategic Considerations

Fire spreads as a result of heat transfer. Heat, like water, takes the path of least resistance. Understanding this principle is instrumental in the prevention of heat build-up in an area. Heat transfer, if left unchecked, will lead to flashover, full involvement, and eventually collapse. It may also lead to a backdraft situation if conditions are right, such as in the case of a tightly sealed-up structure. Heat transfer also causes combustibles to reach their ignition temperature at great distances from the parent body of the fire.

The strategic objective of the incident commander is to inhibit this heat production process and keep it at its present location, or confine it, so that the extinguishing agent in the proper form and quantity can be applied. In other words, the incident commander must forecast how heat production will influence firespread. These firespread characteristics will, to a great degree, depend on the structural traits of a building. Certain buildings lend themselves to specific types of heat transfer. For example, an ordinary-constructed building with an open stairwell invites heatspread via convection currents to the topmost point of the building. Venting over the top of the stairs will tend to localize the fire. All fire extinguishment and support activities, such as ventilation, are aimed at stopping heat transfer, which is, confining the fire. Confinement must come before extinguishment; it does not happen the other way around.

It is important to understand how a building influences firespread. Much of this is discussed in the next chapter. Openings in construction, doors left open, and prevailing winds will all have an influence on heat and subsequent fire travel. Heat will, in most cases, travel upward until it meets an obstacle, where it then spreads out laterally. If not properly controlled through ventilation, extinguishment efforts, or both, heat build-up will cause fire involvement both in areas adjacent to and remote from the original fire.

Understanding heat transfer and its application to specific types of building construction will guide the incident commander in placement of lines, positioning of vent openings, and the decision of whether to first attack a fire or protect the most severe exposure.

Major fires may require the incident commander to focus a major portion of his available resources to protection of exposures and cutting off the fire, while allowing the parent body of fire to diminish to a point where the forces can extinguish it.
(Ron Jeffers, NJMFPA)

The Stages of Fire

The three stages of fire previously known throughout the fire service were 1) incipient, 2) free-burning (steady-state), and 3) smoldering. This book, in some places, still refers to these stages for ease of explanation and understanding. These stages, however, have been changed. There are now five stages of fire.

Ignition

The Ignition Phase is where the elements of the fire tetrahedron (fuel, heat, oxygen, and the uninhibited chemical chain reaction) come together. The fire is small and usually confined to the material of its origin. This stage would be akin to the incipient phase of fire. A fire discovered in this stage is easily controlled with an aggressive attack. Heat transfer is usually not a problem, but if left unattended or improperly handled, the fire conditions will continue to accelerate and grow.

Growth

In the Growth phase, a fire plume begins to form above the ignited material. Rising gases begin to spread upward and head toward the ceiling. Once the gases reach the ceiling, they begin to spread laterally and bank downward, which radiates heat to other items in the area. Initially, the walls and ceiling absorb heat from the fire. Eventually however, a point of heat saturation is reached and the walls and ceiling can no longer absorb heat and begin to re-radiate this heat back into the room. This re-radiated heat is called "thermal radioactive feedback", and if not controlled promptly, it becomes largely responsible for the development of flashover conditions. The process is self-accelerating; the more heat in the room, the more the gases are distilled, which in turn lead to the

This fire is in the growth stage. Heat is being absorbed by the walls and radiated back to the center of the room. The block walls will keep the room well insulated, leading to an earlier flashover. A properly applied fire stream with timely ventilation is the answer to this problem.
(Bergen County Fire Academy)

creation of even more heat. The more heat the room holds, the more gas is radiated back toward the center of the room and the quicker flashover conditions build. A fire in this Growth phase must be aggressively attacked with proper and coordinated support to confine it to the area of origin before the transfer of heat creates a more serious firespread problem.

Flashover

The Flashover phase is the transition between growth and the fully developed fire. The thermal radiation feedback from the walls and ceilings cause the destructive distillation or pyrolysis of all combustible materials in the compartment. These gases replace the oxygen in the room and are gradually heated to their ignition point. Somewhere around 1200°F, which is the ignition temperature of carbon monoxide, the gases ignite. This ignition results in a full-flame atmosphere in the room or compartment. Flashover is the actual ignition of the atmosphere in the compartment. Fire will exist in spaces where there are no combustibles present to feed on, thus, it is the ignition of the combustible gases that have been distilled from the combustibles that fill these spaces and create flashover conditions. Once an area has flashed over, victim survival is not possible. Flashover also begins the destruction of the structural integrity of the building.

Flashover signals the absolute end of potential victim survival in the fire area and the beginning of the structural disintegration stage. A flashover will often break out the windows and vent the fire.
(Ron Jeffers, NJMFPA)

Fully developed

The Fully Developed phase is a post-flashover condition where all combustibles in the room are now in flames. The Fully Developed phase is where the fire begins to attack the structural components and the danger of collapse begins. Adjoining spaces begin to be influenced by the fire and are, in turn, heated to their ignition points as the fire spreads away from the area of origin. In the compartment of origin, the maximum heat output is being released in this Fully Developed phase.

Decay

The Decay phase is when the fire consumes the available fuel in the room and the rate of heat-release begins to decline as the fire runs out of fuel in the area of origin. The compartment becomes a mass of rubble as temperatures also begin to decline. Walls and ceilings, if still intact, can continue to hold considerable heat. This is especially true in buildings with masonry walls.

Smoldering fires would be categorized as a Decay phase fire, however because of the air-tightness of the compartment or some other restricting factor, the heat would not have been dissipated. This high-heat, low-oxygen condition in a sealed-up compartment sets the stage for a backdraft explosion.

This fully developed fire is attacking the structural elements of the building. At this point, even a rapid, well-supported, and aggressive interior attack may not save the structure. *(Louis "Gino" Esposito)*

HEAT TRANSFER

There are three methods of heat transfer that, if not controlled, cause fire extension to various areas of a structure. These are:

Several methods of heat transfer can be seen here. The fire is spreading up the combustible wood siding via convection. Note also the ignition of the power lines. This is due to radiant heat, which may also cause severe damage to the apparatus in this photo. *(Photo by Louis "Gino" Esposito)*

Conduction

Conduction is the transfer of heat from one body to another via direct contact by those bodies. This is the simplest form of heat transfer, and coincidentally, the least common method of fire extension in a building. Conduction follows the law of latent heat flow, which states that heat will travel from a hotter object to a cooler object in an attempt to balance the heat. Nature is always trying to keep things in balance, and conduction is an illustration of this. An example of conduction would be when a metal spoon is placed in a hot bowl of soup. Initially, the handle will not be hot, but eventually the handle will become too hot to touch. Heat has transferred from the hot soup, on a molecular (molecule-to-molecule) level, to the spoon.

A hot pipe in contact with a wooden floor will not cause an immediate problem, but over time as the heat from the pipe dries out the wood around the pipe, the wood may ignite. What has happened is that the heat, through conduction, has driven the moisture out of the wood. This has two consequences. The first consequence is that the drier wood will be more conducive to a temperature rise. The second consequence is that this dried-out wood will have a more reduced ignition temperature than a piece of wood that is "green" and has high-moisture content. If this heating continues over a long period of time, ignition may occur. This ignition is called pyrophoric ignition.

Different materials conduct heat more readily than others. This is called thermal conductivity. Generally, gases and liquids are poor conductors of heat and exhibit low thermal conductivity. Solids, however, range from poor to excellent thermal conductivity depending on their characteristics. Metals such as silver, aluminum, and copper are some of the highest heat conductors. Steel is also a good heat conductor. Unprotected steel has been a factor in the collapse of many buildings. Other solids such as concrete, brick, wood, and glass are relatively poor heat conductors. This is why buildings whose exteriors are constructed of concrete and

These unprotected steel beams are excellent conductors of heat. Conducted heat causes the steel to expand. These steel girders pulled out of the walls and dropped the roof.

brick are excellent in terms of fire resistance. Wood will certainly conduct heat, as mentioned above, but it has to dry out somewhat first, which will slow the spread and the temperature build-up. While glass will allow heat to pass through it, it will not readily absorb it.

Transmission of heat cannot be completely stopped by any insulating material, but can be slowed by certain heat-insulating material. A metal cabinet will conduct heat to its contents much more readily than a wooden cabinet. The best protection against the transfer of heat is an air space. If there is an air space between two materials, then conduction can be arrested as air currents carry the heat away. This, of course, may lead to other problems, but as far as conduction goes, this is the best way of minimizing its effect.

Factors determining heat transfer by conduction. The amount of heat transferred by conduction between two substances depends upon four factors.

1. The thickness of the material. Thinner material absorbs heat and reaches the ignition temperature more rapidly. Thicker material takes longer to heat because there is more mass to absorb the heat. The thicker material is more efficient in balancing the heat absorbed.
2. The temperature differential between the point of contact and the point of departure. A colder object conducts more heat away from a hot object because a balance between the two is sought. The hotter the transferee at the time of initial heat conduction, the less time it takes to reach ignition.

3. The thermal conductivity of the material. As mentioned, some materials absorb heat more readily than others, thereby reaching ignition temperature faster.

4. The total amount of time exposed. The longer the exposure, the more heat is absorbed.

The concept behind the transfer of heat by conduction is a simple one. If the rate of heat conducted exceeds the rate of heat dissipated, then a temperature build-up occurs. If the build-up goes on long enough and is of sufficient temperature, ignition eventually takes place.

Conduction and fire operations. Fire travel by conduction is insidious and will sneak up on you if you are not prepared for it. Thus, the best defense against fire extension via conduction is building familiarization before an incident and recognition via reconnaissance during an incident. The incident commander must consider every way in which a fire can travel. Fire travel by conduction can be from one building to another via steel beams that serve both buildings. This may be the cause for a fire spreading from one basement to another or across drop ceilings that have support members serving more than one building occupancy. Pipes may also serve several basements. One building may have its basement partitioned into sections by concrete blocks, but pipes may pass through the sections. A fire exposing these pipes may conduct heat to the wooden beams that the pipe passes through in adjacent areas. Be certain that in any fire that involves a basement or drop ceiling that adjoining areas are checked for fire extension.

I was involved in an investigation at a fire that occurred in the kitchen of a restaurant on the ground floor of a 10-story, fire-resistive high-rise office building. The first alarm companies responded on alarms activated and an odor of smoke in the building. The alarm panel indicated the alarm was activated on the 2nd floor. Companies investigating the area found smoke on the 2nd floor coming out of the wall. It appeared, at first, that the fire was electrical in nature, but once the walls were opened, no fire or heat source was found, however there was still a good deal of smoke. The company officer stated to me later that there was an odor of burning wood. This seemed a little strange because this was a building that had no wood in its construction and the smoke was emanating from the areas where there should be no wood present.

The companies were then alerted to smoke in the closed bank on the ground floor. The rear of the bank was adjacent to the restaurant, separated by a small service hall. There was a slight haze of smoke in the bank toward the ceiling area. After the bank door was forced and the ceiling opened and examined, there was still no indication of the smoke's origin.

The next place checked was the restaurant because patrons complained that there was a haze of smoke in the area. The manager led the companies to the kitchen of the restaurant. There was a haze of smoke there as well, emanating from the area of the large ovens. The ovens were shut off and checked further. Since nothing in the oven was burning, the determination was made to check the walls behind the ovens. Because these ovens were extremely large and tough to move, the companies decided to breach the wall at the rear. This was a back wall that served the area where the ovens were located in the restaurant that was most easily breached from a small service hallway between the bank and the restaurant. The wall was breached and the fire was found and quickly extinguished. Upon inspection, it was discovered that the ovens were secured to the studs in the wall behind them by large steel bolts and these bolts passed through plywood that apparently was placed there to aid in the anchoring process. Over time, the heat from the oven conducted through the bolt and to the plywood. After time, this conducted heat drove the moisture out of the plywood and caused it to ignite.

The temperature in these ovens averaged about 500-600°F for about 14 hours per day. The maximum temperature that wood should ever be exposed to on a constant basis is 300°F. Plywood will ignite normally at about 650°F, but due to the drying-out process caused by the constant exposure to the heat from the ovens, the ignition temperature of the wood was substantially reduced. Extreme temperature conducted over time caused the ignition of this fire.

Convection

Convection is the transfer of heat through the circulation of heated matter. There must be a circulating medium involved that carries the heat. This medium can be a liquid, but is usually air. Convection heat is a result of a

Autoexposure is caused by convection. Convection currents flowing out the window of this fire apartment caused the fire to spread to the floor above. Open windows in the summer will be especially conducive to this fire spread, especially those located in shafts. Personnel must be assigned to recon the rear as part of the ladder company's initial scene assignments. *(Mike Borelli, FDJC)* fire's influence on the air. Heated air expands, and by doing so, it becomes lighter. The hotter the air becomes, the faster it rises and the more it will expand. Convection causes fire to spread upward through shafts, stairwells, and open stud channels. Products of combustion naturally rise and continue to do so until they either meet an obstacle, such as a ceiling or a door, or cool to the point that they are equal to the ambient air in the area, at which point the products of combustion stay in that area or stratify. This stratification occurs mostly in high-rise buildings.

Convection heat carries with it the products of combustion, such as smoke and fire gases. It is this convection heat that causes havoc in a building by igniting fires away from the area of origin and killing occupants on upper floors. More people die from the toxic products of combustion than do by fire. This is a testament to the fact that heat and the products of combustion carried to other areas of the building by convection currents make this the most common and deadly manner of firespread in a building. The heated smoke and gases serve as the circulating medium for the transfer of heat by convection.

Factors determining heat transfer by convection. Convection heat is the most common cause of heat transfer and subsequent firespread in a building. As the products of combustion rise, their heat is passed to surrounding objects. This heat, if unvented, can raise the temperature of these combustibles to their ignition point. A fire burning on a lower floor that has access to an open stairwell will cause the products of combustion to rise. The upward rise of the convection heat is often retarded by the presence of a ceiling on the top floor. As the upward flow of this heat is blocked, these gases then accumulate

and spread laterally at the top floor. This is called mushrooming and causes the temperatures on the top floor to build to levels that cause the ignition of combustibles on the top floor. It is for this reason, convection, that the vertical arteries at the roof level are opened immediately at a fire.

Generally, the lower the fire is in the building, the more the building is exposed to convection currents. Thus,

Although this fire may never get out of the cellar, the smoke and heated gases can rise to choke the rest of the building causing death on floors not directly affected by flames. Vent this building and things will get better. *(Bob Scollan, NJMFPA)*

the majority of the firespread problem inside buildings can be directly attributed to convection.

The fire mentioned in the next chapter at the Westview Towers in North Bergen, New Jersey, is a prime example of the deadly consequences of uncontrolled convection currents. The doors to the hallway, which were left open, allowed the heated gases and smoke to rise up the now unenclosed stairwell. This caused the death of two residents in the stairwell several floors above the fire. In addition, melted doors in the hallway on the 12th floor, some eight floors above the fire, were further evidence of the severe convection heat generated by the fire on the 4th floor that, incidentally, never really extended out of the apartment of origin.

Direct flame contact is actually a type of convection. Ignition by direct flame contact is, in fact, the result of the heat using the medium of the gases produced by the flame to heat the surface of the exposed object to its ignition temperature. Since there is no space between the object that is burning and the object that is directly exposed, the great majority of the heat energy produced is transferred directly to the exposed object, with little or no heat lost. Direct flame contact is the most efficient form of convection.

Convection and fire operations. The control of convection currents is absolutely critical if the incident commander intends to confine the fire to the area of origin. If not properly vented, dependent on the fire's location and extent, these convection currents can prevent the advance of the hose line on the fire floor. Proper ventilation pulls these convection currents up and out of the building, while at the same time localizing the fire.

Where to direct these convection currents depends predominantly on the location and extent of the fire. Generally, a fire on a lower floor of a residential dwelling with a peaked roof will be vented horizontally on the fire floor, although in the case of balloon-frame construction, it may be necessary to vent the roof. In a multiple dwelling with a flat roof where the fire is not on the top floor, horizontal ventilation of the fire floor and opening and examination of the natural roof openings such as the bulkhead door or scuttles will suffice. Regardless of the type of building, any fire on the top floor or in the cockloft will demand a hole be cut in the roof as directly over the fire as is safely possible.

I once responded to a fire on the second alarm where I was told to check on the roof operation. There was fire on the 2nd floor of a 5-story multiple dwelling. I figured the natural openings would already be opened and my time on the roof would be short. When I got to the roof, I saw a 4' x 4' opening in the roof, which had just been cut. The roof firefighters were in the process of pushing down the top floor ceiling with a pike pole. I inquired as to what they were doing and they replied that their duty was to "get the roof". I commented on the nice job they did of cutting the roof; it was actually a textbook hole, however, I had to inform them that the fire was three floors below the roof and the hole was unnecessary. I then led them down the stairs to the top floor where we entered an apartment that was completely undamaged by fire or smoke. However, there was a large hole in the ceiling. This event made it apparent that some training was in order for this over-aggressive ladder crew because a great deal of totally unnecessary damage was done. If the fire is not on the top floor or in the cockloft, cutting the roof has absolutely no effect on the fire and is not only a waste of time and energy, but it creates grossly unjustified secondary damage. Natural openings, such as the bulkhead door, and the checking of any soil pipes should suffice.

Radiant heat was the reason for this melted vinyl siding. If the distance factor cannot be overcome, only water placed directly on the exposure will protect it.

Radiation

Radiation is the generation of heat energy in the form of electromagnetic waves emitted from the surface of a heated body. Radiation waves travel at the speed of light and receive no resistance from the air. The heat from the sun is an example of radiant heat waves.

Radiant heat waves travel in straight lines in all directions and are not affected by the wind. However, areas downwind (leeward) will still be of critical concern as these exposures will be threatened by both radiant heat and convection heat. Those exposures on the upwind (windward) side will only be affected by radiant heat waves. This does not mean that these exposures on the windward side should be ignored. Radiant heat is insidious in that it may pre-heat an exposure to the point of ignition before the danger is realized. There is usually very little smoldering or incipient (ignition) activity of the exposed material. Items ignited via radiant heat waves usually burst into a free-burning (growth) stage fire, with all of their exposed surface igniting at once.

Heat generated as radiant heat reacts with exposed material in one of three ways.

1. Radiation is reflected by an exposed material
2. Radiation passes through an exposed material
3. Radiation is absorbed by an exposed material

Reflected or passed-through radiation does not cause a heat build-up of the exposed surface. Radiant heat passes through gaseous substances and also through some liquids, especially if they are transparent. The more opaque or non-transparent the substance is, the more heat will be absorbed. Radiant heat that is absorbed creates a heat build-up. This heat build-up increases the temperature of the exposed material, thus creating the danger of ignition.

Factors determining heat transfer by radiation. The degree of heat build-up will be determined by several variables:

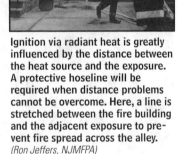

Ignition via radiant heat is greatly influenced by the distance between the heat source and the exposure. A protective hoseline will be required when distance problems cannot be overcome. Here, a line is stretched between the fire building and the adjacent exposure to prevent fire spread across the alley. *(Ron Jeffers, NJMFPA)*

1. *The temperature of the exposing fire.* Certain materials give off more British thermal units (Btu) during the burning process than others. Wood and comparable products give off about 8,000 Btu's per pound, while plastics give off as much as 16,000 Btu's per pound. Flammable liquids and gases give off even more, producing between 16,000 and 24,000 Btu's per pound. This heat-producing quality of the exposing fire greatly influences the exposure. Prior knowledge of the fireload can be of great value when estimating the Btu production and attendant suppression needs of a fire.
2. *The geometry of the exposed surface.* A vertical surface absorbs heat more readily than a horizontal surface because there is more surface exposed to the parent body of fire to heat up. The smoothness or roughness of the surface also makes a difference. A rough surface, such as splintered wood, ignites more readily than a smoothly-polished surface. This is because the wood splinters have a larger surface-to-mass ratio and reach ignition temperature quicker.
3. *The color of the exposed surface.* The darker the material, the more heat is absorbed. This is why on a hot summer day, it is more

Heat from this fire melted the vinyl siding, exposing and igniting the combustible asphalt ("gasoline") siding. Note the path of flame spread across the siding at the front of the building and the heavy fire damage on Side D. The wind here was blowing from left to right as evidenced by the lack of damage to exposure B.

Closely spaced wood frame dwellings are extremely susceptible to fire extension by way of radiant heat. Hose streams placed to keep the walls between the buildings wet may not be enough to overcome the BTU generation from the parent body of fire. Master streams may be required to overcome the radiant heat problem.
(Tony Casteluccio, Newark FD)

This strip mall fire is considered a line source. The flame front is limited in size and height. If left unchecked, it could spread to become an area source.
(Bill Tompkins)

comfortable to stand on the white concrete sidewalk than on the black tar street.

4. *The relative combustibility of the exposed material.* Some materials are highly combustible, and therefore, reach their ignition temperature more rapidly than others. An example of this is asphalt siding, also known as "gasoline siding" because of its highly combustible nature as compared to brick, which is virtually non-combustible.

5. *The thickness of the material.* Thinner material absorbs radiant heat at the same rate as thicker material, however, the thinner material has less mass to distribute the heat build-up and, therefore, experiences a quicker overall temperature-rise and a more rapid ignition.

6. *The moisture content of the exposed material.* For ignition of wood and related combustibles to occur, the moisture must be driven off to a great degree. Wood and similar materials are not easily ignited when the moisture content is greater than 15%. The lower the moisture content, the easier the ignition. Moisture in wood and other combustibles acts as a heat absorbent, reducing the likelihood of ignition.

7. *The distance of the exposed material.* Distance is the most important factor in regard to exposure protection. The amount of radiant heat diminishes with distance. The rate of diminishment depends on the size of the fire and the distance between the exposure and the heat source.

The heat sources that influence exposures are:

a. *The point source.* A small fire can be considered a point source. The radiated heat from this fire diminishes quickly over distance and presents little or no hazard to exposures.

b. *A line source.* A line source is characterized by a flame front in the form of a long line with little height. This type of source emits nearly twice the radiant heat as the point source. An example of a line source is an advancing brush fire or a fire sweeping through a strip mall.

c. *An area source.* A large fire such as a conflagration can be considered an area source. A conflagration is defined by the NFPA Fire Protection Handbook as "a fire with major building-to-building flame spread over some distance." Radiant heat issuing from an area source diminishes very little with distance because the distances between exposures are small compared with flame height and width.

Certain conditions also influence the ignition of an exposure. A high rate of radiant heat transmitted over short period of time is more of a threat to exposures than the same amount of heat generated over a long period of time. Theoretically, the worst case in regard to radiant heat exposure would be a fire exposing a relatively close, dried-out, thin, vertical, dark-colored, rough surface of highly combustible material.

Radiation and fire operations. While distance is the greatest ally the fire department has against the threat of exposure ignition, we sometimes have no control over the distance factor. We cannot move buildings but sometimes we can move exposures, such as exposed vehicles, rail cars, ships, and tanks and barrels of flammable liquids or gases. When the distance factor does not work to our advantage, the application of water to the exposed surface keeps the surface cool and prevents the build-up of heat to the ignition point. Water curtains between exposures are not effective because the radiant heat waves pass right through the transparent water stream. The best protection is by coating the exposed wall or other threatened members with a film of water. This works well with combustible walls.

This lightweight wood truss condominium complex under construction created a veritable firestorm that required hundreds of firefighters to control. It spread to nearby homes and threatened an entire city located on top of the cliffs above it. *(Ron Jeffers, NJMFPA)*

With non-combustible walls such as brick, block, or metal, the exposure threat is dependent upon the amount, size, and area of the exposed windows. This tactic of coating exposed objects with a film of water works with windows as well because it slows the transmission of heat waves to the interior. Firefighters can also help their own cause in this case by closing windows on the exposed side and removing any combustibles such as curtains, drapes, etc. To aid in dissipating any heat build-up, windows opposite the exposed windows can be opened to release any heat already present in the room or apartment. Opening the bulkhead door or a scuttle may also be of benefit in this situation.

Case Study

2-4 Potter place fire

The building. 2-4 Potter Place was a 5-story, multiple dwelling of ordinary construction. At the time of the

Water applied to a combustible surface will prevent heat buildup along that surface. Place the stream on the exposure to be most effective. If possible, alternate the stream between the fire and the exposure. *(Ron Jeffers, NJMFPA)*

fire, the building was vacant and undergoing massive renovations. The interior was entirely gutted down to the studs on every floor, leaving the floors wide open. In fact, the only totally intact building feature was the stairway. It was due to the renovation, open and unenclosed from the ground floor to the roof.

The exposures. Exposure A was a street. Across the street, about 25' away was a sister structure, 3-5 Potter Place of the same height and construction. 3-5 Potter Place was, however, fully occupied at the time of the fire. Exposure B was a 1-story, masonry garage. Exposure B1 was a 3-story dwelling of ordinary construction. Exposure B1 was on the corner of Potter and Clifton Terrace. Exposure C was a rear yard. Behind the rear yard were the rear of the buildings on Clifton Terrace and Liberty Place, most notably #29 and #27 Clifton

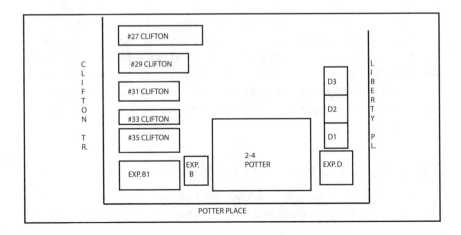

Extensive building renovations permitted an unimpeded avenue for fire spread from the first floor to the roof. It took only minutes for this building to become fully involved. Convection heat was the firespread culprit here. *(Ron Jeffers, NJMFPA)*

Terrace. These two buildings were significant because they were deeper than the otherwise closer buildings on Clifton Terrace and Liberty Place. Exposure D was a 2-story, residential dwelling of ordinary construction. Exposure D was on the corner of Potter and Liberty Place. Exposures D1, D2, and D3 were attached 3-story, residential dwellings of ordinary construction located on Liberty Place.

The fire. This fire occurred on a warm afternoon in April. There was no wind to speak of and, upon arrival, a report of nothing showing was given. A further investigation revealed a moderately-sized fire in progress on the 1st floor, rear. By the time the initial lines were stretched, the building was well on its way to becoming fully-involved because there was no fire-stopping present due to the area left open by the renovation process. Thus, a defensive strategy was employed.

The collapse and subsequent exposure fires

A collapse of Sides 2, 3, and 4 occurred simultaneously after about half an hour. The building collapsed in a curtain-fall fashion, dropping straight down in place. Part of the wall on Side D went astray and crashed into the exposures on Liberty Place, Exposures D1, D2, and D3. This ignited a fire on the top floors and roofs of these buildings.

As the front wall remained intact, the brunt of the heat release with the collapse occurred to the rear of the building, thus igniting the rear top floor of #27 Clifton Terrace, some 150' away. The subsequent fire in #27 Clifton Terrace

Heavy fire erupts from 27 Clifton. Even though there were several structures closer to the original fire building, the open window at the rear and the orientation of the building in relation to the collapse-produced heat release caused ignition. This ignition was a result of radiated heat. *(Ron Jeffers, NJMFPA)*

required what amounted to a second alarm response to extinguish it. The top floor and the attic were lost. All of the other exposures on the Side B on Potter Place and Clifton Terrace were left untouched by fire.

There were several factors that led to the ignition of #27 Clifton Terrace. When 2-4 Potter Place collapsed, the air in the building, as it fell, was compressed and forced out toward the exposures. Along with this compressed air was a tremendous blast of heat. Recall that a high rate of radiant heat transmitted over short period of time is more of a threat to exposures than the same amount of heat generated over a long period of time, thus, this was the time of maximum heat radiation as the walls were no longer confining the fire. The radiant heat could not blow out through the front as the wall was still intact, so it reaped its greatest influence on the rear, the path of least resistance. As the fire expanded from the compressed state, it radiated out and rose as heated air will do. The rear, top floor of #27 Clifton Terrace was slightly deeper than any other exposure in the area. In addition, an open window with a curtain flapping out of it was located on the 3rd floor, rear. This curtain ignited and spread fire into the window and the structure. The vehicle for this heat travel was radiant heat released from the parent body of fire.

Flashover

Regarding firefighter safety and survival, the best prevention against flashover is the recognition of the signs of an impending flashover. Too many firefighters are killed in flashovers every year, especially since every one of

Note the volume and color of the smoke both on the porch and at the window on the right. These conditions should be a cue to firefighters of the possibility of an impending flashover.
(Lt. Joe Berchtold, Teaneck, New Jersey Fire)

Flashover simulators allow students to learn the signs indicative of an impending flashover. Normally, the rear doors are closed and the students sit on the floor in the lower portion of the simulator. The actual flashover takes place in the upper portion, with the ignited gases rolling over the student's heads.
(Bergen County Fire Academy)

these deaths is preventable. The root problem is training and recognition. If you ask the average firefighter what are the signs of flashover, you will get varied and widespread answers. That's because many firefighters, who actually witness a flashover first-hand, never get to tell about it or are too busy getting out of the building to notice.

Flashover has been traditionally defined as the simultaneous ignition of all combustibles in an area due to heat build-up. In essence, the definition says that all combustibles in an area are raised to their ignition temperature and burst into flames at the same time. This definition is not 100% accurate.

Flashover is actually the ignition of the superheated atmosphere that consists of flammable gases, which have been distilled from all the combustibles in the area. Heat from the parent body of the fire raises the temperature of the other combustibles in the area, including the floors, walls, and ceilings if they are combustible. These combustibles, in turn, begin to emit flammable gases, mostly carbon monoxide, through the process of destructive distillation. The heating of other combustibles becomes self-accelerating with the heated combustibles, floors, walls, and ceilings, all radiating heat back to one another. As the temperature in the room rises to the ignition temperature of carbon monoxide, about 1200°F, these gases ignite. This explains an atmosphere of full-flame involvement in areas where no combustible material exists. For anyone who has trained in flashover simulators, or witnessed a flashover travel down a hallway, this is a more apt definition.

Indications of flashover are observable on both the outside of the structure and on the inside of the structure. Sometimes, what is seen on the exterior is not evident on the interior and vice-versa. For this reason, all members on the fireground, both inside the building and outside, have a responsibility to continuously size-up the fire conditions for flashover potential, as well as for other safety hazards such as backdraft and collapse.

Flashover variables

Room size. The build-up of heat in a small room is much quicker than in a large room. This is because there is less area to heat, and thus, decreases the time for occupant survival and development of flashover conditions.

Size or number of openings within a room or space. The more compartmentalized or sealed up the room is, the more prone to early flashover. This is because the accumulated heat has less of a chance to escape and causes a more rapid heat build-up. The more opportunities for heat to escape, the longer it takes for flashover conditions to develop.

The rate and amount of heat release (fireload). This has a direct bearing on the amount of time it takes for the room or area to flashover. The heavier the fire load, the quicker the heat accumulates. A room heavily loaded with plastics releases at least twice the amount of heat (Btu) than ordinary combustibles. In addition, quantity also affects the conditions. If there is more fuel, there will be more heat, it's as simple as that.

The insulation qualities of the structure itself. Attempts to conserve energy leads to a virtual sealing-up of buildings. Thermopane windows, more efficient insulation, and the presence of drop ceilings not only reduces the size of the compartment but also disallows the heated products to escape, thus causing a rapid build-up of temperatures. Cocklofts in older buildings are also prone to flashover and backdraft because they are relatively sealed-up until either a firefighter or the fire releases the flammable products of combustion from them. Released improperly, they

Large homes that have 2-story atriums and cathedral ceilings can result in the same misleading flashover size-up as can commercial occupancies with high ceilings. Raise the flag of extreme caution when a black smoke layer is above your head, but the heat level is still tolerable.

kill firefighters and destroy property. In addition, renovations that add drop ceilings and extra partitions allow these new concealed spaces to develop flashover and backdraft conditions. Extreme caution should be used when-ever opening up these spaces.

Ceiling height. Ceiling heights have a direct bearing on room size. Rooms that have very high ceilings result in misleading flashover size-up information. The smoke layer conceals the high ceiling, causing a tenable situation at the floor level, all the while a life-threatening situation develops above the heads of the firefighters. Large homes that have 2-story atriums and foyers can have tenable situations on the ground floor, while a flashover occurs on the 2nd floor where firefighters are searching bedrooms. In the entrance hallway, the layer of smoke may be far above the heads of the firefighters, but due to the open construction, this layer of heat and smoke may be ignitable on the 2nd floor. Many times this traps firefighters. A second means of egress, such as ground ladders raised to 2nd floor windows, may be the only way out of a room when the upstairs hallway flashes over.

The heavier the smoke, the more fuel involved. The more fuel involved, the more heat production. Heat production leads to flashover. Take notice of smoke color, pressure, and volume as exterior cues of an impending flashover.
(Newark, New Jersey FD)

Gases above their ignition temperature burst into flames, causing full involvement of the top two floors of the building. Recognition of the signs of flashover, both on the interior and exterior, is the best method of preventing rapid fire involvement casualties.
(Newark, New Jersey FD)

These variables may also have a synergistic influence on the fire, exponentially decreasing the time to flashover.

Signs of impending flashover – exterior

Smoke color. Fires in ordinary combustibles produce smoke of a grayish to brownish color. If plastics and other petroleum-based products are involved, which is almost always the case, the

Due to the lack of fire at the window on the left, this ignition at the doorway may be a vent point ignition. This ignition is a definite indication that the superheated gases in the building are at or above their ignition temperature; they just lack oxygen.
(Lt. Joe Berchtold, Teaneck Fire)

smoke appears darker and closer to black. In fact, the closer to flashover the conditions become, the darker the smoke will be.

Smoke movement. Smoke carried on convection currents naturally rises away from the area of origin. The hotter the smoke becomes, the faster the movement. Smoke that is pushing and billowing rapidly out of an opening is a sign that flashover conditions are approaching.

Smoke volume. Larger volumes of smoke almost always mean more fire involvement. More fire involvement means higher temperatures. Heated smoke expands as well as rises. When the windows and/or doors are showing large volumes of pressurized, heavy, black smoke over much of their entire opening, flashover may be imminent.

Vent-point ignition. There are times when the smoke traveling down a hallway or in a room is above its ignition temperature, but because there is not enough oxygen on the interior, no ignition takes place. You can be sure, however, that when the superheated smoke finds an opening, such as a door or a window, the oxygen-to-gas ratio will balance properly and ignition will occur. This ignition around doors and windows, usually at or near the top, is called a vent-point ignition, and is one of the last exterior signs that flashover is about to occur on the inside, as the smoke and gases are already above their ignition temperature.

Any or all these signs may be present from the exterior, or they may not. They must be watched for because companies on the interior may not notice them until it is too late.

Lean flashover can be seen at the ceiling level in this room. Convected heat is distilling gases from the ceiling, which in turn ignite briefly. This ignition will be more sustained as the fire grows, heating other combustibles, releasing still more gases.

Signs of impending flashover – interior

Lean flashover. As the fire gases rise away from the original fire, they spread across the ceiling and heat the upper portions of the room. As the ceiling area heats up, flammable gases are distilled from the ceiling material. As these gases become distilled, small flashes of flame away from the parent body of fire are produced. This is especially noticeable in corners because this is where pockets of heat accumulate and are trapped. This is a noticeable sign that ceiling temperatures are beginning to get very hot. Water application at this point should solve the problem and destroy the chances of a flashover occurring. Unfortunately however, this phenomenon of lean flashover, is many times obscured by smoke.

Blinding, thick, dark smoke. This occurs because of large amounts of fire gases being produced and an extraordinary amount of incomplete combustion. These gases, if confined, bank down to the floor. While not always a reliable factor, the presence of large volumes of smoke must be a reason for caution. As the ceiling temperatures get hotter and hotter, these gases and smoke reach their ignition temperature. The hotter the smoke is, the closer to flashover the conditions are becoming.

Heat build-up. Hot smoke that forces a firefighter down to the floor is a definite signal that flashover may be imminent. This heat has banked down from the ceiling because of thermal radiation feedback and is now close to the floor. In fact, the lower you are forced to the floor, the greater the chance of flashover.

Rollover. This is one of the last signs before a flashover. Flashes of flame produced by sporadic ignition of the superheated gases ignite at the ceiling level. These are sometimes called "snakes" or "dancing angels" and are mixed with the smoke. Before entering a room that is heavily charged with smoke, check the smoke emanating from the door, especially at the top, for signs of rollover. Unfortunately, rollover, like a lean flashover, is often masked by smoke.

Test the atmosphere. Firefighters searching a room should periodically look up for signs of rollover. This is often difficult while wearing an SCBA, but it can be done. Another technique is to reach up into the smoke to feel for heat build-up. You do not need to take your glove off because if the heat is that intense, it will be felt through the glove. Depending on the conditions, the firefighter will need to make a decision on whether to continue the search or back out.

Firefighters advancing a hoseline can also test the atmosphere. While it is usually taboo to discharge water on smoke, a small amount of water can be used to test the atmosphere for flashover-type heat. This is accomplished by shooting a quick dash of water toward the ceiling. If the water comes back down, the temperatures are not hot enough to produce a flashover and the advance can continue. If, when the water is discharged at the ceiling, a sizzle is heard, then the temperature is approaching the boiling temperature of water or even hotter, and more caution should be used. If, however, when the water is directed at the ceiling and there is no sound and nothing comes back down, beware. The ceiling temperatures are already so hot that the water was instantly vaporized, which means that the ceiling temperatures are extreme. A decision will then have to be made about whether to continue the advance or to back out. Remember that water is your greatest weapon against an impending flashover, but if you cannot find the seat of the fire (the source of the greatest heat) it may be wise to back out of the area.

Flashover prevention

It has been said that water kills flashover. This is absolutely true. Water is the only effective weapon that prevents a flashover if it is applied properly and in a timely fashion. The overhead must be cooled to a point where the temperatures of the superheated gases are cooled below their ignition point. This cooling does not take a lot of water because the hotter the gases are, the more they affect a stream of water. The method is called "penciling". It consists of applying short bursts of water toward the ceiling to cool the overhead while the line advances. Water is at its most efficient when it is turned to steam. Using this principle, the superheated gases overhead can be used to convert water to steam, thus cooling the atmosphere well away from the nozzle. It is critical that this water be applied in short bursts on a tight fog, at about a 30° angle. The water must be applied from outside the compartment, such as a doorway or the entrance to a hall. It is absolutely critical that ventilation be accomplished on the side opposite the attack team, otherwise the developed steam may envelop the attack team and burn them. Using this method with proper and timely ventilation, it is possible to "pencil" the ceiling as the line is advanced down the hall toward the fire apartment or room.

Tactics to delay flashover may be executed by the ladder company on the roof to draw a flashover condition away from searching firefighters, who do not, as of yet, have hoselines available. This ventilation must be done vertically. This draws out the hottest gases and delays the development of flashover. Note that the ventilation is vertical. This is because it is sometimes impossible to predict the effect horizontal ventilation has on a fire, especially before a hoseline is in place. Once a line is flowing on the fire, the chances of flashover are drastically reduced and horizontal ventilation can be accomplished with more confidence.

I once saw a video of firefighters on a porch of a building that was showing signs of flashover. There was heavy, black smoke violently pushing out of all the windows on the 1st floor and flowing under the porch roof. The firefighters on the porch were bunched up and obviously waiting for the line to be charged. As the camera shifted to the left, Side B, there was a firefighter with a Halligan hook breaking a side window. Almost instantaneously after he broke the window, the entire 1st floor lit up, practically engulfing the firefighters waiting for water on the porch. Uncontrolled tactics almost always cause something to go wrong, which can have disastrous consequences. Flashover conditions are nothing to fool with if you don't know what you're doing.

Flashover safety precautions

Remember that when the threat of flashover exists, firefighting problems increase. Firefighting operations such as search and line advancement become more dangerous. The first and most critical safety precaution the firefighters should be aware of is to recognize the warning signs of a flashover. This alone saves more firefighters from getting burned and killed than any other safety-related factor. To be aware is to be alive.

Mentioned above is the testing of the atmosphere, either with a gloved-hand or with a hoseline. These furnish reliable indications that the situation is deteriorating.

Listening to recon reports of those in other parts of the building, especially on the exterior, may also cue the interior firefighter in to a potential problem.

A firefighter recognizing the signs of an impending flashover must make the crucial decision of whether to continue the search or find an area of refuge until a hoseline can be brought up to attack the fire. A limited search can be executed, which will quickly assess the area of egress to the room while keeping the firefighter in a relatively safe position. A firefighter can hook his foot around the doorframe and stretch himself and a tool just inside the

Once the fire flashes over, the area becomes fully involved. This is generally the time that the building demolition process begins. Structural members are beginning to be compromised. In a commercial occupancy such as this, heavy streams will be required. (Bob Scollan, NJMFPA)

opening to probe for a victim. In doing this, the hooked foot acts as a lifeline, which allows the firefighter to maintain contact with the exit point. In a room that is about to flashover, this is about the limit of the search to be extended. If no victim is found, the door is closed and the firefighter moves to a more tenable area or exits the building, depending on conditions.

Flashover and occupant/building survival

After a flashover, a full-flame environment exists in the compartment because all the gases will have now ignited and are not only feeding off all the combustibles in the room, but are flowing out of the room, thus heating combustibles and spreading fire and smoke to other areas of the building. These areas will, if no fire department intervention occurs, reach their flashover points and continue to conquer the building. Occupants caught in a flashover cannot survive and will be dead in a very short time, usually less than one minute. Firefighters cannot survive this environment either and should not attempt the search of an area that has already flashed over. It is a totally unacceptable risk.

Once a hoseline is stretched and the fire is being attacked, aggressive search techniques can once again be performed under the protection of the line and with adequate ventilation to clear the way.

Flashover also signals the beginning of the end in regard to building stability. Before flashover, generally only contents are involved. After flashover, the structure, itself, becomes involved and potential collapse increases as the fire reaches the Fully Developed Phase. Potential collapse is even greater when the fire reaches the Decay Phase because the structure will, most certainly, have been compromised. All personnel on the fire scene must perform continuous evaluations of the structure.

Conclusion

As stated throughout this chapter, the incident commander, as well as all other personnel on the fireground, should have a thorough understanding of heat transfer and how it spreads both fire and the deadly products of combustion throughout the building. Hoselines must be stretched and rescue operations must be initiated in anticipation of this spread. Suppression efforts and support operations, such as ventilation, should focus on minimizing the effects of heat transfer.

Questions for Discussion

1. Discuss heat conditions present in each of the five stages of fire.
2. Discuss the variables involved in the transfer of heat by conduction.
3. What are some of the methods of preventing the build-up of convection heat?
4. Discuss the variables involved in the transfer of heat by radiation.
5. What is the difference between the three types of heat sources that may cause an exposure fire?
6. What causes a flashover?
7. What are the variables that contribute to the creation of flashover conditions?
8. What are some of the signs, both interior and exterior, that a flashover may be imminent?
9. What are some of the ways that firefighters can test the atmosphere in a room for heat build-up?
10. What safety precautions should a firefighter take when attempting to enter and search a room displaying flashover conditions?

CHAPTER THREE
Building Construction & Firespread

This chapter addresses the problems inherent in the different building construction classifications. In the last chapter, heat transfer and fire behavior were addressed. Understanding fire behavior and heat transfer is similar to knowing the rules of the game. Understanding building construction is akin to being familiar with the field the game is played on.

One of the most critical responsibilities of all officers is to be proficient in the study of building construction because it is this knowledge of the field where the game is played that will allow them to best protect their crews. Without this knowledge, major mistakes can be made, which can have tragic results.

Building construction will greatly influence fire spread. Although this fire was through the roof on arrival, the presence of fire walls in the fire building and a non-combustible exterior wall on the rear exposure allowed companies to confine the fire to the area of origin. *(Ron Jeffers, NJMFPA)*

I stated earlier that it is more important to know the building that is on fire than it is to know the fire that is in the building. This is because building features influence firespread. The ability to forecast this spread will, to a great extent, be determined by the incident commander's grasp of building construction. In fact, the incident commander's knowledge of the strengths and weaknesses of each classification of building construction, coupled with the application of the Point of Entry Rule of Thumb (safest, most effective path of least resistance), will lead to the best decisions regarding line placement, access, ventilation, and extension prevention, to name just a few.

The five building classifications according to the NFPA are:

Class 1: Fire-resistive construction
Class 2: Non-combustible/Limited-combustible construction
Class 3: Ordinary construction
Class 4: Heavy-timber construction
Class 5: Wood-frame construction

Class 1
Fire-Resistive construction

Fire-resistive construction consists of walls, columns, beams, floors, and roofs made of non-combustible materials or limited-combustible materials. The building's structural elements do not add to the fire load of the structure. Build-

ings of fire-resistive construction are designed to allow the contents within the structure to burn without causing a massive failure of the building's structural components. The principal intention of fire-resistive construction is to inhibit the spread of fire outside the compartment of origin and to expedite occupant escape.

Fire resistance is applied to building elements, such as steel, to increase its resistance

Two types of fire-resistance are used on this column. The lower portion is encased in concrete while the upper portion has been sprayed with fireproofing material. The horizontal structural members in the ceiling have also been sprayed.

to the ravages of heat, which causes it to break down and lose strength. Encasement of steel structural elements in fire-rated sheetrock or concrete gives it a fire-resistive rating. In addition, spraying the steel with a fire-resistant material also attains a fire-resistive rating.

The greatest ally to the fire department in buildings constructed of fire-resistive materials is that the fire is usually compartmentalized, that is, each unit of construction, especially in residential high-rises, is essentially self-contained. Fires in these buildings usually do not spread beyond the apartment of origin. In commercial buildings of fire-resistive construction, floor spaces are usually more open in nature. As a result, firespread is a greater problem. However, due to the fact that a wet, automatic sprinkler system is usually mandatory, the fire is generally contained to the area of origin by virtue of sprinkler activation.

Firespread problems of fire-resistive construction

Extensive firespread beyond the area of origin is rare, however, there are several building features that compromise the complete compartmentalization inherent in this type of construction. These are listed and explained below.

HVAC system. The first and most important firespread element in fire resistive buildings is the presence of a central heating, ventilation, and air-conditioning system (HVAC). This system may serve the whole building or just a segment, such as a bank of floors. If left operating at a fire, this system may allow the products of combustion to spread beyond the area of origin, causing toxic smoke and gases to permeate uninvolved areas. One of the first actions for the incident commander to take is to order these systems shut down as soon as possible. Additionally, the incident commander, or his designee, should consult with the building's maintenance personnel as to how this system can be utilized to assist in the confinement and/or ventilation of the fire.

False floors and drop ceilings. Another feature found in buildings of fire-resistive construction are concealed spaces either above or below the floor. Those above the floor can be found in almost any occupancy but are usually found in commercial occupancies. Suspended drop ceilings usually house the building's HVAC system components that serve that area. There are usually no fire-suppression systems in this area because the sprinklers and smoke detection systems run throughout the ceiling are designed to protect the contents below.

Concealed spaces below the floor, created by the installation of false floor slabs, are usually found in areas where computers are located and are filled with a maze of wiring that links the computer systems. These areas are

This HVAC system will be hidden above a drop ceiling and penetrate every space in this building, weakening the intended compartmentalization. Sprinkler systems, intended to protect the life hazard, are usually not designed to protect this ignition-source rich environment.

usually protected by a complete, automatic-extinguishing system, utilizing a clean agent such as CO_2 or halon. An agent such as dry chemical will create a huge mess that tends to be difficult and expensive to clean up and may cause damage to sensitive computer equipment.

Elevator shafts

A feature common to high-rises is the presence of elevator shafts. These shafts, if not properly constructed, can spread fire, heat, and smoke to other floors. Cars moving in the shaft can also cause smoke to travel with the car to areas below the fire. Furthermore, if not utilized properly during a fire, they can cause death and injury to those inside the car.

Compactor and incinerator shafts

Chutes serving compactors and incinerators will rise through the entire building and penetrate every floor. Fires originating in or near building compactors and/or incinerator shafts can cause fire and smoke to spread to upper floors of a building. Even a relatively minor fire somewhere in the shaft can cause hot products of combustion to mushroom at upper floors and ignite nearby combustibles. Fire may also spread into adjacent areas via conduction if adequate clearance from combustibles around the chute was not provided or if the integrity of the chute has been compromised due to improper construction or lack of maintenance.

At a fire in a 34-story residential high-rise, made of fire-resistive construction, a small fire in the compactor shaft on the 10th floor caused a fire to ignite combustibles in the top floor penthouse. Reconnaissance of all floors, especially those at the top of the shaft, was necessary. This required a considerable amount of manpower. Wet sprinkler systems and automatic smoke dampers that trigger instantaneous shutdown of the equipment should be installed in these systems.

Access stairs

Access stairs are open stairways, sometimes of the winding type, that connect two floors rented out to a single occupant. These open passageways between floors should be protected by an automatic sprinkler system because there is nothing to stop a fire from extending from one floor to the next. Firefighters searching the floor above the fire can find themselves trapped by a rapidly extending fire via the access stair opening. The discovery of an access stair should be immediately communicated to the command post. Preplanning and building familiarization are the only ways to ascertain this information before a fire.

Occupant indifference

Probably the greatest factor in regard to firespread and the subsequent havoc reeked, on both the building and its occupants, is the indifference displayed by these same occupants in regard to fire safety. This is most often caused by laziness first and indifference second.

Occupants often alter and, subsequently, negate fire containment features of the building. The self-closing device on some apartment doors will often be removed by the tenant because the automatically- or self-closing door will often inconvenience the resident, locking them out or being a hassle when hands are full of groceries, etc. If a fire occurs in the apartment, any doors left open quickly turn the hallway into an oven, filling it with extreme heat and dense smoke. This not only

Blocking open hallway doors turns potential evacuation stairwells into chimneys. Convection-driven products of combustion will render these stairwells unusable.

hinders our efforts to locate the fire apartment, but makes the stretch down the hall a punishing affair, especially if the wind is blowing into the apartment from the exterior.

Additionally, hallway doors are often blocked open to allow a breeze to get to the floor from the stairwell. This also allows fire and other products of combustion to vent into areas where people may be trying to escape. Occupants have been overcome and killed in stairways because residents left the doors on the fire floor open.

It is imperative that occupants, in all buildings, be taught safety procedures in the event of a fire.

Case Study

Granton Avenue high-rise fire

This fire occurred in a fire-resistive, high-rise on a very hot summer day. Although the building did have a standpipe, it was not equipped with a sprinkler system, having been built before the code required one. Due to a grandfather clause, the sprinkler system did not have to be installed. The ceiling and floor of the apartments were reinforced concrete, while the walls between the connected apartments as well as between the apartments and the hallway were constructed of double sheetrock on aluminum studs. Four people were killed and many firefighters and occupants were needlessly injured. The need for and value of fire safety education can be no clearer than the evidence produced by this fire.

The building. 6115 Granton Avenue, also known as the West View Towers, is a 22-story, fire-resistive high-rise located in North Bergen, New Jersey. There are 296 apartments in the building, which is approximately 230' high. It was also set back from the street about 125' and there is a downgrade from the street to the front of the building. Because of this downgrade, the 4th floor was actually about 6 stories above the ground.

The fire. The fire began in Apartment 4E on Side A. The occupant of the fire apartment, who lived with her sister, was a cancer patient, and as a result, was on oxygen a good deal of the time. There were three oxygen canisters in the apartment, two small pony bottles of aluminum construction, and one 4' bottle of steel construction. Although she was very ill, the resident was a heavy smoker. She would often take her breathing apparatus off and place it on the couch to have a cigarette. This act caused the cushion to become oxygen-enriched, thus lowering the ignition temperature of the cushion. When a lit cigarette was accidentally dropped onto the couch, it readily ignited and quickly spread, severely exposing the aluminum oxygen cylinders.

The attack. Firefighters stretching lines down the hall were met with an extreme heat condition, characteristic of this type of construction along with the oxygen-fed fire, which was as yet unknown to the incident commander and to the attack forces. As the attack lines were advanced down the hall, an explosion occurred and blew a large hole in the wall between the fire apartment and the hallway wall. The ensuing fireball engulfed the hall, severely burning the firefighters on the attack lines. It was later determined that the explosion was caused by the rupture of one of the pony aluminum oxygen tanks in the area of origin.

The rescue attempt— fatalities #1 and #2. The residents of the fire apartment, having been cut off from the exit to the hallway by the fire on the couch that had now extended to other combustibles, sought refuge on the terrace. As the aerial was unusable because of the setback and overhead wires, firefighters attempted the rescue using an extension ladder and a roof ladder. Due to the downward grade, the extension ladder did not reach the objective, so a roof ladder was extended up to the 4th floor balcony. Before the hooks could be set, one woman jumped onto the

Looking down the hallway. The fire apartment is on the right. Note that while the explosion caused the wall to blow out into the hallway, the steel door remained intact. The ensuing fireball burned firefighters advancing down this hall. *(North Bergen FD)*

Due to the setback, a ground ladder and roof ladder were used in an attempt to reach the fire apartment balcony. *(North Bergen FD)*

ladder. She could not maintain her grip nor could firefighters hold the ladder. As a result, she fell approximately 50' to her death. Her sister, a very elderly woman, was still on the balcony. As firefighters attempted to reach the balcony via the roof ladder, the terrace doors blew out as a result of the same explosion that blew out the hallway wall. The terrace became instantly untenable and the woman succumbed to the products of combustion on the terrace.

Preventable deaths. Obviously, the fact that smoking materials were routinely used while the woman was on oxygen displayed an indifference and lack of awareness of fire safety behavior. This led to the ignition of the couch. In addition, statements taken from their neighbor indicated that the women had a good deal of time in which to escape via the front door. She stated that three times she attempted to get the occupants to leave the apartment and each time the conditions were worse. Finally, the smoke and heat was so severe, she had to abandon the attempt.

Fatalities #3 and #4. A middle-aged couple was found overcome in the stairwell. Attempts to revive them were unsuccessful. Upon learning of the fire on the 4th floor, they attempted to leave their apartment on the 10th floor and escape via the stairwells. They were found in the stairwell between the 6th and 7th floors. These deaths were preventable as well. These victims would have been safer in their apartments. Many occupants chose properly to remain in their apartments, which, in most cases, if their location is not on the fire floor or directly above, is the safest course of action to take at a fire in a fire-resistive building. Protection-in-place is usually the best way of safeguarding large numbers of occupants in high-rise fires.

Directly contributing to the deaths of the couple was the fact that residents had blocked many fire doors open to get a breeze in the hallway. The stairwells quickly filled with heat and smoke turning the now-unenclosed stairwells into vertical death chambers. Door numbers, many floors above the fire, were melted as a result of this convection heat. Again,

This statement is somewhat misleading. These instructions may lead occupants in otherwise safe areas to enter stairwells that may become filled with smoke and heat.

lack of responsibility and indifference on the part of the residents greatly added to the problem at this fire.

One other theory came from this fire. It is a widely known fact to firefighters and occupants alike that normally operating elevators should not be used in the event of a fire. The fact that the couple chose to enter the stairwell may have been as a direct result of instructions furnished by the same sign that urges the avoidance of the elevators.

The sign says,

> "IN CASE OF FIRE, DO NOT USE ELEVATORS.
> USE STAIRWELLS INSTEAD."

The decision to flee and subsequently to use the stairwell, put them in the path of the heat and smoke from the fire. This, again, was because the stairwell doors on many floors, including the fire floor, were blocked open. If they had remained in their apartment, they would have survived this fire because the fire floor was six floors below their floor.

Since this fire, officials are attempting to have the signs, through legislative channels, reworded to better guide occupants on all floors. It may say something to the effect of:

Note the spalling of the concrete and the damaged rebar in the ceiling above the point of origin, the couch. Note also the huge hole in the wall caused by the rupture of the oxygen cylinder. The pike poles leaning up against the wall are against the opposite hallway wall. Again, note that the steel door is still intact. The blast took the path of least resistance, through the wall. *(North Bergen FD)*

> "IN CASE OF FIRE, DO NOT USE ELEVATORS.
> USE STAIRWAY ONLY AS DIRECTED BY THE FIRE DEPARTMENT.
> OCCUPANTS ARE TO REMAIN IN APARTMENTS AND AWAIT FURTHER
> INSTRUCTIONS FROM THE FIRE DEPARTMENT."

Of course, this is wordy, but these comprehensive directives may be necessary to accomplish the objective of protecting the life hazard as effectively as possible. Nothing can take the place of an effective fire safety education program, but, in the confusion and sometimes panic at a high-rise

Rupture of this small oxygen cylinder caused this fire to escalate beyond what was normally expected at this type of structure. Oxygen-fed fires will burn more intensely and with tremendous heat. *(North Bergen FD)*

fire, instructions in key places, such as elevator lobby areas and stairway doors, may do what the program could not: give direction when the incident is actually occurring.

Investigative findings. Investigators found that the area of origin was indeed the couch in the fire apartment, and the aluminum pony cylinder of oxygen had indeed ruptured. This caused both the terrace windows to blow out as well as the blast that blew a hole in the wall that led to the burn injuries suffered by the attack team. The larger, steel cylinder had not ruptured but activated the relief valve, causing oxygen to be expelled in the apartment, feeding the fire. This caused severe spalling in the area where it was located as well as intensified the heat condition faced by the attack and search teams.

Conclusion. The fire in this building and the subsequent fatalities and injuries could have all been avoided. The fire and the subsequent extension could be directly attributed to:

1. Carelessness and indifference of the occupant in regard to smoking materials in the proximity to a highly flammable gas, in this case, oxygen.
2. The lack of awareness on the part of the occupants as to the rapidity at which a fire can spread and what to do in the event of a fire.
3. The blocking open of the doors in the hallway that led to heat and smoke extending to the stairwells and upper floors.
4. The lack of knowledge on the part of the occupants on the upper floors who attempted to leave their apartments and escape down the stairwells.

Structural problems of fire-resistive construction

These buildings are the ally of the fire department in regard to building stability during a fire. The most serious collapse hazard in fire-resistive buildings that are fully constructed is the spalling of concrete. This occurs when moisture trapped in the concrete expands, due to heat, and breaks off in pieces or sometimes in chunks. Spalling usually only occurs as a result of direct flame contact. Therefore, the only weapon the fire department has against the spalling of concrete is an effective hose stream to cool the area and reduce temperatures.

While these buildings are the most resistant to collapse at their finished and occupied state, they are extremely susceptible to structural failure in the construction phase. In this construction phase, wet concrete is poured into wooden molds called formwork. It takes concrete approximately 28 days to harden to the point where collapse is not a threat. The construction process, however, moves rapidly as floors are erected, and the concrete is supported by the formwork (sheets of plywood being held up by a series of 4" x 4" posts). Uncured concrete floors are literally a wood-frame building holding up a concrete building. The chance of a catastrophic collapse is substantial if a fire in the formwork destroys the wood's ability to support

This concrete has spalled to the point where the steel rebar has been exposed. The steel has sagged toward the heat source. Spalling results from direct flame contact. *(North Bergen FD)*

Here, metal takes the place of wood as a vertical support for formwork. Wood is placed above to hold the wet concrete. This metal is unprotected, thinner in dimension than the wooden 4" x 4" posts, and is likely to fail earlier.

the weight of the concrete. Sometimes, instead of the 4" x 4" posts, the formwork is supported by steel screw jacks. The problems of fire attacking unprotected steel apply to these jacks. These jacks fail before the wood posts do. Use extreme caution when operating around these weak supports. Firefighters should be aware of the myriad problems they will face when attempting to fight a fire in a fire-resistive building under construction. Fires in fire-resistive buildings under construction will be discussed more in depth in a later chapter.

Class 2
Non-Combustible/Limited-Combustible
Construction

This type of building construction is similar to fire-resistive construction, but is not as protected. Walls, floors, and the roof support system (including the roof deck) are all made of non-combustible material. The walls are usually constructed of concrete block or metal.

One of the main differences between fire-resistive construction and non-combustible construction is that in fire-resistive construction, the steel structural elements are coated with a fire-resistive material such as fire-rated sheetrock or concrete. Non-combustible construction provides none of this fire-resistive protection, thus the steel is exposed to a fire in the contents below.

Unprotected steel and block walls are indicative of non-combustible construction. The great majority of the fire load will be in the contents. The unprotected steel will not stand up well to fire exposure.

Firespread problems of non-combustible/ limited-combustible construction

There are several major problems regarding firespread in non-combustible/limited-combustible construction.

Contents (fire load). The content of the structure determines the fire load and, consequently, the hazard. Buildings of this type often house everything from supermarkets and retail stores to storage facilities and office buildings. In addition, most modern strip malls are usually built of this construction. Fires that are extensive upon arrival should cause the incident commander to consider a defensive-offensive strategy, defensive in the store or area of origin and offensive in threatened exposures in an effort to pinch off the horizontal firespread. This strategy is often used as the potential for the complete involvement and destruction of an entire row of buildings may outweigh the value of saving just one store. By the time the lines are in place, the fire may have already extended beyond the store of origin and it will be time to play catch up.

Metal deck roof fire. The roof in this type of construction, while not combustible in and of itself, has been altered to make it highly susceptible to fire ignition, spread, and failure. The corrugated steel decking is laid over lightweight steel parallel bar joist trusses. Above the steel decking, combustible felt and insulation is added to make the roof weather-resistant. The whole roof is then covered with tar that is hot-mopped in place. A fire originating in the contents below will heat up this tar to its ignition point. This causes a larger fire problem as the roof becomes involved. In addition, flaming molten tar balls often drip through the seams in the roof deck, igniting additional fires in the contents below. All of this reduces the safe operating time for firefighters on the roof and inside the structure.

The only structural fire load the building will contribute will be the ignition of the roof covering. The corrugated steel roof deck will be sealed with hot-mopped tar or neoprene rubber. Once ignited, the combustible roof represents another incident command problem—firespread across the roof and flying brands. *(Bill Tompkins)*

Another way of sealing this roof involves covering the roof with a rubber or neoprene covering. This rubber is also combustible and can spread a fire across a roof. Both the roof and the drop ceiling must be examined to determine the extent of the fire. In fact, at an advanced fire in one of these structures, the only safe strategy to employ is a defensive mode.

Large, open floor areas. Many of these structures are wide-open and large in size, covering many thousands of square feet. A large structure housing a single occupant may have few, if any, partitions to contain a fire. Incident commanders should consider using 2½" lines with solid bore nozzles whenever attacking a fire in these large structures. These large, open areas should raise a flag of suspicion because they are usually indicative of truss construction.

Drop ceilings/cockloft (hanging ceiling space). These structures are usually not the residential type, thus they may be occupied by any other occupancy classification. To increase the aesthetic value of the building, drop ceilings are usually added. The addition of drop ceilings has virtually added a cockloft beneath the metal joist roof. This cockloft, also called a hanging ceiling space, allows fire to travel its width and heat the unprotected steel above. Above these drop ceilings are

Before the drop ceiling is installed, great quantities of wiring and HVAC equipment will be run beneath this corrugated steel deck, creating a cockloft through which fire can travel and destroy the lightweight steel trusses.

literal highways of wiring for everything from computers and telephone systems to fire protective systems and light fixtures. The components of the HVAC system are also located in this drop ceiling. All of the fire protective systems are usually positioned to protect the contents below. A fire starting in these spaces can spread extensively before the fire is discovered.

Many times, these structures are occupied by several businesses such as in a strip mall. Between each store, there may or may not be fire stopping above the ceiling. In fact, the partitions that serve as the dividing walls between stores may only go to the level of the drop ceiling, leaving the entire space above the ceiling open above all the stores. A fire in one store will expose them all and will require a significant amount of manpower and equipment to prevent full involvement of the row. Even if there is fire stopping between the stores, wiring and HVAC systems will often pass through each partition wall, creating a space for fire to travel. At fires in these structures, the incident commander should never expect that the fire stopping is sufficient to contain the fire to one store. Lines should be stretched to each immediate exposure in anticipation that the fire will spread and firefighters with pike poles should be committed to pull the ceilings to check for extending fire.

The HVAC system itself may present another problem. I once responded with the ladder company to a report of smoke in a store that was attached to several other stores. There was a slight haze at the ceiling level and a definite odor of something burning. However, no source could be found. The solution was found when the crew on the roof opened the HVAC motor compartment and found a belt that had been burning. It, subsequently, spread the smoke from the burning belt throughout the entire HVAC system, which exhausted to the entire row of stores. This widespread area caused the delay in finding the source of the odor. When you are at a building with an HVAC system and have an odor of smoke throughout the premises and cannot pinpoint the source, make sure the roof is checked to see if the problem is in the fan motor housing.

Structural problems of non-combustible/ limited-combustible construction

Non-combustible construction is the least stable of all construction types in terms of susceptibility to collapse. This is due mainly to the fact that unprotected steel is a principle construction material, often supporting the roof. Steel is a poor performer in fire. Non-combustible construction often has a metal deck roof supported by lightweight steel trusses of the parallel-bar joist type. In addition, unprotected steel lintels often span the area above show windows and support free-standing parapet walls. This steel is unprotected and will be exposed to the heat from a contents fire below.

This unprotected steel I-beam serving as a lintel will be exposed to fire blowing out of the show windows that will be installed below it. Above this lintel will sit a block parapet wall (installed on the left, but not yet in the center). Heat-induced twisting of the unprotected steel lintel can cause this entire parapet wall to fail. The horizontal collapse zone should be the entire width of the wall.

Characteristics of steel under fire conditions:

- At 400°F, steel begins to lose strength.
- At 1200°F, steel will have lost 60% of its strength.
- Steel can be expected to fail at temperatures in the proximity of 1000°F.
- Steel elongates when heated. A 100' steel beam heated to 1000°F will expand 9½".
- Expanding steel, if restrained, pushes out walls or drops roofs.
- The failure rate of steel exposed to fire depends on several variables:

  1. The size of the steel. Lightweight steel will fail much more quickly than heavier steel. For example, a lightweight bar joist will fail much quicker than a steel I-beam.
  2. The load the steel is subject to. There are often heavy HVAC units on the roof, weighing several tons. This concentrated load over the fire area will dramatically decrease the time for collapse.
  3. The temperature and the distance from the steel to the fire. The higher the temperature and the closer the member is to the fire, the greater the chance for failure.

These variables have a synergistic effect on the steel's ability to remain in place. For example, a very hot fire in close proximity to the steel bar joists, which are holding up a large HVAC unit, fail quicker than if the HVAC unit was in some other area. The presence of any heavy roof objects should be part of the roof size-up and relayed to the incident commander.

Unprotected steel will sag, warp, and twist when exposed to fire. This lightweight steel roof collapsed when the roof supports failed. Note the humps where the walls and heavy objects are located. In a collapse, this is where victims may be found. *(Pete Guinchini)*

It should also be mentioned that the lightweight bar joist truss is, like all trusses, an open system. This characteristic exposes all components of the truss to the fire and heat simultaneously. Also, since these trusses are open, this allows fire to spread across a ceiling, thus weakening adjacent trusses.

These roofs, constructed of lightweight steel bar joist trusses, can be expected to fail in as little as five minutes. Unlike their wooden counterparts, lightweight steel sags before it fails, giving warning to roof firefighters. The operation can continue, if conditions permit, in a less-exposed area.

Steel is a poor performer in fires. Unprotected steel twists and warps under the assault of fire and heat. How it fails and the extent of the damage depends on whether or not the building is attached to other structures. If the building is unattached, the expanding steel may push out the walls and collapse the roof.

If the building is attached to other structures or even to different parts of the same structure, the steel may be unable to expand and may then expand downward, twisting and warping. This action may pull walls inward if the steel is restrained. If the steel is unrestrained, it may cause the roof to collapse with the warping steel while the walls remain intact. Notice that the word "may" is used when addressing how the steel fails. This is because building failure cannot be predicted, accurately, by any one person at any one time. Some structures fail in predictable

The elongating steel was restrained by the attached exposure building on side B, but was able to cause the collapse of wall on Side D. This led to the subsequent collapse of the non-bearing wall on Side A, crushing the cars in front of the building.
(FF Jeff Richards, NHRFR)

ways; some do not. Intelligent incident commanders do not take chances when predicting how and when a building will collapse; they play it safe and establish a proper collapse zone.

Other than applying fire-resistive material to unprotected steel, the only way to prevent failure of an unprotected, steel structural member is to keep it cool. An automatic sprinkler system strategically placed to protect the steel is the best method of ensuring that collapse due to heat exposure does not occur. If this feature is absent, water from hose streams can cool steel to a point where the failure hazard can be reduced. These streams should be of heavy caliber applied from a distance. Any heavy fire in this type of construction should prompt incident commanders to conduct operations in a very cautious and pessimistic manner. If the fire is of large proportions and initial operations are not having any success, consideration should be given to withdrawing the forces and pursuing a defensive strategy.

I remember being at a fire where steel I-beams were used as main roof supports. The fire strategy was defensive and master streams were in use. However, because of the depth of the building, the master streams could not reach the farthest I-beam from the street. As a result, the beam twisted and pulled the walls in with it, collapsing the roof at the rear. The I-beam, about 20' away and closer to the front of the building, was kept cool with the master stream and did not collapse.

It should be noted that these failure temperatures apply to the temperature of the steel itself and not the temperature of the fire. However, temperatures in fires can rapidly reach over 1000°F at the ceiling level in minutes. Lag time should not be trusted. Get some cooling water on the steel or anticipate its swift collapse.

Class 3
Ordinary Construction

In ordinary construction, often called "brick and joist" or just "brick", exterior walls and structural members are constructed of non-combustible material, such as brick or masonry. Side walls are usually, but not always, the bearing walls. Front and rear walls are usually non-load bearing. Bearing walls carry themselves and the weight of the structure, whereas non-load bearing walls usually only carry their own load.

Ordinary, sometimes called "brick", or "brick and joist", construction has masonry exterior walls and combustible interiors. If they are more than 25' wide, steel columns will be used to reinforce the floors. Buildings of ordinary construction are rarely erected today.

Interior structural members such as walls, columns, beams, floors, and the roof are constructed primarily of wood. Older buildings of ordinary construction have wooden wall and ceiling studs that are covered with plaster on lathe, plaster on wire mesh, or both. These are very tough walls to open, especially the wire mesh type. Fire traveling in these areas often spreads quickly.

Wooden roof beams are often large, constructed of

2" or 3" by 10" or 12" joists covered by wood planking that serves as roof boards. This wood planking is then covered with roofing paper or felt, with maybe a layer of tin in between. The roof is then covered with hot tar. Over the years, the roof may have been repaired many times, each time adding a layer of paper and tar to the roof. Over the years, the roof may become very thick. Trying to vent this type of roof is very difficult. These built-up layers of roofing material can bind up the saw, especially when a fire below softens the tar. After the cut is made and the material removed, it often resembles a seven-layer cake, sometimes as thick as 8".

Rather than add "layers to the cake", building owners can address the problem of the leaky flat roof in a couple of ways. First, they may build a completely new roof over the old, porous roof. Called a rain roof, it will be raised several feet over the old roof and be pitched for drainage. Usually constructed of lightweight materials such as wooden trusses, this roof, in effect, adds an additional cockloft over the old roof, one that will conceal fire and may collapse quickly. Firefighters venting

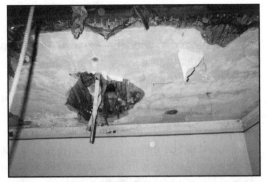

This ceiling is constructed of plaster and wood lathe with wire mesh sandwiched between them. While these are extremely difficult to open, they must be examined, as fire travel in concealed spaces above ceilings and behind walls is common in buildings of ordinary construction.

A new layer of tar and tarpaper is being added to this roof. The tar is hot-mopped to the old roof, the tarpaper is laid out, and the hot tar is then again added to seal the tarpaper. Note the propane cylinders and tar-cooking kettle in the background. The day before this picture was taken, the tar kettle overturned, igniting the roof. The fully involved roof caused the propane cylinders to reach BLEVE (Boiling Liquid, Expanding Vapor Explosion) level, rocketing them off the roof like missiles.

this roof must be aware that in order to vent the original cockloft, they still must also get through the original roof, a now almost impossible task.

The second method may be to rip the old roof off and add a new roof, this time constructed of lightweight wooden trusses or lightweight-steel parallel-chord bar joist trusses. In either instance, a roof that was once a stable work platform has now been compromised to a degree where early failure will occur.

If you observe a new roof being installed on an older building, it is wise to inquire as to its construction. This may serve you well at another time. Fire-fighters should be aware of both of these altered roof situations through building familiarization and pre-fire planning tours. Without this information, sending a team to perform a routine roof vent may have deadly repercussions.

Fire problems of ordinary construction

The exterior masonry walls are intended to limit the spread of fire from one building to another. However, wooden interior structural members have little or no designed fire resistance. The combustible interiors of these buildings are built to burn. The older they are, the more dried out the wood is, and the more easily ignited it becomes. The major firespread problems in these buildings are in concealed spaces that allow uninhibited firespread. These concealed spaces spread fire both vertically and horizontally.

Open stairways. This is the major vertical artery (read "path of least resistance") in which fire will travel in the building. It is also, unfortunately, the primary escape route for occupants and the attack route for the fire forces. Controlling this route is critical in controlling the fire and its spread in the

This stairwell is open from the ground floor to the bulkhead area. The products of combustion will travel rapidly up this stairwell, also the main route of egress. It is imperative the fire strategy focus on controlling this artery.

building. If fleeing occupants leave the door to the fire apartment open or if the fire burns through the door, the stairwell quickly becomes untenable, forcing those occupants above the fire to seek escape routes elsewhere. This complicates the operation because victims will now be on the fire escapes, on the roof, and at windows. Any time occupants seek escape routes other than the usual means *(i.e.,* the interior stairs) the time it takes to remove them, the additional manpower and equipment required, and the accompanying danger to both occupants and rescuers increases dramatically. At the top of the stairway, there is usually a bulkhead door or scuttle that will have to be opened to allow the heated products of combustion to escape and clear the stairway.

Other vertical shafts: pipe chases, channel rails, and dumbwaiter shafts. There will also be other vertical shafts that must be checked for fire travel. These shafts may be interior or exterior.

One interior shaft where fire can travel is in the pipe chases, most often located in the bathrooms or the kitchens. In buildings of ordinary construction, especially apartment buildings and tenements, the kitchens and the bathrooms are often stacked one above the other and back-to-back in adjacent apartments. A rapidly extending fire can go straight up the chase and expose all floors. A fire located in or in the vicinity of these areas should cause a very quick commitment of companies to the floor above the shafts to check for fire extension. Ladder company personnel on the roof should also be examining these shafts and opening up as necessary. Also, don't forget that fire can drop down in these shafts as well, so a surveillance of these areas should also be maintained.

Bathrooms and kitchens will often be stacked in buildings of ordinary construction. This will provide a vertical artery by which fire can travel from floor to floor. These areas must be examined.

Buildings of ordinary construction more than 25' wide will have built -in steel I-beams to aid in handling the compressive load of the building. These I-beams will be concealed inside walls where they will be boxed-in by sheetrock and wall studs. Called a channel rail, the area around the I-beam is a vertical artery for fire travel, one that is not as readily observable as the pipe chase. Companies opening up walls must communicate the presence of channel rails and any other vertical artery to the incident commander as soon as possible so the shafts can be opened above. Other shafts, such as dumbwaiter shafts, may conceal and spread fire to upper floors. These shafts may have been closed-up for years, but that does not mean that fire cannot get

The presence of steel columns and girders are readily visible in this building under demolition. Note how they are enclosed in partitions, making them difficult to find in an intact building. Note also the dumbwaiter shaft, another area for vertical fire spread.

This shaft, only visible from the roof, can cause fire to spread from building to building via the windows on the shaft. Ladder companies on the roof must relay the presence and conditions of the shaft to incident command immediately.

into them. Therefore, they must be checked. There may also be elevators, compactor chutes, and incinerator chutes present. All of these shafts need to be examined because they will be open, thus allowing firespread to all floors.

Light and air shafts. While the interior shafts are a primary concern, do not let fire traveling up an exterior shaft surprise you. Many times, these shafts are only visible from the roof. Not only may there be fire traveling in them, but there may also be victims hanging out of windows that border on the shaft. These shafts are capable of spreading fire to upper floors of the fire building as well as to the adjoining building via the shaft. Many times, windows facing the shaft are directly in line with one another, or worse yet, the exposed window is slightly above the fire apartment window. In the summer, windows left open can allow the upward flow of gases created by the fire extending into the shaft to ignite curtains, spreading fire into exposed apartments.

Recognition of the presence of shafts between buildings is vital. Here, the open end of an attached fire building shows one-half of a shaft. From this cue, incident command must assume that there will be a fully enclosed shaft between the attached structures and seek confirmation of this from crews on the roof and inside the building. *(Bob Scollan, NJMFPA)*

For this reason, the building and any adjoining exposures must be constantly evaluated to ensure no fire travel is undiscovered.

Combustible cockloft. In terms of horizontal firespread in buildings of ordinary construction, the largest culprit is undoubtedly the combustible cockloft.

This access opening leads into the cockloft, the largest concealed space in buildings of ordinary construction. While this bulkhead landing area will not be an ideal place to be in the fire building, it may be used for recon and stream penetration from an adjoining building.

Cockloft fire tactics require roof ventilation above the fire as well as aggressive pre-control overhaul on the top floor. This type of fire is extremely manpower intensive, especially if the building is attached.
(Ron Jeffers, NJMFPA)

The cockloft is the space between the ceiling of the top floor and the roof boards. This space can be several feet high. Fire originating in or extending to the cockloft space creates the need for an immediate opening of the roof to stop the horizontal firespread. These cocklofts are frequently open over a whole row of buildings or over the top of a large apartment building, sometimes referred to as H, E, T, I, or U-type buildings. This corresponds to the way the building looks from the aerial view. Fires that are allowed to travel in the cockloft will burn the top floor and the roof off of the building, and if the building is attached, all the buildings in the row down the entire block. In fact, this type of fire destruction is usually indicative of poor fire containment tactics, misplaced priorities, a failure to request proper support in a timely fashion, or a combination of the three.

Strong control and coordination must be present when a fire is suspected in the cockloft. If the gases in the cockloft are above their ignition point and have not ignited because of the lack of oxygen, extreme caution must be exercised. The roof will have to be opened first in order to allow the products of combustion to leave the building. The ceilings may then be pulled. Opening the ceiling on the top floor before the roof is open may cause a backdraft in the cockloft, causing the now-ignited products of combustion to envelop the top floor and firefighters working under the opened ceiling. Many firefighters have been killed when uncoordinated cockloft venting has been conducted.

Structural problems of ordinary construction

These structures usually are not prone to early collapse, but are more susceptible to burn-through, which will eventually weaken the building. Buildings that are subject to abuse through neglect and due to previous fires should also red flag the incident commander that the collapse of this building may occur sooner than normally anticipated.

Parapet walls. A parapet wall is a continuation of the brick wall from which the exterior wall is constructed. It is the free-standing wall that sits atop the roofline. The parapet may be located only at the front of the building, but in most cases, it encircles the roof. It may be waist high or it may be six or more feet high. The danger is that as the parapet wall is a freestanding wall; it is prone to collapse. It may collapse due to a various number of causes: as a result of fire exposure, as a result of uneven expansion due to either ice accumulation or heat on one side and cold on the other, when struck by master streams, or by being hit with an aerial device (such as a tower ladder basket or an aerial ladder, just to name a few).

When a parapet wall is in danger of collapse, the building should be cleared for a distance equal to the entire height of the wall on which the parapet rests (the height

This parapet wall of these two attached buildings not only encircles the perimeter walls, but also the enclosed shaft at the center. Roof firefighters must ensure that they cross over the parapet in a safe area. Just because the building is attached at the front does not mean it is attached at the rear or center. Always probe before you cross a parapet wall.

of the building). However, this vertical collapse zone will not be enough. The parapet wall will often be reinforced laterally by steel reinforcement rods; a collapse of one section can collapse the wall for its entire length. Therefore, a horizontal collapse of at least the width of the wall must also be established. This is critical when the fire is in a row of old-style taxpayers, where the parapet extends over the entire row of stores or buildings. But it must also be enforced at the apartment house fire where the parapet wall may not be as wide as at a taxpayer.

Performing recon at the rear of this building reveals the height of the parapet, not apparent from the front or side. This side view condition is more prevalent in wood-frame multiple dwellings.

Because water will seek the path of least resistance downward, the location of a downspout or scupper must be at roof level. This is an indication of parapet height.

There is an additional danger with these parapet walls. They are often topped with a copingstone, which can weigh as much as 50 lbs. These copingstones are often not bonded to the parapet, as the bonding material may have deteriorated over the years and has not been reinforced or replaced. The only thing holding the stone in place is gravity. It can easily be knocked from the parapet by a master stream, by firefighters hoisting or lowering tools in the area, or when a firefighter attempts to lean over the roof edge to perform a task or size-up the rear or sides. Use caution when working around these copingstones.

Recognition of the height of the parapet wall may be determined in several ways. The first and most effective way is prior knowledge of the building and its roof. Without this, the firefighter may have to wait until he is at the roof edge to ascertain the height of the wall above the roofline. There are a few ways to get this information from the ground before the firefighter attempts to get to the roof. The first method is to check the sides of the building. Sometimes the parapet is only on one or two sides of the building, which is especially true on corner buildings. Looking at the end of the parapet from this area will tell you how high it is. Another method is to check for the presence of scuppers or downspouts used to drain water from the roof. These are usually even with the roofline. If the downspout is 3' below the roofline, it is a good bet that the parapet wall is 3' high. Firefighters should always attempt to gain information about an area before it is entered or accessed. Surprises are no fun on the fireground.

The decorative metal cornice. Also called an extended cornice, this purely aesthetic roof element, found mostly on buildings of ordinary construction, is an extension of the roof boards covered with decorative sheet metal to give an overhang appearance from below. From a roof operation standpoint, this is a very dangerous area of the roof to work on if you are unfamiliar with its presence or characteristics. Not only does this building feature allow fire to spread from one area to another via its hollow interior, the decorative metal cornice may be unstable and dangerous due to the fact that it is likely to be as old as the building itself and, being more exposed to the elements than most other parts of the building (top, bottom, and side), may be in a seriously weakened condition. This type of roof feature is found all over the Northeast and Midwest, usually on old, ordinary-constructed buildings that are usually 3 to 5-stories high, but I have also seen them on old, ordinary-constructed buildings of both high-rise proportions and of 2-story heights. They are also present on wood-frame, old-law tenements (more on these later).

The extended cornice may be found on any attached building at the front to add to the aesthetic value of the building. Buildings built of ordinary construction 5-stories and higher usually have a parapet instead of an extended cornice, but there may still be one at the front of the building and it is usually raised above the rest of the roof level to the height of the parapet. This serves to tip off the roof firefighter as to its presence. A serious fire that has originated in or extended to the cockloft may quickly weaken this structure to the

This cornice may receive nothing more than a coat of paint when this building is repaired. This may hide any weaknesses caused by the previous fire.

point of failure. In addition, the age of the building and the subsequent years of neglect may have caused this cornice to deteriorate to a point where any load placed upon it will precipitate a collapse of the cornice. Previous fires, which have exposed these structures and then been renovated, may also contribute to the weakness of the cornice. The weakened cornice may receive nothing more than a fresh coat of paint to hide the weakness caused by the fire.

This offset chimney along with the raised lip at the edge of the roof indicates the presence of an extended cornice below. Do not operate between the chimney and the roof edge.

There are few indicators to tip you off to the metal cornice from the roof level and they are not always present or reliable. First, the chimney may be offset from the roof's edge for a distance equal to the width of the extended cornice. The chimney, if built into an exterior wall and not run through a shaft somewhere else in the structure, is usually flush with the exterior wall on most buildings of ordinary construction. This is not the case if a decorative metal cornice is present. If you notice that the chimney is several feet from the roof's edge, beware. The area between the chimney and the roof's edge may be unsupported.

Second, there may be a raised or depressed lip on the roof from where the cornice is attached to the rest of the building. This is a good indicator, but

The best way of determining the presence of an extended cornice is to size up the building before you go to the roof. This cornice is clearly present from street level.

This is the same building as in the previous photo. From the roof, the only indication that the extended cornice is present is the location of the chimney. The unsupported roof edge looks just like the rest of the roof surface.

is usually only present at the front of the structure in attached buildings and at the front and side of buildings that are unattached or that are located on a corner. This lip may diminish or rise as the roof's edge runs from front to rear. During night operations or where smoke obscures vision, getting too close to the roof's edge could place roof personnel in more danger than they bargained for. If these buildings are located on a corner, the sides may not be raised up and will look like any roof without a parapet.

The most reliable way to recognize and point out this type of roof is during preplanning inspections. Be familiar with the types of buildings in your response area. But, even in an unfamiliar response area, doing a proper size-up from the ground level before ascending to the roof can tip off the firefighter of the presence of the decorative metal cornice. The cornice is plainly visible from the street level if the smoke does not obscure it. The cornice extends out from the walls of the building at the roof level for a distance of about 2' or more. Having this information before you attempt to make the roof is vital for firefighter safety on the roof.

Safety tips for the roof firefighters:

1. Preplan the roof types and hazards in your response district.
2. Be a student of roof construction.
3. Size up the roof before you go there.
4. Work in pairs and look out for each other.
5. Be sure that where you are about to step will hold your weight.
6. Use extreme caution when working near the roof's edge, especially on any flat roof without a parapet.
7. Never operate between your cut and a roof edge as the fire may suddenly erupt out of your vent hole and knock you off balance (and off the roof).
8. Be wary of the "Circle of Danger". If a man operating a tool can spin around in a 360° rotation and touch you with his hands (or worse, with the tool he is operating), then you are in the Circle of Danger (which can quickly become the "Circle of Death"). Take steps either to protect yourself by letting him know you are there or (better yet) move out of this theoretical circle.
9. Be careful of the saw – never let a live, spinning saw operate more than 6" above the roof deck. Place it on the roof as soon as it is taken out of the cut.

Rear roof view of shaft. Not only should you be sure an area will hold your weight, but you must make sure there is something to step onto. In this building, you can cross over the parapet wall to the adjoining building at the front of this building, but not at the rear. Use extreme caution, especially in heavy smoke.

This building has been exposed to the elements for so long that there are actually trees growing on the top floor. It is certain that structural elements have compromised beyond the point of safety. No interior operations should take place at this building.

Buildings that have had previous or repeated fires. Buildings of any construction that have had previous fires pose a problem, but none worse than buildings of ordinary construction. For some reason, these buildings are often not immediately demolished after a serious fire. Usually a wood-frame building that has suffered a serious fire is bulldozed the next day, but the Class 3 building lives on, possibly because the walls are still intact. Renovations may be planned, but they certainly do not happen overnight. The interior is still as wooden as a frame building. As a result, these buildings sit for long periods of time open to the weather and vandals. There may be holes in the floors and roof, missing steps or entire stairways, and many other factors that make the building dangerous to firefighters, including reduced stability. Efforts must be made to ensure responding firefighters are aware of building weaknesses due to both firefighting and neglect. These awareness methods are discussed in chapter 12.

Class 4
Heavy-Timber Construction

In regard to the main structural members, heavy-timber buildings can be likened to Class 3 buildings on steroids. In heavy-timber construction, exterior walls and parapets are made of non-combustible material, such as brick or masonry. Interior structural members are made of wood of larger proportions than ordinary construction. Columns must be at least 8" x 8" in diameter. Girders or joists, such as those used for flooring, must be at least 6" x 10" in diameter. Placed on top of these girders is usually 3" planking laid on its side covered with a 1" top plank for flooring. This is a very substantial building. Buildings of this construction, also called "mill" construction, are often very old, and, generally, are only located in the Northeast. They were built to house textile mills and other business activities popular during the Industrial Revolution. In addition, many old (and sometimes new) churches are also constructed with heavy timber.

This and many other heavy timber and "mill" constructed buildings are still in service today. These buildings represent a heavy structural fire load. A wet-pipe automatic sprinkler system is the best protection.

The steel rods, as evidenced by the decorative stars, which tie the walls of this heavy timber building together, were part of the original design. They will be affected by heat in the same manner as any unprotected steel element.

Today, many have been renovated to house such occupancies as condominiums and museums. Others are left to rot.

The "star" spreader plates in this building are placed in an arbitrary manner and are evidence of a weakened wall. Expansion of the steel rods will likely allow the wall to collapse. Use extreme caution when this condition is encountered.

Often, steel rods or cables tie the walls together. These rods usually terminate at the outside wall with a spreader plate. This plate usually takes the form of a decorative metal star, a circular, square, or diamond plate, or a plate that resembles the letter "S". Metal bracing is also used. Generally, if the spreader plates are symmetrical in nature, they were probably built into the original design of the structure. If they are placed in an arbitrary, seemingly haphazard manner, they were probably placed there to tie up a weakened wall. Extreme caution must be exercised around this type of wall. It is important to remember that the heat of a fire, whether built into the original building or added later, adversely affects all steel rods. Firefighters must be aware of this building condition before the fire.

Firespread problems of heavy-timber construction

Massive structural fire load. Although the contents of the building may contribute considerably to the fire problem in these buildings, the main problem of heavy-timber construction in regard to fire is the massive amount of fuel in the form of wooden timbers that make up the structure's interior and roof.

These buildings are sometimes referred to as "slow-burning". This is because the rate of burning of any material, especially wood, is a function of the ratio of the surface area to the mass of the structural member. Heavy-timber structural members are huge; therefore, the surface-to-mass ratio is relatively small. Because of this factor, heavy timber usually maintains its structural integrity for long periods of time before failure. In comparison, a lightweight wood truss has an extremely large surface-to-mass ratio. This is one reason why failure occurs in as little as five minutes. Wood burns through at a rate of 1" every 40 minutes. A 2" thick piece of wood being attacked by fire on both sides will burn through completely in 40 minutes. It will have lost its ability to support any weight well before that complete burn-through time is reached.

Hazardous contents and operations. Hazardous operations and contents in the structure may negate the resistance to easy ignition of the structural members. Processes that involve plastics, flammable liquids, and flammable gases create hot fires, which cause the tim-

The structural fire load in heavy timber buildings may be substantial, however the small surface-to-mass ratio of the timbers make them less susceptible to easy ignition. Note the lack of concealed spaces, also a characteristic of this type of construction.

bers to ignite quicker than if they were exposed to ordinary combustibles. In addition, oils and other easily ignited fluids may have soaked into the wooden floors over the years. This allows a fire to spread rapidly, often beyond the control of a sprinkler system (if there happens to be one installed).

Renovations. One advantage to this type of construction is the absence of concealed spaces. All areas are open to potential stream penetration. This advantage is being annulled in recent years as these old heavy-timber buildings are converted to condominiums and other occupancies. Often, materials such as a lightweight wood truss or wooden I-beams are used in the renovations, virtually constructing a lightweight, collapse-prone building inside a heavy-timber structure. Partitions are added, large, open floor areas are being divided in half to make two floors where there was once

Renovations rarely make fire operations safer or easier. Here, windows have either been partially or completely bricked up. This will cause ventilation difficulties and heat retention problems that will lead to earlier flashover and collapse.

This converted condominium, once a commercial occupancy, is heavy timber construction. What was once a building without structural voids is now permeated by concealed spaces where fire can attack structural members. Although sprinklers protect the living areas, the renovation-created voids are unprotected.

Some heavy timber buildings are located in the heart of congested urban areas. Heavy fire in these buildings will create enormous exposure problems. Even though the structural members are substantial, prolonged heavy fire will cause the wood floors to fail and collapse in a pancake fashion.
(Newark, New Jersey FD)

one, and drop ceilings and other renovations are being added, which create a multitude of concealed spaces. These new void spaces should be protected by sprinklers, but are usually not because the codes most likely do not require it. As a result, a fire extending into these spaces can eat away at heavy timber materials, which, due to renovations, are now supporting more than their own weight and may collapse sooner than before.

Lack of or inadequate sprinkler systems. The only tool in the arsenal of the fire service against the infernos occurring in these buildings is the installation of a sprinkler system to protect the building. However, most of these old structures were built prior to the mandatory installation of sprinkler systems and are still not protected. Once ignited, these buildings radiate incredible amounts of heat, creating an enormous exposure problem. Heavy fire in these types of buildings are beyond the capabilities of most fire departments to handle. The best action to take is to establish collapse zones and protect exposures.

Structural problems of heavy-timber construction

The main advantage of heavy-timber construction is the structural integrity of the members. As mentioned earlier, the surface-to-mass ratio of these buildings make them both harder to ignite and more resistant to collapse. However, like ordinary construction, previous fires, years of rot-causing neglect, and renovations that negate the benefits of heavy timber cause these buildings to be more prone to collapse.

Once the floors collapse, the enclosing walls essentially become free-standing. Here the walls have collapsed in a curtain fall manner. The blast of radiant heat created by the collapsing walls can ignite or scorch anything in its path, including apparatus and firefighters. *(Mike Borrelli, FDJC)*

It must be mentioned that a collapse of one of these mammoth structures that is heavily involved in fire will ignite combustibles well away from the parent body of fire, sometimes as far as 1000'. Incident commanders must be ready for this collapse-created exposure problem.

Class 5
Wood-Frame Construction

Wood-frame construction consists of exterior walls, interior walls, floors, roofs, and other structural members and supports made completely or partially of wood. Wood-frame buildings are usually residential occupancies, but may house businesses, churches, and hotels. Wood-frame structures are rarely higher than 3-4 stories. Literally, the entire building is combustible.

As all wood-frame structures are combustible on the interior and the exterior, the likelihood of spread from one building to the next is a possibility that the incident commander must consider. Many wood-frame buildings are built with very little space between them, sometimes only a few feet. Others are attached and have a common cockloft, as in the case of row houses. Fires in these structures require a large commitment of manpower operating in an offensive-defensive mode to confine the fire to the building of origin.

A proactive exposure protection strategy is required to prevent ignition of closely spaced wood-frame dwellings. A line must be directed into this alley to wash the wall of the exposure. Lines must also be stretched into the exposure. *(Ron Jeffers, NJMFPA)*

Siding on the building also plays an important role in the firespread profile. Siding made of highly combustible material, such as asphalt or wood, acts like a fuse for fire extension from one building to the next. These exposed structures may have to be kept wet as an initial measure when fire extension is a serious threat.

The combustible siding not only plays an important role in spreading fire to adjacent structures, but it also allows fire to spread upward along its combustible exterior, causing spread to the upper floors of the fire building via the exterior. As many wood-frame dwellings are covered by asphalt siding, also known as "gasoline siding", this upward exterior spread may be extremely rapid. This spread of fire from floor to floor via the exterior is called "autoexposure".

These structures, like ordinary construction, also have unenclosed stairways that channel heat, smoke, and flame to the upper floor(s). However, unlike ordinary construction, the doors to the compartments (rooms) will not be solid, but are usually of hollow-core construction. They usually serve bedrooms on the upper floors and will, unlike an apartment door, usually be left open, especially at night. For this reason, it is imperative at a fire in these wood-frame structures to attempt to gain access to the upper floors in as many ways as possible to conduct search and rescue operations. Line placement must also focus on protection of these vulnerable vertical arteries.

Wood-frame buildings are also susceptible to collapse. Not only may the building be consumed by fire to the point of failure, but there may be some alterations made to the building that will decrease the time it takes for a collapse to occur.

On all types of wood-frame buildings, structural additions that create an eccentric load on the walls increase the chances of collapse. An eccentric load is a load imposed on a member that is perpendicular to the cross-section of the structural member, but is offset from the center, creating a tendency in the member to bend. Veneer wall coverings and fire escapes are two examples of additions that place an added strain in the form of eccentric load on exterior walls.

Veneer walls are a single thickness of masonry, usually brick or stone, decorative in nature, designed to improve the appearance of the building. Veneer walls are dependent on the wooden wall they cover for stability. Also, fire may travel in the spaces behind a veneer wall, which

Fire venting out of this first floor window spread fire into the second floor. The vinyl siding melted away, exposing the combustible asphalt siding. The fire, using the combustible siding like a fuse, burned up the wall and entered the dwelling through the second floor window.

further weakens the bond between the veneer covering and the wall. These veneer walls may be found on all types of wood-frame buildings. Veneer walls

Note the space that exists between the original wood frame wall and the brick veneer. A fire can travel in this space and weaken both the wood exterior wall and the metal bonding elements holding the veneer wall in place.

Corner buildings may take the path of least resistance when collapsing, leaning over into the street. When heavy fire is present in a corner or otherwise unsupported building, be prepared for a lean-over collapse. *(Bob Scollan, NJMFPA)*

often fail in what is called a "curtain-fall" collapse, which is the wall falling straight down as if a curtain had been cut at the top.

Wood-frame buildings may collapse in several ways. The front or rear wall may fall as one piece for its full height. Called a "90° angle collapse", it covers the most area with collapse rubble and is the primary reason for collapse zones being established at least the height of the wall.

Buildings located on a corner often fall into the area of least resistance. If they are attached at one side and not at the other, such as on a corner or in the situation where a lot or driveway is next to the building, the building may collapse in what is called a "lean-over". The whole building may list to one side until gravity does the rest of the job of collapsing it into the adjoining yard, lot, or street. These areas should be kept clear if a heavy body of fire is present.

A two-story addition to this wood frame structure collapsed in an inward/outward manner. The enclosing walls fell outward. The roof fell inward into the center. This collapse occurred without warning. No one was injured.

The most deadly of the wood-frame building collapse is the inward-outward collapse because it often occurs without warning. The top two floors fall inward while the bottom floor collapses outward. All four walls may collapse in this manner simultaneously. If there is a large body of fire in one of these buildings, this may be the only sign that an inward-outward collapse is imminent.

Firespread in these buildings as well as collapse characteristics often is influenced by the way the building was put together. This is discussed in the following section.

The three basic types of wood-frame construction

Braced-frame construction. These buildings are very old, many built before the turn of the 20th century by expert craftsmen. Also called "post and girt" or "post and beam" construction, they are held together by vertical posts

The mortise and tenon connection is clearly seen in this braced frame structure. This connection contains the least amount of wood and will likely be the point of failure.

and horizontal beams or girts. These posts and beams (girts) are held together by a connection called a mortise and tenon connection. The connection is somewhat like a square peg in a square hole, which is then pinned together with a wooden peg called a "trunnel". The wooden beams are, generally, 4" x 4" or 6" x 6". The walls are not load bearing because the load rests on the posts and girts, so proper joint connections are critical to stability.

Firespread problems of braced-frame construction. Braced-frame buildings, like balloon-frame buildings, are likely to be more than 100 years old. Thus, their ability to spread fire throughout the building is great. The wooden members are extremely dry and may have split due to age, thus making them easier to ignite. This fire will attack the mortise and tenon joints that are holding the building together. Once a fire in this type of building gets a stronghold on the building's structure, the firefight may be doomed to failure, and collapse may be imminent.

Buildings constructed of braced-frame construction often fail in an inward-outward fashion, with the two top floors collapsing inward while the bottom floor collapses outward onto the street.

Structural problems of braced-frame construction. The weakness inherent in this type of wood-frame construction is the failure of the mortise and tenon joints that hold the walls together. Like a truss, the point of connection is the weakest area in terms of resistance to fire. This is because the tenon joint has a smaller dimension of wood at the end, which is mated with the hollowed-out mortise. These smaller areas are more vulnerable to destruction by fire.

A problem inherent in all wood-frame construction, but even more critical in braced-frame construction, is the fact that there is no compensation on the lower floors in regard to the added load the upper floors place on the weight-bearing structural elements. For instance, in a 3-story, braced-frame building, the top floor will hold only itself and the weight of the roof. The 2nd floor will hold its own weight, plus the weight of the top floor and the roof. The most weight and the weakest point will be the ground floor, which will hold the weight of itself plus the weight of all other floors and the roof. The dimensions of the timbers on the 1st floor will be the same as that of the timbers on all other floors. This places the greatest structural load on the ground floor. That is why, when there is

Note the pegs or "trunnels" holding the horizontal members to the larger vertical members. The size of the wood members on the lower floors is the same as those on the upper floors without any compensation for the increased weight. This connection is the weakest area of the structure and usually the first to fail.

heavy fire on the ground floor of a braced-frame building, collapse should be anticipated. Because of this added weight, 3-story, braced-frame buildings are much more susceptible to collapse than 2-story, braced-frame buildings. The difference is the added, uncompensated-for load of the additional floors.

When braced-frame buildings fail, they often collapse without warning in an inward-outward fashion. The bottom floor falls outward, while the top two or three floors collapse almost straight downward in an inward fashion. If the building stands alone or is located on a corner, it is even more likely to collapse, as the sidewalls do not receive the same support as an attached building. This is still another indicator of collapse potential in an old, heavily involved building.

Balloon-frame construction. Balloon-frame construction consists of a wood framing system in which all of the studs are continuous for the full height of the building, from the foundation to the eaves line. There is no inherent fire stopping between floors. These are also very old buildings, often what we see as 2½-story frames, with the top floor being the attic or, as is usually the case in the inner cities and now in the suburbs, an additional apartment. Sometimes there will also be dormers on the top floor.

Firespread problems of balloon-frame construction. As balloon-frame buildings, unlike the other types of wood-frame construction, have open stud channels from the foundation to the eaves, the possibility of rapid firespread from floor to floor via the exterior walls

Heavy fire on the lower floors caused the inward-outward collapse of this braced-frame building. Ensure collapse zones are established on all exposed sides when these buildings are heavily involved. *(Ron Jeffers, NJMFPA)*

is a real threat and cannot be ignored. A fire in the basement may quickly break out in the attic. In addition, floor joist channels may be attached to "ribbon" boards, a horizontal board attached to the interior of the vertical stud, or attached to the vertical stud itself. This condition leaves an open horizontal channel in proximity to the open vertical channel, allowing fire to spread in many directions simultaneously. This spread is often undetected until the fire exposes itself.

Firefighting is these buildings is also very manpower-intensive because the fire forces struggle to get ahead of the fire. Pre-control, as well as post-control overhaul, must be extensive to assure the fire has been cut off, extinguished, and has not extended via the stud channels.

Lines must be placed in anticipation of the firespread. If the incident commander waits for fire to show in an area, it is often too late to stop it if there is no hoseline waiting for it. In this case, opening exterior siding and sidewalls to let the fire vent may be an acceptable alternative. This requires close coordination between interior forces and exterior forces.

Structural problems of balloon-frame construction. There are several factors that contribute to the failure of balloon-frame buildings. First, these buildings are usually built with 2" x 4" vertical studs. These studs hold up larger wood members that serve as floor joists and the roof rafters. While the floor joists may be the same size as the vertical studs, they also may be larger. Roof joists will sometimes be 2" x 10". Smaller vertical members holding up larger horizontal members are a definite disadvantage when fire is attacking these stud channels.

Secondly, the non-bearing front or rear walls may fall out at a 90° angle collapse.

Balloon framing allows unimpeded vertical spread to all floors of the building. Note the absence of fire stopping in the stud channels leading into the cockloft. In old wood frame buildings, the building must be considered to be balloon frame until proven otherwise.

Although the top floor looks like it may have been an addition, the two lower floors appear to be balloon frame. It is almost impossible to stop a well-advanced fire in one of these structures. *(Newark, NJ Fire Dept.)*

This may leave the floors intact, and due to the now-open front, the radiant heat problem may intensify.

Other times, either the bearing walls will fail, causing the floors and roof to collapse simultaneously, or the floors may fail and bring the bearing walls they are set into with them.

A platform-frame home under construction. The first floor and the second floor are built as separate platforms. The advantage of this firestopping quality will usually be negated by pipe chases, wiring, and ducts for central HVAC systems.

Platform-frame construction. Platform-frame was the last type of frame construction in the industry if you don't count lightweight truss. It is constructed of a wood-stud framing system, in which all studs are 1-story in height and the floors provide some inherent fire stopping. Also, usually limited to three stories, platform-frame construction is built so that the 2nd floor is another platform built right on top of the 1st floor studs. The 3rd floor is then built atop the 2nd floor the same way.

Firespread problems of platform-frame construction. Platform-frame buildings were built in an attempt to confine the fire to one floor. Each floor is a platform unto itself. Thus, there is a good possibility of confining the fire to the area and/or floor of origin. However, the utility systems in the building such as plumbing, electric, and, when present, HVAC systems, negate the effects of the platform frame. Holes are drilled and wires, piping, and ductwork are run from area to area and from floor to floor. These areas must all be checked for fire extension.

Other areas to check in all wood-frame buildings, but mostly in platform frames, are soffits. Soffits are framed-out areas in a building that are usually above kitchen sink areas and cabinets. These areas are probably not firestopped and allow fire to spread from area to area in a blind space. There may also be an open soffit adjacent to the soffit above the kitchen cabinets in the adjoining apartment, allowing fire to spread laterally unnoticed. These areas must be opened if a fire has occurred in proximity to them.

Structural problems of platform-frame construction. Platform-frame construction fails in much the same way as a balloon-frame building does, especially when the bearing walls are being destroyed. Bearing walls, with windows in them, also weaken the wall. Walls with windows in them are more likely to collapse than walls that are unpierced for their full height and width. When these walls fail, the floors and roof go with them to the degree that the failure occurs. The lower the bearing wall destruction is, the more extensive the collapse. Also, floors may burn through and cause localized collapses while the rest of the floor remains intact. This intact floor should not be trusted to hold a firefighter. Collapse of any kind in any building should cause an immediate re-evaluation of the current strategy. Buildings can

Pipe chases and duct channels will negate the intended compartmentalizing feature of platform construction. This is the path of least resistance for vertical fire spread in an otherwise compartmentalized structure.

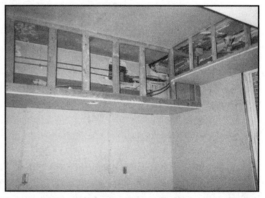

These soffits are not only an area for fire spread, but also a potential ignition area. Note the wires being run through the soffit as well as the penetration in the wall that could spread fire to adjacent rooms or apartments.

always be rebuilt. Firefighters who stay in a building that is losing the battle against gravity cannot be rebuilt.

Lightweight-wood construction

Lightweight-wood construction, which includes both lightweight 2" x 4" wood truss and laminated wooden I-beams, are the latest of the wood-frame

Note the trusses used in the floor construction and the sheetrock used to separate each unit. The open peaked roof trusses are likely to be installed above the top floor ceiling without any firestopping, allowing fire to spread across the roof area from one end of the complex to the other.

construction methods. Nowhere else in building construction is the statement, "a structure is only as strong as its weakest link" more appropriate. Large and small private dwellings, condos and townhome complexes, as well as structural renovations of older homes make use of this type of construction. This construction is found either as parallel-chord trusses and laminated I-Beams in floor and flat-roof construction or as peaked (triangular) trusses found in roof construction. Lightweight-wood structures are usually no more than a 4-story building.

Firespread and construction problems of lightweight-wood construction

1. *Minimal dimension of the wood.* A fire attacking this wood will be able to attack all sides at once. Wood burns at a rate of 1" every 40 minutes. At this rate, fire attacking both sides will burn completely through in 40 minutes. The wood will have lost its ability to support the roof or floor long before that. Some of this construction is not even 2" x 4" anymore. Many are now 1½" x 3" or less. There is no way to predict when this roof or floor will collapse, however, lightweight-wood truss construction can be expected to fail, without warning, in as little as five minutes of fire exposure. Remember that this timeframe begins not when companies arrive on the scene, but the moment the fire begins to attack these structural members.

2. *Questionable connection methods.* Probably more dangerous from a firefighting standpoint than the minimal dimension of the wooden structural members is the method used to attach the trusses together. A sheet-metal surface fastener, also known as a "Gusset plate" or "gang nail" is a piece of light-gauge sheet metal with small nailing points protruding from it. It penetrates the wood for approximately

Five lightweight wood members are joined at this sheet metal surface fastener. The failure of this connection can be catastrophic. Note the minimal depth of the penetration points of the connector into the wood.

These lightweight trusses will expose the entire floor area simultaneously In addition, the HVAC equipment snaking through the truss is likely to allow fire spread from floor to floor. Expect early failure.

¼" and readily pulls free when exposed to heat as it curls away from the wood. These fasteners are used throughout the truss to hold it together. Usually, these trusses are delivered to the construction site as a completed unit. If they are handled roughly or left out in the elements, the sheet-metal surface fasteners' ability to hold the truss together may have been compromised before it is even in place in the building.

3. *Failure to completely compartmentalize each dwelling unit.* The fire-rated sheetrock used to compartmentalize each dwelling unit may only extend to the ceiling of each unit. Fire that burns through the ceiling in any one unit will rapidly feed on the truss loft in each successive unit. If there is any wind to speak of, this firespread will be rapid, exposing many trusses almost at once. This makes operating on the roof too risky in adjacent dwellings. If the fire is not controlled while it is still small, the building is doomed.

4. *Open construction of the truss.* One of the most significant characteristics of the truss is that each piece of the truss will be simultaneously exposed to the fire. This open construction spreads the fire from one truss to the next, and may expose them all in a short amount of time. The openness of the truss area also provides for lateral firespread

under the roof that is above the truss. Failure of one may trigger a domino effect in the trusses, causing a total collapse of the entire roof and supporting assembly.

5. *Laminated I-beams.* Formed with 2" x 4" wood beams used as the top and bottom chords and laminated plywood as the web member, this member will also fail without warning in as little as five minutes. Adding to the fire susceptibility of the plywood I-beam is the fact that the glue used to laminate the plywood is combustible and will add to the vulnerability of the plywood. It will also cause the beam to burn faster and hotter.

Once the building has flashed over, the only acceptable method of fire attack is establishing and maintaining defensive positions outside an established collapse zone, using the reach of the master streams to extinguish the fire and prevent extension to adjacent exposures.

Conclusion

It is not the intent of this book to explore the building construction subject in depth. There are many excellent texts that cover this area more comprehensively. The intent here was to provide a thumbnail rundown, or tickler file, if you will, on the inherent dangers in each type of construction.

All fire personnel, officers and firefighters alike, should seek out as much information as possible in the area of building construction. Surprises caused by building-related factors are rarely favorable on the fireground. Preplanning and building familiarization, along with a solid education and experience, will reap the most benefits when addressing this problem.

These laminated plywood I-beams are being used to support the floor of the condominium complex. They will fail rapidly, causing a floor collapse.

Questions for Discussion

1. Name the characteristics of each of the five construction classifications.
2. Discuss examples of how occupant indifference has led to fire fatalities.
3. What are some of the ways that the fatalities in the Granton Avenue fire could have been avoided?
4. What are some of the inherent problems of steel when exposed to fire?
5. Discuss the dangers of the decorative-metal roof cornice and how to best go about recognizing it.
6. Discuss how the paths of least resistance causes the spread of fire in each of the five construction classifications.
7. Discuss why it is critical to protect the stairway in ordinary and wood-frame buildings.
8. Discuss the problems created by light and air shafts and how to best address them on the fireground.
9. Discuss some of the ways that firefighters can operate safely on roofs of burning buildings.
10. Discuss the differences in the four types of wood-frame construction.

CHAPTER FOUR
Strategic Modes of Operation

Fireground strategy is viewed by many to be simple; there are only two ways to address a fire, offensive and defensive. Fire strategy is just not that simple. The complexity of the fireground is such that incident commanders must not get hamstrung into just two opposite ends of the spectrum. There are gray-area situations in fireground strategy where certain elements of both of these extremes enter into the actual fireground experience and are used in varying proportions to address present and future fire problems. These strategies are referred to as "transitional" strategies, namely the offensive/defensive and defensive/offensive modes.

To determine which operational strategy to pursue, the incident commander must address several factors. These factors,

Comparing current/forecasted fire conditions and the resources available to control them determine to a great degree the operational strategy that will be chosen by incident command. *(Ron Jeffers, NJMFPA)*

alone or in combination, are instrumental in the strategic mode decision. These questions are:

1. What is causing the main problem? (fire, Haz Mat, technical rescue, etc.)
2. What is the life hazard? (firefighter/civilian)
3. How much of the building is involved at this time?
4. How much of the building is uninvolved and threatened?
5. Has structural stability been compromised?
6. What is my manpower and apparatus profile?
7. Can I get the building vented?
8. How threatened are exposures?
9. How quickly can additional resources get to the scene?
10. What, if any, special circumstances must be considered at this incident?

Addressing these questions help the incident commander arrive at a decision based on a realistic forecast of what the initial arriving companies can safely accomplish during the early stages of the operation.

There are six basic classifications of firefighting strategy. In addressing these strategies, firefighter intervention ranges from total to non-existent.

The Basic Classifications of Firefighter Strategy

Generally, an incident escalates in magnitude as conditions deteriorate, causing the strategy to move from totally interior (offensive) to totally exterior (defensive). The opposite progression may also occur as additional manpower, apparatus, and/or additional water sources are brought into the game. No attack (non-intervention) is gaining acceptance in certain types of operations, while the indirect method of attack, although rarely used, can be effective if the conditions are appropriate for its use. In addition to these six strategies, strategy modification, as well as the multiple strategy incident, must be considered and will be discussed.

Offensive

When the offensive strategy is employed, the fire forces are operating inside the building. Lines are stretched to the seat of the fire, thus placing the attack team between the fire and the means of egress (stairwells, for example). Support activities such as forcible entry, ventilation, and overhaul (checking for fire extension) are carried out in coordination with the attack operation. A thorough primary search is also extended to remove any victims in danger.

This is the most common strategy because most fires are offensive upon the arrival of companies. By initiating a fast, aggressive, interior attack, all problems are solved at once. This is the greatest advantage of the offensive attack. Exposure problems are eliminated and the fire is usually localized in this mode of strategy, thus maximizing chances of success.

At offensive fires, lines must be aggressively advanced to the seat of the fire while providing protection for the paths of egress. Proper and timely support from ladder company personnel will greatly assist in making this possible. *(Pete Guinchini)*

During the offensive fire situation, the command post must be continuously monitoring conditions and reports from the interior, rear, sides, and roof. Information gained from these areas, as well as the view from the command post, drives the decision to either remain in the offensive mode of operation or begin the strategy modification process. It is imperative that the incident commander be sensitive to cues that conditions are changing because these are cues that will be the impetus to strategy change. Some of these cues include, but are not limited to:

- Failure to locate the seat of the fire in a timely manner
- Evidence of smoke getting darker and more voluminous even though water is being applied
- Forcible entry difficulty
- Any difficulty in ventilation
- Problems with the water supply
- Indications of flashover or structural compromise
- An operation that "eats up" a good deal of manpower (i.e., a difficult rescue)
- Fires in attached buildings that meet any of the preceding criteria
- Fireground experience or gut feelings

Removal of victims from anywhere other than the normal means of egress (i.e., interior stairs) will require a greater amount of manpower and may alter a strategy that may otherwise be strictly offensive. *(Ron Jeffers, NJMFPA)*

Offensive/defensive

This strategic mode is utilized when there are more problems at the initial stages of the operation or when the firespread has the potential to rapidly out-flank the current compliment of apparatus and manpower. This is routinely a second alarm or "all-hands" fire, depending on the manpower levels of the first alarm assignment. The chances for successfully stopping the fire in the area of origin are good, however, there may be factors that, if not considered at this time, may cause the withdrawal of forces and start the operation sliding into the defensive direction.

A fire in an attached or (in this case) in an immediately adjacent building, will demand early and aggressive fire containment operations in the fire building while additional companies operate in adjacent areas to stop fire extension. This may require early additional alarms. *(Bob Scollan, NJMFPA)*

In the offensive/defensive mode, the major portion of the fire forces operate on extinguishment of the main body of fire, while a smaller compliment, which may need to be reinforced by the additional alarm or later-arriving first alarm companies, operates on exposure control and reconnaissance. A primary search of all areas must be initiated immediately. This includes threatened exposures, which must be evacuated before any real threat of firespread materializes.

The offensive/defensive strategy is often employed at working fires in buildings that are attached, such as row houses or taxpayers, where the common cockloft and other areas are vulnerable to horizontal fire extension. The strategy used here is aimed at confining the fire to the area of origin with the exposure forces, while extinguishing the main body of fire with the main compliment of personnel.

Offensive/defensive operations are the beginning of what some authors call the "marginal" mode, for chances of success of the offensive attack are not assured and steps are actively being taken to cut off the fire before it chases the fire forces out of the structure.

Defensive/offensive

This strategy is used when the companies on hand are not sufficient to handle the main body of fire via an offensive attack. They, therefore, operate in a "holding" action, essentially a defensive posture, to keep the fire confined until reinforcements arrive. A limited primary search in the vicinity of

To keep the fire manageable in this mixed-use occupancy, it may be necessary to initially pursue a defensive attack on this fire while lines are stretched to the upper floors to protect the primary search. Allowing the parent body of fire to burn may endanger upper-floor operations aimed at life safety. *(Bob Scollan NJMFPA)*

the egress points may be the only search attempted due to the conditions. When there is sufficient manpower to operate in a safe manner, then the mode may be switched to offensive. This operational mode may also be utilized when at a fire that, due to the volume of fire upon arrival, would normally be a defensive-strategy fire. However, due to a life hazard, quick offensive tactics are employed to cover the rescue. These tactics may include line placement and operation that may push fire into uninvolved areas and venting to draw fire away from a victim (which may also spread the fire). Once the rescue is made, companies are withdrawn and a defensive strategy is pursued.

A defensive/offensive strategy may also occur when a heavy body of fire is present upon the arrival of one or two companies. Due to the lack of manpower, the only safe action may be to protect exposures until sufficient forces arrive to bolster the initial attack. This may also have to be the mode selected in the event one engine company is on the scene of a heavy body of fire, in a store for example, and a victim is showing at a window on the lower floor. In this case, a deck gun may be placed in service to knock down or hold the fire in check while the other personnel make the rescue. Another defensive/ offensive option presents itself where only one line is initially available when fire is found traveling up the combustible exterior wall of a wood-frame dwelling, threatening to spread to upper floors. In this case, the initial line may have to be utilized to knock down the exterior fire (a defensive action) before being advanced into a structure to pursue a offensive attack. An alternate strategy here would be to use the deck gun to extinguish the combustible wall fire while the interior line is being stretched. The deck gun is shut down when the interior line is place and ready to begin the attack. Proper operational control and communication between interior divisions and the firefrighter controlling the exterior master stream is essential. A defensive-offensive strategy may also be utilized based on the availability of a sufficient water supply. The operation may have to initially operate in a defensive mode, protecting exposures and alternately hitting the main body of fire until an adequate water source can be secured. Proper and effective pre-fire planning should have taken this into account before the fire.

Indirect method of attack

This method, while it is technically a defensive/offensive strategy, is rarely utilized, as the conditions required for its use are seldom present. It requires several factors to be present.

1. Fire is in the Decay or smoldering phase (oxygen below 15%).
2. No live victims in the involved area.
3. A relatively small, sealed-up area in which to operate.
4. A high-heat condition present.

The indirect method of attack, developed by Lloyd Layman of the United States Navy for use aboard ship fires, takes advantage of the phenomena that occurs when water turns to steam. Water is the most efficient heat-absorbing agent known. When it turns to steam, it is at its most efficient, expanding more than 1700 times its original volume. Thus, one gallon of water will convert to 1700 gallons of steam at 212°F. The higher the temperature, the more pronounced, and often violent, the reaction that takes place. Also, the more effective the results.

Fires in the third stage often reach temperatures in excess of 1000°F. The area is full of flammable gases above their ignition temperature. The only missing element is oxygen in sufficient quantity for combustion to continue. Introduction of this oxygen in the wrong place at the wrong time, such as the opening of a door or window, can a have disastrous effect, which results in a backdraft explosion that can not only kill firefighters, but can also destroy equipment and apparatus, which in turn, causes the building to collapse. Any venting must occur at the highest point in the building to release these superheated gases.

When this third stage fire exists, we can take a little time to attack it properly. There are two schools of thought here. The first school of thought is, as mentioned before, venting at the highest point and allowing the building to release the superheated gases to the atmosphere. Flaming combustion will resume and the fire can then be totally vented and attacked by using large diameter handlines. This method will be discussed later in this book.

The second school of thought is the indirect method of attack. The principle of the indirect method of attack is the transfer of heat from combustible materials above their ignition temperature to non-combustible material, in this case, water in the form of fog. Again, all of the conditions must be ripe for a backdraft. In the indirect method of attack, a fog stream or two, dependent on the area, is introduced through a small opening from the exterior. Using fog instead of a straight or solid stream is essential to the success of the indirect method of attack. In comparison with other types of streams, water discharged as fog will expose the most surface-to-mass ratio for the water droplets, therefore allowing for rapid expansion to steam, and absorption of great quantities of heat.

It is most desirable that the indirect method of attack be conducted from the exterior of the structure, although, in rare cases and under very strict control from the incident commander, it may be conducted from the interior of the structure. To be most effective, the application of the fog stream is introduced in the area of principal involvement. The injection of the fog stream should be made at the upper levels of the area if possible, which is where the heat concentration is highest. If this is not possible, it will still be effective at lower levels because of the almost uniform levels of heat in the area. As the fog streams are injected, the water expands and, in doing so, is instantaneously vaporized. This vaporization absorbs heat and causes steam to be exhausted to the outside of the building. This allows cooler air from the outside to flow to these areas. This displacement of the interior atmosphere will immediately start upon the injection of the water fog. From the exterior, evidence of effectiveness can be gauged by observing the following:

1. Smoke being violently driven from the building.
2. Smoke and condensing steam simultaneously released from the building.
3. Condensing steam and lesser volumes of smoke as the fire is smothered.

At this point, the building should then be completely vented and lines stretched in to finish extinguishment and check for extension.

It should be noted that this indirect method of attack is effective only on the fire floor and possibly the floor above. Beyond that, it is unlikely that the heat condition will be sufficient to create enough steam for this attack to be effective.

The indirect method of attack does have a place on the fireground. Strict control and discipline are necessary ingredients. It is the observant incident commander who uses this strategy when the conditions warrant it.

Defensive

All fire forces are committed to exterior operations in the defensive strategic mode. This strategy is employed when the extent of fire and nature of exposures are such that the fire forces are overwhelmed and must apply streams from a distance or outside established collapse zones. In this attack method, exposure protection is the priority while the fire building is given a lesser priority because of the size of the fire and/or the dangers present.

When the incident is defensive upon arrival, the incident commander must announce this in the initial radio report. Part of this report must state that a primary search will not be extended.

Generally, only about 1% of all fires require a strictly defensive strategy upon arrival. About 5% of all fires escalate to the point where the strategic mode is forced to the defensive end of the spectrum. If a great majority of your fires are being fought in the defensive mode, it might be time to evaluate your operations.

Few fires are defensive upon arrival. The incident commander must make a decision whether to attack the parent fire or protect exposures. This decision is based upon, among other things, the proximity of the exposures, manpower on-hand, and the water supply. *(Ron Jeffers, NJMFPA)*

Interior Defensive Operations

When we think of defensive operations, we generally visualze batteries of master streams deluging a structure from outside of the collapse zones. In the purely defensive posture, this is usually the case. However, there are many instance when defensive-oriented operations take place from the inside the structures. A positive risk versus gain analysis in regard to a potentially threatened area will enable incident command to establish a "stop" point inside a building. This decision should be made after careful analysis of the fire situation from the vantage point of the command post and from reports generated from the roof, rear, and interior (both inside of the fire building and inside the exposures). Consultation with division and group supervisors as well as with the safety officer may assist in determining not only the feasibility of such an operation, but also exactly where and when these operations should take place.

Interior defensive tactics usually occur when offensive/defensive or defensive/offensive strategies are being pursued. Thus, the fire is doubtful in regard to the present alarm assignment and the success of the stop may make the difference between a successful operation and a major loss.

Some of the situations where defensive operations take place inside a structure:

*   Lines are placed on the safe side of fire walls to keep the fire from extending into uninvolved areas of commercial building.

- Lines and support companies are positioned on the top floor of a multiple dwelling on the safe side of trench cut operations to assist in the "stop" from the floor below.
- Lines are placed and ceilings are pulled on the top floors of attached building where common cocklofts are either present or suspected.
- Lines are placed, ceilings pulled, and combustibles cleared from the top floor and the attic area inside peaked-roof townhouses and garden apartments.
- Lines are placed and ceilings pulled in adjacent exposures of strip malls.

These defensive-oriented operations demand that strict command and control exist on the fireground. First and foremost, the action plan and how it will be implemented must be communicated to all personnel. It is best to assign a chief officer to supervise the particular areas of operation, as a considerable amount of danger will exist. Safety must always be the overriding concern of all fireground operations. If there is any doubt as to the safety and tenability of the area, withdrawal and surrender of the area must be the only alternative.

No attack (non-intervention)

This is the mode of operation when the hazards are so great, that intervention may create more problems than it attempts to solve. This is the case at fires involving certain chemicals, such as pesticides; where the runoff water from hose streams may do more damage to the environment than letting the fire burn out. Companies should position uphill and upwind and protect exposures.

This is also the strategy to employ when the incident is beyond the scope of fire department capabilities, such as hazardous materials, building collapses, terrorist incidents, confined spaces, etc. The best action here is to protect life (including the life of personnel) and call in the experts to mitigate the incident. Collect as much data about the incident as possible in order to provide an effective briefing for the technical experts when they arrive. The role of the fire department may become one of support in these types of incidents. Fire department operations may be limited to the evacuation of exposures, providing water supplies, assisting in the decontamination process, and other tasks that support the incident's action plan.

Another such incident is when explosives are involved or are seriously exposed. Attempting an attack on a building containing explosives is a loser.

Operating in the non-intervention mode takes a great deal of discipline on the part of operating forces, especially when life is involved. Statistics show that over 60% of all victims are would-be rescuers; rescuers who intervened without appropriate knowledge about the incident. Incident commanders should not put forces in this position. The line between the acceptable risk and the unacceptable risk is a thin one, so training, experience, good judgment, and sometimes luck on the part of the incident commander is what it takes to make the difference. It is the mark of the true professional that can realize that the incident may be best handled by someone else.

Strategy Modification

We must not think of fireground strategy as a closed-end, one-time decision. Utilizing a snapshot mentality such as this can only lead to problems. Fireground management must be open-minded and able to adjust to the situation. Thus, modification of the fireground strategy should not be a crisis on the fireground. Properly coordinated and implemented, it is an effective method of bringing the situation under control in a disciplined and systematic manner. If improperly coordinated, injuries and chaos are often the result.

Strategy modification is based on operational control. The ability to modify the strategy from one mode to another is directly proportional to the control the incident commander has over the fireground.

Strategy modification becomes a reality as a result of continuous evaluations of the fire situation and the willingness to be flexible and adapt to a changing situation.

The incident commander must always consider the present strategic mode and the mode that will be pursued if conditions warrant a change. This "Plan B" must always be in his pocket and ready for use. A "what-if" mentality allows the incident commander to stay one step ahead of the situation. The decision on whether or not to modify the current strategy will be greatly influenced by reports generated from those areas unseen by the incident com-

Attached buildings often force incident command to consider strategy modification. Incident command must maintain a "what-if" mentality until the fire is definitely contained. *(Ron Jeffers, NJMFPA)*

mander, namely the interior, roof, and rear. Using the information gathered from these reports and the progress, or lack of it, will guide this strategic decision.

Case Study

Kennedy Boulevard multiple dwelling

The fire building was a 6-story, Class 3 building, in a "project" or "low-income" complex. Several apartments in the fire area were in the process of being renovated. There were fire escapes in the rear of the building. The building, which faced Kennedy Boulevard, a wide four-lane, two-way street, was setback approximately 100'. To the rear of the building was a steep decline that led to a brush area and was basically inaccessible from that direction for apparatus positioning. Also, because this building was in the center of the complex, and all the buildings were connected, access for ground ladders at the rear was limited as the ladders would have to be carried a great distance to get them in place.

The wind was unusually strong, blowing out of the west and into the rear of the building at a steady 30mph with gusts up to 50mph. This wind condition is typical of buildings that are located at the top of hills or cliffs.

Upon arrival, fire was venting from two windows on the 3rd floor, Side A and threatening to spread via autoexposure to the 4th floor. Also, the fire was located behind a partition wall that hindered access to the fire from the front door of the apartment.

Fire has extended to the fourth floor. Wind coming from the rear of the building was gusting up to 50mph, preventing advance. The Telesquirt is being positioned. *(Ron Jeffers, NJMFPA)*

The offensive attack. As the fire was venting out the window, it seemed that an aggressive interior attack would quickly solve the problem. This was the proper strategy at the time. However, companies trying to make an attack from the front door were met by an extreme heat condition. This was due to the wind and the layout of the apartment. To get to the fire area, the line had to cross the living room, be stretched around to the right (into the wind), and down the hallway, where the line had to turn left 180° to attack the fire room. The stretch was extremely difficult because of the wind condition that was blowing toward the crews attempting to attack from the doorway.

As it became apparent that there was too much heat for one line, an attempt to make the apartment was tried again, this time with two lines in tandem. This fared better, but could not make the final push into the apartment because the wind was even stronger at the last bend. It was later discovered that nearly half of the floor in the fire room had been removed to facilitate the renovation, and had the line been able to get to the area from the interior hallway, it could have led to a firefighter falling through the floor. One thing to always remember is that gravity never takes a day off. This open-floor condition pulled additional wind into the fire floor and as a result, the lines could not advance in the face of this heat.

The defensive operation. The incident commander, as a result of the untenable condition, ordered that the lines back out of the apartment and that all personnel evacuate the fire floor. A safe haven was established in the hallway on the floor. A master stream from a Telesquirt was put into operation to knock down the heavy body of fire. Although the fire had extended to the 4th floor, the master stream from the Telesquirt did quick work in knocking down this heavy fire. While this was happening, the incident commander was already putting his next strategy into action.

The Telesquirt operates in an attempt to diminish the parent body of fire as lines are being positioned to attack from the rear. The progression of the strategy from a frontal attack to a defensive attack to rear attack was successful at this fire.

Change of attack direction. The incident commander, realizing that the major limiting factor was the wind blowing into the apartment from the rear, decided to change the direction of attack. The lines that had attempted to attack the fire via the interior hallway were proven ineffective. The decision was made to stretch 2½" handlines to the 3rd and 4th floors. These lines were stretched via adjacent apartments and out onto the rear fire escape where the attack could be made, not only, with larger diameter handlines, but also with the wind at the firefighter's backs. At the same time, companies were stretching a large diameter supply hose and a manifold to the rear via the side of the building to supplement the water supply at the rear. This was a time-consuming and manpower-intensive operation. This hydraulic tactical reserve was not used, but provided a viable alternative in case there was a problem with the interior lines stretched through the building. These tactics were successful at bringing this fire under control due to the always-thinking incident commander being one step ahead of the operation.

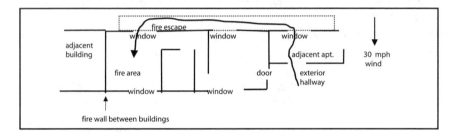

Conclusion. When the current strategy is ineffective, the incident commander must evaluate why this is so and make adjustments to the action plan to solve the problem. In this case, the initial attack lines were unable to penetrate the heat barrier from their position. Because of the wind direction and velocity, the correct direction of attack was from the rear, using large diameter lines. To implement this new plan took time. The fire would have definitely extended in a major way to the uninvolved areas of the building and the battle might be lost before the new attack positions and handlines could be in place. Therefore, it was necessary to use a master stream in a holding action, a defensive move, while the next offensive position was being set up. When these lines were in position, the master stream was shut down and the lines advanced.

Lines are repositioned to attack from the rear fire escape–the windward side. Once the Telesquirt had accomplished the objective of knocking down the heavy body of fire, the master stream was shut down and lines were advanced from rear to front.
(Bob Scollan, NJMFPA)

Fires that require strategy modification also require a large commitment of manpower. This is the most important aspect of strategy modification. Additional plans require additional resources. To attempt this with an inadequate level of manpower is dangerous and irrational. The prudent incident commander forecasts this and summons help early. In this case, the incident commander requested additional alarms in order to have the manpower on scene to stretch the large diameter lines while the master stream was knocking down the heavy fire in the front of the building. If the same crews were used to back out the initial lines and stretch the larger lines into position on the fire escape, the fatigue factor would surely have overwhelmed the forces to the point that needless injuries would have been suffered and the success of the operation would have been greatly jeopardized. The concept at this type of fire, as in high-rise fires, was to throw plenty of resources at the fire. There must be a strong tactical reserve to replace fatigued personnel who have the fought the initial battle.

Managing the Multiple-Strategy Fire

There are times when, due to inherent conditions such as location and extent in relation to both interior and exterior exposures, weather, especially wind conditions, and manpower on hand, different strategies must used at the same time to fight the same fire. Also coming into play will be the size and layout of the fire building, as well as any built-in firebreaks such as firewalls.

Fire breaks through the roof of this contiguous townhouse complex. A defensive strategy was exclusively pursued in the fire area while aggressive offensive operations were taken in the exposed attached dwellings to contain the fire to the dwelling of origin. Without strict control over these types of operations, success is unlikely. *(Bob Scollan)*

This is not to say that both an offensive and defensive strategy may be pursued in the same area of the fire building, but that different strategies can be pursued in different areas of the same fireground, as long as they can be done safely and in a controlled manner. Utilizing a multiple-strategy attack requires strict discipline on the part of the forces as well as an organized command structure. The best way to accomplish this without sacrificing scene safety is by reducing the span of control. This allows the decentralizing of the fireground by assigning chief officers to the various areas of operation, especially those out of direct sight and control of the incident commander, and by demanding that regular progress reports be transmitted to the command post by those assigned to supervise the various operational areas.

Case Study

The Certified Bakery fire

The building and fire condition upon arrival. This fire occurred at rush hour in a heavily congested area of Union City, New Jersey. The fire building was a U-shaped, 1-story, Class 3 building that had multiple occupying businesses. To the right (Exposure D) was a 1-story homeless shelter that was attached to Exposure C, which was a 3-story homeless shelter. Exposure C, originally an embroidery factory, faced 36th Street. These two exposure buildings abutted the rear and west side of the fire building.

The fire occupancy, Certified Bakery, was broken into three sections, the production section that occupied the rear of the building, the delivery section that faced 37th Street and was served by two rolling steel doors, and the sales portion that faced Park Avenue. A firewall separated the production section from the delivery section. There were two openings in the firewall. At the time

Aerial view shows the production area (main fire area), the delivery area to its right, and the sales area is in the foreground (darker roof). The Park Avenue stores are directly attached to the sales area. The exposed three-story shelter is at the left.

of the fire, these doors were open. To the left, toward the east and on Park Avenue, the sales portion of the bakery connected to several businesses that faced Park Avenue. A beauty salon abutted the bakery sales portion. Adjacent to the beauty salon was a furniture store, which was then connected to a hardware store at the end of the row.

Between the two sections of the "U" was a service alley that was 5' wide. With firewalls on each side, this separation was instrumental to the strategy pursued at this fire and would play a key role in the saving of the businesses on Park Avenue. The fire originated in the production section of the bakery in the area of the chimney seen in the diagram. I responded with four engine companies, two ladders companies, one deputy, and one safety officer on the first alarm. As I was leaving the station, which was a good half-mile away, I could see a large column of black smoke spiraling up from the area. The fire had already gotten a good head start. It turned out that the employees had tried to fight the fire for quite a while before summoning the fire department. The fire started in the area of the ovens and spread into the ceiling, thus feeding on the dust from the production process that had been going on for more than 40 years. In fact, it appeared that there might have been a dust explosion or at least a dust-fed traveling flashover prior to our arrival.

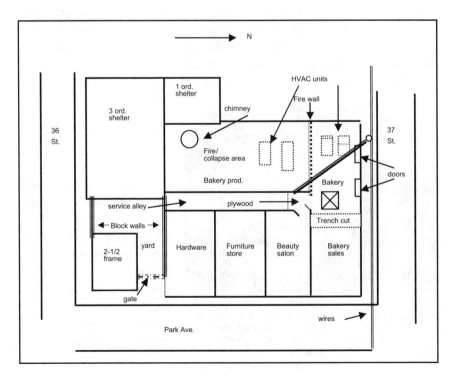

The service alley in between the main fire area and the Park Avenue stores was an area of major strategic importance. Although narrow, it was instrumental in keeping fire from spreading to the entire complex.

Fire issues from the roof of the production area and severely exposes the three-story homeless shelter on Side C. When faced with this situation, be prepared to fight fire on several fronts (think manpower here). *(Ron Jeffers)*

Conditions on arrival were a heavy fire through the roof in the area of the ovens (production area) and a heavy smoke condition in the delivery area where the fire doors had been left open. It was also apparent that the homeless shelter was severely threatened by the fire issuing through the roof.

Fire area operations. I established incident command on the 37th Street side, requested a second alarm, and had the first alarm companies engage in a cautious offensive attack and primary search in the delivery area. This was due to the fact that the business was still open and employees were still coming out of the overhead doors upon our arrival.

The first-arriving ladder company initiated the primary search with the first two engine companies stretching a 2½" line into the delivery area. The third engine company began stretching an additional 2½" line into the delivery section while the second arriving ladder company laddered the roof on the Park Avenue side for recon and ventilation. Reports from the interior confirmed an "all clear" on the primary search and that the fire had not passed the firewall between the delivery section and the

Companies struggled to maintain positions on the safe side if this fire wall. The collapse on the fire side was total. Damage on the safe side was relatively minimal. This fire wall, along with the narrow alley, was a structural feature that helped save the complex.

View of Side C. Fire had ignited these trees and was brushing up against this three-story exposure very early in the operation. Lines stretched to the roof of the hardware store assisted in protecting the exposure.

production section, but the roof of the production section was collapsing due to the heavy HVAC units on the roof. The reports from the roof confirmed this as well as the presence of the service alley between the fire area and the Park Avenue occupancies. Thus, it looked like the only way the fire could spread directly to the Park Avenue occupancies was via the bakery delivery section. It would have to wrap its way around the "U" to get there.

With this knowledge, I ordered the lines to remain on the safe side of the firewall and direct the streams into the burning production section. The ladder company pulled all of the ceilings on the safe side of the firewall to ensure the integrity of the wall. This now turned into an interior-defensive operation on the safe side of the firewall. The objective here was to keep the fire out of the delivery section.

The deputy arrived on the scene, and after a briefing, the command was transferred. A third alarm was requested. I was ordered to recon the area and report on conditions, so I walked around the corner to the Park Avenue side. The fourth engine company had arrived. It was a Telesquirt and was positioning on the south side of the Park Avenue stores between the end of the row and the 2½-story frame dwelling. Looking into the yard behind the fire building, I could see that the fire was licking up the side of the 3-story, homeless shelter. It had ignited several trees and was probably already inside the exposure. At that point, the chimney on the roof of the fire building collapsed, causing the roof to open further and exacerbating the exposure problem.

As the exposure problem was severe, I ordered 2½" lines stretched to the roof of the hardware store via the yard on the south side of the fire building. These lines were stretched and manned by second alarm companies. They operated in a defensive fashion from the safe side of the service alley to darken the fire issuing from the roof. The Telesquirt was put into service, having received a water supply from one of the second alarm engine companies, but the setback prevented it from reaching the main body of fire. The second alarm ladder company was ordered to enter and recon the stores on Park Avenue. I then ascended the aerial to the roof and took command of the roof division. The ladder company on the roof had already cut a sizable hole in the roof of the delivery section and had opened up the large skylight. Only smoke was issuing from this area and because of the firewalls, the fire had not penetrated into the open service alley. So far, so good.

Exterior exposure operations. The reports regarding the severe exposure problem prompted the incident commander to send the entire third alarm compliment to the 36th Street side with orders to protect the 3-story, homeless shelter. An additional battalion chief, responding on the third alarm, was placed in charge of the operation. Fire had broken through the rear windows and had already entered the exposure. The ladder company evacuated the building while an aggressive offensive attack was initiated to push the fire back out of the building. Lines were stretched to the 2nd and 3rd floors and ceilings pulled in a feverish effort to stop the firespread into the building. Luckily, the 1st floor was unpierced at the rear. Once the fire was knocked out of the building, the lines were directed at the main body of fire via the windows facing on the bakery. The operation in the exposure now turned defensive. Additional lines were also stretched to the roof of the exposure via the aerial to add to the defensive fire stream volume from Side C. While it may seem like a long time to the reader, the time it took from my arrival on the scene until the third alarm companies were in place and initiating offensive operations in the exposure was within 20 minutes.

Once this vent hole over the delivery area was completed, companies were withdrawn from this area. The roof strategy was to keep the fire from spreading to the left, toward the stores on Park Avenue. The main fire area can be seen here as well as Exposure C, the homeless shelter.

Roof operations. Once on the roof, I was able to observe the service alley and determined that firespread across the alley was not a threat at this time. The real threat was the area where the bakery delivery section met the sales section, which, if the fire spread to this area, would threaten the stores on Park Avenue. The vent hole had been completed on the delivery area roof by this time and the skylight had been vented. The cockloft below this area was found to be about 10' deep, a very large area for firespread. No other operations were permitted on the roof of the delivery area due to the heavy roof load.

The electrical service connection was located at the center of the roof, attaching in the service alley. The area on the far side is the "U" where the two portions of the complex were connected.

This master stream was shut down when it became apparent that the operation was not having the desired effect. Command must remain flexible, making a constant evaluation of both fire conditions and operations in order to safely adjust the strategy. (Ron Jeffers, NJMFPA)

Another problem affecting to the roof operation was the presence of an overhead, electrical service connection. This connection originated at a utility pole adjacent to the fire building and was strung at head level across the roof where it entered the service connection to the building at the service alley. This presented some difficulties to the roof teams.

Due to heavy fire and the collapse in the production area, the operation in the delivery area had shifted to a totally defensive attack. The lines had been backed away from the open firewall and were now operating from the sidewalk. A ladder pipe was also placed into service to knock down the heavy fire venting from the burned-away roof of the production area.

Fire was now issuing from the vent hole in the delivery area, a sign that it had passed the firewall on the interior and heavy smoke was also issuing from the skylight. The decision was made to create a trench-cut between the delivery area and the sales area. The thinking was that this would keep the fire out of the Park Avenue stores. Four companies utilizing three saws quickly made the cut. It was decided not to use the skylight as part of the trench, a tactic normally used, due to the fact that it was too close to the fire area. The skylight would be utilized as the indicator of when to pull the trench open. Just adjacent to the skylight, the roof pitched slightly upward. Figuring this would provide a path of least resistance for the fire, it was decided to make the cut at this pitch.

As the trench was being cut, I was constantly evaluating the ladder-pipe operation. It seemed the stream was pushing the fire into the delivery area. I requested it to be shut down. This seemed to work because the fire in the delivery area abated once the master stream operation had been terminated. The interior lines also made another advance into the delivery area and attempted to maintain the interior position on the safe side of the firewall. Most of the production area roof had now collapsed and the interior positions were easier to maintain.

The trench did not have to be opened because the fire never reached the skylight. Companies on the roof continued to monitor conditions from a safe area.

Interior Exposure Operations. Another problem was observed both from the roof and from the recon operation in the Park Avenue stores. An open door was discovered leading from the bakery delivery section to the service alley. This door directly exposed another open door that served the sales area and a closed wooden door that led to the adjacent beauty salon on Park Avenue. These openings were covered over at the roof level by a sheet of plywood, so they were not evident from the roof. Fire was evident below the plywood and when it was opened, the threat to the exposures became apparent. There was a threat of fire entering the Park Avenue stores via this opening. From there, it would have access to the drop ceiling space and could spread to adjacent stores.

The command post was notified of this situation, in turn, they ordered lines stretched into the beauty salon and the ceiling pulled at the rear. The plywood covering in the service alley was also completely removed to keep the fire from spreading laterally. This was essentially an offensive operation. The incident commander also assigned a battalion chief to this area of operation.

As the operation on Park Avenue depleted the tactical reserve, the fourth alarm was requested to provide relief of the operating companies and ensure the manpower pool was replenished. This fire was subsequently contained to the area of origin, the production area of the bakery.

There were several key factors that allowed the incident commander to keep this fire from spreading not only to adjacent interior exposures, but also to exterior exposures. These were:

1. The decentralization of the command structure, reducing the span of control. Chief officers were assigned to oversee operations in each major area of operation.
2. The willingness to request additional alarms early in the operation. Foresight is critical to the incident command game.
3. The ability to pursue several strategic modes at once. Control of manpower through delegation allows the complex action plan to succeed.

4. A willingness to modify the action plan and strategy based on reports generated from assigned operational areas.

Conclusion. Identification of the strategic mode is critical in controlling the fire scene. The strategic mode chosen by the incident commander is a decision that should not be taken lightly. All personnel on the scene should be aware of the strategic mode chosen and know their role in the action plan set by the parameters of that mode.

Conclusion

Incident commanders must remember that evaluation regarding the effectiveness of the current strategic mode must be continuous process. The incident commander who cannot effectively change the strategic mode does not have control over the scene and operating personnel. This is a dangerous position to have. An out-of-control fire scene will most certainly compromise the safety of personnel.

Questions for Discussion

1. What are some of the factors on which the strategic mode decision is based?
2. Describe the six basic classifications of firefighting strategy.
3. Give some examples of when an offensive/defensive strategy would be utilized.
4. Give some examples of when a defensive/offensive strategy would be utilized.
5. What conditions must be present to utilize the indirect method of attack?
6. Name some of the situations where the non-intervention mode would be appropriate.
7. Discuss some of the tasks firefighters can accomplish during the non-intervention strategy to keep the situation from worsening.
8. What are some of the ways the incident commander can establish control when modifying the chosen strategy or when directing the multiple strategy fire?

CHAPTER FIVE
Private Dwelling Fires

The Occupant Factor

Private dwelling fires account for a great majority of fire deaths in the United States each year. While the majority of these are one and two-family structures of wood-frame or ordinary construction, and the fireload is relatively light, the fact that most fatal fires occur here can be attributed to the following five factors.

Unsafe sleeping habits

Many people in private dwellings sleep on the floors above grade and bedroom doors are often left open. This condition invites the products of combustion into the room, killing occu-

Private dwellings account for a great majority of fire-related deaths each year in the United States. The fire service must seek to make up for the lack of code enforcement capability in private dwellings with an aggressive public fire education program. *(Ron Jeffers NJMFPA)*

pants via smoke inhalation before heat and flame have permeated the area. Closed doors save many lives because they act as an initial barrier to the products of combustion. Fortunately, most people have installed smoke detectors, and if they have been maintained properly, their chances of being awakened and escaping a fire in the early stages are greatly enhanced.

Carelessness and indifference

Unfortunately, smoke detectors do not work if not properly maintained. Many people disconnect them in cooking areas because they keep activating and they never reconnect them. Also, the batteries are usually allowed to die. Smoke detector batteries should be changed twice a year; a good schedule to keep is when the clocks are to be changed. This is usually not done, and these silent sentinels are rendered useless.

In much the same vein, no matter how much is written to warn people of the dangers of smoking-related fires, smoking is still going on in bed. The same can be said for overloaded electrical outlets, especially during the Christmas season. Many people are also killed each year because of fires caused by electrical carelessness.

This indifference also carries over to understanding fire hazards. A fire investigator relayed a story to me about a fatal fire in his jurisdiction. It was a bitterly cold night. A family in a row frame had no heat and none of the electrical outlets worked. The tenant had a space heater and by cutting off the plug, he stripped the wiring, and pushed it through the floorboards into the cellar. From there, he attached it directly to the leads where the circuit breaker is supposed to be attached to the electrical panel. It was being used to warm three young children huddled together on a pullout couch. The heater must have been working because the family went to sleep, with the parents sleeping in the back bedroom. A fire caused by the improper wiring ignited the bedding. The children never got out of the bed. They were killed right where they were sleeping and burned beyond recognition. These were preventable deaths.

Improper storage and housekeeping

Basements, attics, and garages can sometimes be described as the "overflow drain of the soul". Items that people have accumulated over the years and that they no longer need, but cannot part with, find their way into the these areas. This "packrat mentality" has caused many fires and has also caused firefighters to become lost, trapped, and subsequently killed in these areas. I have been in basements that were piled so high with "stuff" that hoseline advancement was virtually stopped at the foot of the stairs. Search operations here should be conducted using a lifeline and thermal imaging camera. If there is no life hazard, using the reach of the stream may be the safest action to take in an excessively cluttered cellar or basement.

In addition, many hazardous materials are stored in garages and basements, turning them into a veritable minefield for the unsuspecting firefighter. Firefighters should take notice of certain indicators that could present the potential for a problem involving unusual storage. Signs on the building or a commercial vehicle routinely parked in the driveway can be tip-offs to unusual dangers in the premises. Exterminators, pool maintenance contractors, and landscapers can store a bevy of hazardous chemicals in these areas. A hunting enthusiast may have live ammunition stored anywhere in the house or garage. If you see a lot of stuffed animal heads or rugs, think live ammunition and take steps to ensure you don't become the next trophy.

These commercial vehicles, indicating that a pool maintenance service is run out of this home, should be a signal to operating firefighters that hazards out of the ordinary for this type occupancy are likely to be present.

A homeowner with a pool, whether it be built-in or aboveground, is likely to store chlorine and other pool maintenance chemicals somewhere on the premises. These items may be stored in an outside shed or garage. If there is no outside storage area, think basement storage. It is best, when arriving at a fire in a home with a pool, to inquire as to where these chemicals are stored and take actions to safeguard firefighters from this non-routine exposure. Demand that firefighters wear SCBA at all times in all areas.

In addition, during the winter months, many homeowners store their barbecue equipment, including propane tanks in the garage. Due to the possibility of the presence of any or all of these items, a routine fire can easily turn into a disaster when contents not normally expected to be involved are ignited or subject to extreme heat.

Firefighters conducting search and reconnaissance missions should be cognizant of items that may cause a major problem if they were to become involved in fire. The incident commander must be immediately notified of any unusual or dangerous storage. A change in strategy may be the result of these discoveries.

This home actually doubles as a dentist's office. A fire could cause exposure to such hazards as oxidizing and flammable gases, radioactive materials (x-ray equipment), and bloodborne pathogens. The only indication of this problem is the small sign out front.

Complacency

It is a fact that most accidents occur in the home. Well, hostile fires are usually accidents. It has also been said that three things start all fires: men, women, and children. One of the reasons for this is that people are comfortable in their homes and many times display a "this can't happen to me" attitude. This attitude, which causes people to drop their guard in respect to fire safety, can have deadly consequences. Fire personnel should convey a strong message of fire safety whenever in contact with the public. Public education is probably our greatest ally against fire and subsequent death in all residential dwellings. Remember that the primary mission of the fire service is fire prevention. Fire suppression activities are necessary when this mission fails.

Home repair specialists

Amateurs making home repairs and renovations have always been a problem for the fire service. It seems that when the work of these geniuses goes awry, we are called to take care of the problem, placing ourselves in danger to do it. How many times have we seen fires caused by homeowners who have ignored or been ignorant in respect to building codes regarding proper wiring, chimney repair, and heating equipment? Shoddy workmanship has been the cause of many fires in private dwellings as home electricians and plumbers/welders attempt to do the job of the professional. This problem is often compounded when these specialists also become do-it-yourself firefighters and try to put out the fires that they have started, causing alarm delays and unnecessary damage and injury.

I recently investigated a fire that originated in the basement of a private dwelling and spread to the 1st floor. The owner had installed his own HVAC system. The ductwork was run under the floor in the horizontal stud channels. At a point on the other side of the room, the floor was breached to run the duct up through the wall to the upper floors. Proper heat-retarding material was never added to the horizontal duct run and it was installed flush with the basement ceiling. Over time, the constant heating and subsequent drying of the ceiling boards reduced the ignition temperature of the wood. Ignition occurred and the fire followed the duct channel right to the vertical opening where it broke out on the 1st floor. Pyrolysis of the wood, due to conduction of heat from the duct to the wood, was the culprit. The owner had installed the system himself, used no permits, and never had it inspected.

Another time, a woman claimed she was an expert at electrical wiring. This was after the outlet she intended to use for an air conditioner was wired with the wrong size of wire and protected by a breaker with too high of an amperage. The result was an ignition inside the wall that nearly burned the entire backside of the house away. Only an aggressive, pre-control overhaul saved the building. Then, she had the nerve to complain about the damage.

Recently, four 3-story braced-frame buildings were seriously damaged by a fire that was found to be electrical in origin. The fire broke out at about 9:00 p.m. Hours earlier, the fire department responded to the building of origin for a reported smoke condition. The owner was doing electrical repairs in the 2nd floor ceiling. The problem was isolated, the electricity to the area shut down, and the owner directed to get a qualified electrician. It was discovered later that he continued to work on the problem and then turned the power back on. A fire broke out in the ceiling space and traveled horizontally in the floor/ceiling space until it found a vertical chase between the chimney and the wall. It then spread vertically past the top floor and into the cockloft. When the companies arrived this time, the fire was already showing at the roof level. It subsequently spread horizontally through the open cockloft to involve four buildings.

These amateur home-repair jobs have caused countless fires that have burned inside walls and under floors until they were discovered. These same homeowners are usually the ones who dismantle the smoke detectors because they make too much noise. It seems that lousy home repair experts are usually lousy cooks too.

Years past, there were not as many do-it-yourselfers because it was difficult to get all the materials to do a complete job. Today, the rise of the do-it-yourself superstores have spawned a new generation of home electricians, plumbers, and welders, who, in their valiant efforts to be handymen, wind up burning their own homes down.

Firefighting Problems in Private Dwellings

Firefighting problems usually result from something out of the ordinary occurring. Generally, due to the relatively light fireload, these fires should be routine. However, because of such factors as the presence of open stairways, basements, garages, and attics overloaded with stock, setbacks, and extremely combustible exteriors on some homes, the firefight may be tougher than anticipated.

Open stairways

Open stairways act like a chimney, allowing flame, smoke, and gases to spread up to floors above the fire. The only saving grace occupants above the fire may have are properly operating smoke detectors to warn of a fire, or doors that are closed to act as a barrier to the products of combustion. In multiple dwellings, apartments are usually only on one level. Apartment doors exposed to vertical fire, heat, and smoke travel via the common stairwell are usually left closed because they are usually the main entrance to the apartment. In private dwellings, multiple levels are more common. Doors to rooms on the upper floors are usually left open, allowing unimpeded upward spread of the products of combustion.

Excessive and improper storage and debris

Excessive storage and debris in any area of the home adds to the fireload and makes access to the seat of the fire difficult and potentially dangerous. Exterior debris and outside

Open stairways are common to private dwellings, especially those with two-story foyers. A properly placed smoke detector located at the top of the stairs, such as the one at the entrance to the second floor hallway, will be the only warning of a fire originating on the ground floor.

storage also cause an exterior fire to spread to the interior or vice-versa. Fires follow debris accumulation like a fuse to adjacent properties. This exterior storage can also hamper firefighting access to the rear and sides of the building as well as completely render the area useless for ground ladder placement. Inspection visits must ensure that all areas, both interior and exterior, are kept clear of fire hazards and allow unimpeded access for fire department operations at all times.

Excessive rubbish on the exterior of a building such as this renovation debris can act like a fuse and ignite several structures at once. Wood frame buildings are especially susceptible to this type fire spread.

Setbacks and landscaping obstacles

Setbacks on private dwellings usually cause a delay in getting the first line into operation, disallow aerial operations, and/or delay the raising of ground ladders. In addition, a preconnected hose may not reach the building or the fire area, resulting in longer stretches, increased friction loss if excessive small diameter hose is used, and a general delay in getting to any trapped occupants. Decorative landscaping and large trees may also hamper the ability to raise ladders and gain access to the rear and sides of the structure.

Combustible exteriors

Combustible exteriors are a major firespread problem, both inside and outside the fire building. Fire extending to any combustible siding, especially extremely combustible asphalt siding, will extend quickly into upper floors, sometimes more quickly than the open interior stair.

Companies on the interior may not notice the problem of autoexposure, the extending of fire from floor to floor via the building's exterior (also known as "leapfrogging"). A line properly placed to protect the stairs and the members extending primary searches on the upper floors may be outflanked by a fire extending into upstairs windows by way of the combustible exterior walls.

Combustible exterior walls are often responsible for extremely rapid fire spread from floor to floor and from building to building. Old buildings with dried-out wood walls beneath asphalt siding are particularly susceptible to this threat. *(Louis "Gino" Esposito)*

An exterior line was used to stop the exterior fire spread while companies were pursuing an offensive attack inside the building. The exterior line must never enter the interior of the building during this operation.

Incident commanders and firefighters on the exterior must be cognizant of this fire-spread hazard and communicate any exterior-related fire spread immediately to interior crews.

To halt this exterior spread, it is necessary to place and apply a hose stream directly to the combustible exterior. It may also be possible, if the fire is in close proximity to the door being used for the attack, to sweep the ignited exterior before advancing into the structure. If the fire extends to the siding after lines are already working in the building, it may be necessary to use a line from the exterior to extinguish any fire spreading via the combustible exterior. Those operating these exterior lines must take great care to avoid directing the stream inside any opening where interior crews may be operating. This can cause serious injury and push the fire and the products of combustion back into the building toward advancing firefighters. Instead, the line is used exclusively on the outside wall to arrest any exterior extension. Where asphalt siding is present, this may be the most important tactic that prevents spread to the upper floors and to adjacent structures.

A larger problem with combustible exteriors is the possibility of firespread from building to building via radiant heat. A hose stream may have to be placed to cut off this extension and protect the combustible exterior of exposures.

HVAC Systems

Also worth noting is that many of these private dwellings have a central HVAC system installed. Fire spreads rapidly via this system to other areas of the house. This system should be shut down as a matter of procedure and checked for its entire length for signs of fire.

The presence of two HVAC units indicates that a two-zone system is in use in this dwelling. They must both be shut down to prevent spread of fire via this artery.

Commercial occupancies usually position HVAC units on the roof of the structure. In the private dwelling, the HVAC unit is usually somewhere on the perimeter of the building at ground level. If there is more than one unit present, expect two or three zone heating. Ensure that all units are shut down to prevent firespread into other areas of the dwelling via these avenues. HVAC duct work that runs throughout the entire house negates the advantages of platform construction and essentially creates a large

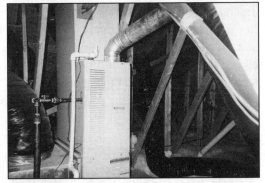

Attic-mounted HVAC units create a concentrated load above the top floor ceiling. This further weakens collapse-prone lightweight wood truss construction.

void in which fire can rapidly travel, especially if the unit is running. The force of the fans will spread fire like lightning.

Some newer private dwellings, many of those built of a lightweight wood truss, have the HVAC unit located in the attic. This creates a concentrated load that will lead to an earlier than usual collapse in a building whose construction is already inherently collapse-prone. In addition, the vibration of the unit may cause the sheet-metal surface fasteners holding the roof trusses together to gradually pull out of the wood, further weakening the connection. Early size-up of this condition is the key. In a newer private dwelling, especially a large home, where there is no HVAC unit on the perimeter, expect it to be in the attic. If available, the owner is the best source of information regarding the location of the unit.

Many older homes were built with the ductwork in place, but the HVAC system was never installed, leaving a highway for fire travel.

The home I grew up in has registers in the walls in virtually every room; however, we did not have a central HVAC system. The duct terminated in the cellar. If a fire was to start in the cellar, the duct could have served as a thoroughfare for vertical fire travel to each room in the house.

Voids in construction

These voids, especially in newer wood-frame construction, often take the form of soffits found in kitchens and bathrooms. Soffits are boxed-out areas that are used to house wires, ventilation systems for stoves, and piping. These are found above kitchen cabinets and bathroom sinks as well as around bathtubs. In older buildings, the pipes that ran from floor to floor were run in the corner of the room and were out in the open. In newer construction, these pipes are run in a blind channel, which is more aesthetic, but they allow fire, smoke, heat, and gases to spread undetected throughout the structure. They may also be hidden in closets to keep them from view. These areas must be opened and examined whenever they are found in or above the area of involvement.

The area above these kitchen cabinets may be a maze of piping that will invite both vertical and horizontal fire spread. Kitchen fires, especially those originating on stoves, demand that this area be opened.

Soffits may also be present as decorative overhangs where the roof meets the top floor. Many times, there are vents in the underside of the soffit to allow air to circulate throughout the attic. A fire issuing form a window can easily enter the attic via these openings

Other voids that may be avenues in which fire may travel are the sleeves that hold sliding "pocket" doors. This framed-out area may not be completely enclosed and once a fire enters this space, it may be free to extend upward through the walls.

Dangerous peaked roof conditions

A good majority of private dwellings have peaked roofs. These roofs may be gable, hip, mansard, shed, gambrel, or a combination of these. Fire-fighters get killed and injured by falling through roofs, falling off roofs, and being struck by roof components. It is imperative that firefighters understand roof construction and roof coverings. Roof slope is also a major consideration during roof operations. The incident commander must always consider this pivotal area of operation in the most cautious manner possible, always erring on the side of safety.

Roofing material involved in fire not only gives off huge amounts of thick, black smoke, but also may be the cause of additional fires in areas remote from the fire building. Brand patrols may have to be established.
(Ron Jeffers, NJMFPA)

The gambrel roof has an extremely high pitch. It is safer and more effective to vent horizontally via the top floor windows than to vertically vent the roof. If absolutely necessary, roof ventilation should be performed from an aerial device.

Roof construction

Peaked roofs are usually constructed in one of three ways. In the first way, which consists of the oldest roofs, the roof construction was made of narrow furring strips spaced several inches apart, upon which the roof covering was nailed. There is virtually no roof deck. This is the weakest of the non-truss roof types. This type of roof construction is not likely to hold the weight of a firefighter.

The second way is one in which roof rafters are utilized. Constructed of larger pieces of lumber and spaced anywhere from 12" to 24", these roofs are more stable to work on, usually covered with wood planking or plywood. When using an axe, the planking is easier to cut and remove during firefighting operations. Plywood offers resistance to chopping because of its resilient qualities, which cause the axe to "bounce" prior to getting a bite on the materials. It is also more conducive to splintering, thus making it harder to pull than the smaller, separated wooden planks.

The third way is the peaked truss roof. This roof construction offers the least amount of support in regard to spacing and is usually sloped at least 45°. The trusses are usually spaced 4' on center and eliminate

This lightweight wood truss roof will be covered over with plywood and is prone to early failure. As there is no ridgepole, roof ladder operations are not recommended. Venting from an aerial device is the only method of venting this roof.

the requirement for a ridgepole. This also eliminates the safety provided by a roof ladder since the hooks, which are usually secured over the ridgepole, will be unsupported. Peaked truss roofs should only be cut from an aerial device. In addition, the truss roof will likely be covered with plywood, which is best cut with a saw. In addition to the minimal support, the peaked roof is constructed of lightweight 2" x 4" wooden members that are joined together by a sheet-metal surface fastener. The limitation of this type of construction has already been discussed at length.

Roof coverings

Peaked roofs may be covered with a variety of materials. Some are easy to breach, others are not quite so. Asphalt shingles are relatively easy to penetrate but are combustible, allowing the fire to spread quickly across their surface. They also produce large quantities of thick, black smoke. In addition, while they are not as slippery as some other types of roof coverings when wet, they can be extremely slippery and begin to melt when exposed to heat, causing a firefighter not properly secured to lose his balance.

Wood-shake shingles create a flying brand problem and have been the culprit of many conflagrations. These types of shingles, while relatively easy to cut through and remove, are prone to decay and mildew, making them slippery.

By far, the most dangerous type of roof coverings are those constructed from more substantial material such as

This mansard roof is covered with asphalt shingles. Fire will spread rapidly over this surface and emit huge quantities of thick black smoke. It is also likely that asphalt shingles are present below the vinyl siding on the walls. *(Newark, New Jersey FD)*

This slate roof covering may conceal an unstable roof. Falling debris is heavy and may be razor-sharp, and when wet, can be extremely slippery. Operations on this roof should be conducted from an aerial device. *(Lt. Joe Berchtold, Teaneck New Jersey Fire)*

tile, slate, and clay. Not only are these materials smooth and prone to slipping when wet, but they can also be very brittle, causing them to break under the foot of a firefighter on the roof. Roof ladders or aerial devices should always be used when operating on these types of roofs. These roofs offer two other dangers. The first danger is that they may hide their instability during a fire. Much like a terrazzo, or concrete floor laid on wood supporting members, these roofs may remain intact even after supporting members have burned away. As soon as the firefighter steps on this surface, he may plunge through into the burning room below. The second danger is the weight and razor-sharpness of this material. Slate and other heavy tiles have been known to slide off the roof and strike firefighters causing deaths and serious injuries. Especially when using master streams in and around these roof coverings, all area around the building should be cleared to protect personnel from these falling guillotine-like roof materials.

Firefighters should also be aware of any foreign material on the roof such as ice, snow, or even rain. Ice may be difficult to see at night and may also be so thin that it is not noticed until the firefighter is already on the roof. Be especially cautious where roof coverings are in close proximity to trees. Sap, wet moss, and leaves from trees can make the roof as slippery as if oil had been poured on it.

In addition, do not trust the chimney as a place to get a handhold when working on the roof. The effects of weather may have caused the mortar between the bricks to erode. Some chimney members are being held in place by gravity only. They will remain stable until either a firefighter grabs them or the impact of a master stream dislodges them. Always use caution around chimneys.

Roof slope

The degree of roof slope is directly proportional to the caution and apparatus support required when accessing and operating on it. A roof with a slope of 30° or less is relatively safe to operate on without a roof ladder. Any roof more than 30° and up to 45° should man-date the use of at least a roof ladder. A roof with a slope greater than 45° should not be walked on at all. The use of an aerial device is mandatory to conduct roof operations. These are for ideal conditions. Any condition that causes a reduction in the safety of the roof operation should man-date a roof ladder or aerial device be used. The incident commander must not allow or accept unsafe roof operations. Again, gravity never takes a day off.

Aerial use is mandatory at steep slope roof operations. If aerial access is not possible due to setbacks, wires, or unusual roof orientation, allowing the fire to vent itself may be the only course of action available to the Incident Commander.
(Ron Jeffers NJMFPA)

Life Safety Problems in Private Dwellings

As far as occupant numbers are concerned, the problem is not as severe as other types of occupancies; however, the ability of firefighters to get to the occupants may be another story. Access to the upper floors is usually limited to the open interior stairs that may quickly become untenable prior to or after firefighters have entered these areas. For this reason, separate means of access, as well as egress, should be established. Ground ladders should be raised to bedroom windows and porch roofs. Attempts should be made to reach the upper floor bedrooms from these areas as well as from the interior. This access from the exterior is the job of the person assigned

Upper floor primary search must be attempted from as many avenues of access as possible. A ground ladder raised to the upper floor is not only a way in for search teams, but also a secondary means of egress. *(Newark, New Jersey Fire)*

Multiple mailboxes, doorbells, gas meters, as well as the presence of a fire escape are all tip-offs of a private dwelling that has been converted into a multiple dwelling.

the Outside Vent (OV) position. Many times, the only way into (and out of) the upper floor is by way of ground ladders. Ladders should be raised at strategic locations at every serious fire that involves this kind of structure.

In buildings that should house only one or two families, the attic and the basement may be used as a living space for a teenager, in-law, or illegal tenant who rented space from the owner to offset bills. Since this is classified as a private dwelling, there is no code inspection that can uncover this. That is why these areas must be searched. In this case, there are usually no means of secondary egress. The only egress to the exterior, especially in non-basement areas, will be via the interior stairs. During exterior size-up, firefighters should look for curtains on attic and basement windows and/or extra doorbells, gas meters, and mailboxes that may tip them off as to the presence of an occupant. The same |is true in areas (potential living spaces) over a garage. Ensure these areas are searched and question homeowners and occupants as to their use.

The last, and certainly not the least, of the life safety problems in these structures is to the firefighter. Since most fire deaths occur in this type of occupancy, it can be reasoned that many firefighters who are killed at fires are also killed in these types of occupancies. Since they are small in area, firefighters often operate alone. However, many private dwellings are customized to fit the taste of the owner, which results in such firefighter traps as dead-end halls and unusual passageways, double and triple-sized rooms, room-size closets, 2-story rooms, and atriums. When a firefighter finds that he or she is in trouble, there is often no one who knows his or her whereabouts until it is too late. No firefighter should ever enter a building alone without someone knowing where he or she is going and what he or she is doing. A working radio and a working PASS device are absolutely mandatory for these types of operations, as is a lifeline and a thermal imaging camera if the basement must be entered. And remember, a PASS device not turned to "arm" is like not wearing one at all.

Critical Interior Control Factors

There are five critical, interior control factors in this and every building fire. They are especially crucial in residential occupancies, but are also a major factor in other occupancies such as commercial, assembly, and the like. These factors follow the acronym CRAVE. They must be considered a system because they reinforce and support each other and the operation. As such, they are not meant to stand alone.

CRAVE

Command. Command can be thought of as the linking element between the other four factors, sort of the glue that holds it all together. Command must be a strong and consistent factor at all fire operations, whether they are offensive or defensive in nature. Due to the inherent danger of working inside a burning structure, offensive-interior operations will benefit most from a solid incident commander presence. To be effective, the incident commander must provide:

- Leadership
- Support
- Reinforcement
- Relief

The incident commander must be an information gatherer. All of this information must be utilized to complete the picture of what is actually occurring on the fireground. *(Ron Jeffers, NJMFPA)*

These factors are nurtured by timely and accurate status/progress reports from the companies on the interior charged with the responsibility for carrying out the other four fire control factors.

It has already been stated that 99% of the problems faced by the incident commander are from places that cannot be seen from the vantage point of the command post. For this reason, the reliance on reports from the interior are critical as they will assist in completing the fireground puzzle and allow the incident commander to match strategies and accompanying tactics with current and forecasted conditions.

For virtually all emergency situations, the incident commander must use a system of problem-solving steps to identify the problem, develop a solution, and put it into effect. The problem-solving process can be thought of as a loop, with the last step leading back to the first until the problem is eliminated. The steps are:

1. Identify the problem—this is basically size-up
2. Gather information on the problem—remember that some problem-solving attempts may have to be made without complete information.
3. Develop several solutions to the problem
4. Choose the best solution given the situation
5. Implement the solution
6. Evaluate the effect of the solution and adjust as necessary

If the problem still exists, the solution may need fine-tuning or a complete reworking. Go back to step #1 and repeat the steps again. Not all problems go away immediately. Continue to size-up the problem, reinforce it with current data, and continue to persevere. All problems go away eventually. The aim is to control the problem and not the other way around.

Rescue. Generally, these dwellings have the sleeping areas on the 2nd floor, but sleeping areas may be found on any floor. Rooms may be found in the attic and basement, as well as at the rear of the 1st floor. Primary search efforts should be focused on removing those in the areas of the fire first (fire floor), then checking all other areas, beginning with the floor above the fire. As stated before, the open, interior stairway will quickly render the main artery for escape unusable. For this reason, it is critical that ladder company SOPs address as many avenues of access into these structures, especially the upper floors as possible. These multiple avenues of access are also available as secondary egress points for interior teams searching on or above the fire floor.

Attack. A primary and continuous water supply must be established. If a secondary supply is possible with the initial compliment of manpower, it is a good idea to establish it.

A hose line with a minimum diameter of 1 3/4" must be utilized for the attack. It must enter the building via the safest most effective path of least resistance, usually the front door. It must go to the seat of the fire while at the same time providing

The initial attack line must be stretched via the safest, most effective path of least resistance to place the line between victims and the fire. This route should also attack the fire from the unburned side. A frontal attack will usually meet these criteria. *(Bob Scollan NJMFPA)*

protection for the main stairwell. Often, this can be accomplished simultaneously. If not, dependent on the fire's location and intensity, a decision must be made on whether to protect the stairs first and attack with the second line or vice-versa. Usually, the quicker that water is applied to the fire, the quicker the conditions will improve. It goes without saying that attack operations must be coordinated with support operations.

A second line, at least equal to the diameter of the initial attack line must be stretched, usually via the same access point, to back-up the initial attack line. This line may, if not required in the main fire area, advance to the stairwell and up the stairs to protect the operations on the floor above the fire.

If a third line must be stretched, and this is usually the case in a serious fire, it must access the building by a different route than the first two lines. This line may be assigned to the attic or to the top floor if a basement fire is in progress. It is best to hoist it up the exterior of the building. It may also be stretched via a ground ladder. Avoid using the aerial for this purpose. Another option may be via a rear or side door provided it will not be used in opposition with any of the initial attack lines. Proper communication and coordination on the part of incident commander with interior companies will avoid these problems.

Water supply considerations. Water supply SOPs dictate how the water is supplied, either as a forward lay, (hydrant-to-fire) or a reverse lay (fire-to-hydrant). If the first-arriving engine decides on an in-line attack, stretching a preconnected line from the engine without securing a water supply, they will certainly require a supply from the second-arriving engine company. This must be coordinated via radio upon scene arrival. A radio transmission such as "Engine #1 is on the scene, Engine #1 has the building" or "Engine #1 has the attack" should cue in the next-arriving engine company that a feed is required. In this case, the second-arriving engine company should state, upon arrival, "Engine #2 on the scene, Engine #2 has the water".

Narrow streets where the first two engine companies arrive before the ladder company are an ideal situation for "bumping". Here, the first engine becomes the water supply while the second engine takes the attack position. The front of the building must be left open for the ladder company.

There may be times on narrow streets that the first and second-arriving engine companies arrive in close sequence before the ladder company. In this case, if the first engine company hasn't begun to stretch a line, the second-arriving engine company can "bump" the first-arriving engine company. The first-arriving engine company moves up and becomes the water supply, while the second-arriving engine company takes the building and becomes the attack engine company. Their position will be past the building, leaving room for the ladder company.

In contrast, a radio report of "Engine #1 is on the scene, Engine #1 has its own water" should cue other arriving engine companies that a feed will not be required. The second or third arriving engine companies should position to secure a second water supply. When the first engine company has established its own water supply, the second engine company entering the fire block before the ladder company is improper because the second engine company will likely block the ladder company's access at the front of the fire building. Accurate and clear radio reports from first-arrivers, recon of the block before committing the apparatus, and simply being aware of what is happening on the fire scene should prevent improper positioning and resulting operational problems.

If a forward lay method is the department's hose lay of choice; the first-arriving company should initiate communication. If the procedure is to "drop and wrap" with the second-arriving engine company operating at the wrapped hydrant, the first engine company (attack pumper) must identify which hydrant they are wrapping so that the second-arriving engine company can establish the water supply in ample time.

Third and later-arriving engine companies may take positions as dictated by conditions, but if nothing unusual presents itself, the third engine company should back down to the ladder company. This will ensure a water supply for a ladder pipe operation or additional attack lines at the front of the building if they

Backing down the block provides tactical flexibility. It is always easier to pull out than to reposition in reverse. Make sure that the position of the engine leaves room for ground ladder access at the back of the ladder truck, usually about 25'.

are required. It is also a small insurance policy if the attack engine malfunctions. As stated, the best position is backing to the ladder company. Nosing in to the ladder company is always a mistake because the engine company is now hemmed in and any repositioning will present a major problem, causing a delay. If the apparatus has to be re-positioned, it is easier to pull straight out than it is to back out.

Apparatus positioning must be practiced and radio communications between engine companies must be a standard part of the response. It must be done at all incidents, so that at major fires, the chances of a mistake are minimized. It takes coordination, a good sense of judgment of the part of the engine company personnel, and a good knowledge of hydrant locations and the water supply profile of the area.

Ventilation. Ventilation of private dwellings can best be accomplished by breaking or opening windows and doors. It is generally not necessary to cut the roof of these structures unless the fire is in the attic, in a balloon frame, or extending up the exterior wall. Even a fire in an attic can be more efficiently and quickly vented by taking out the attic windows opposite the attack line advancement. Roof operations, such as cutting the roof, usually are the duty of a second or additional alarm ladder company.

Before committing to the cutting of a peaked roof, horizontal ventilation must be performed. If the wind condition is favorable and the fire is aggressively attacked from opposite the vent, further roof venting may not be required. *(Bob Scollan, NJMFPA)*

Venting opposite the attack line requires both recon from the exterior for the fire's location and proper and accurate reports from the interior. Venting for life in these situations is most likely accomplished by personnel accessing upstairs bedrooms via ground ladders or low roofs. . During this operation, every attempt should be made to confine the fire to as small an area as possible. This may be accomplished by closing doors. This not only tends to confine the fire, but it also buys some extra time for the search team operating above the fire. Personnel operating in these areas should be cognizant of fire conditions and attack operation status. This allows informed decisions to be made regarding the risks versus tenability profile of the area.

Once tasks are completed, it is critical for personnel to report the results to the incident commander. Likewise, any areas unsearched should be reported as an exception report so that the incident commander can devise an alternative way to get to these areas.

Extension prevention. Extension prevention takes two forms: pre-control overhaul and post-control overhaul. Understanding building construction as well as the concept of the path of least resistance is critical to both pre- and post-control overhaul success. The purpose of pre-control overhaul is to expose and cut off any extending fire. This can be extremely manpower-intensive and often means opening walls, drop ceilings, and cocklofts as well as exposing pipe chases and duct runs. Charged lines in the areas of these operations are critical.

Post-control overhaul will be conducted once the fire is under control. It usually begins where fire control operations end because manpower and lines are already in this area. The primary mission of the tasks associated with post-overhaul is to check for any fire still existing in the structure and ensure it is extinguished. Openings should be made in conjunction with the location and potential spread of the fire. Existing openings in construction should be checked first as fire will also follow the point of entry rule of thumb, the path of least resistance. These areas include light fixtures, pipe chases, electrical outlets, and switch plates. Additional openings may have to be made around door and window frames as well as directly over the fire area if there are no readily-available

Many older buildings have openings under the floor that can spread fire horizontally. This fire travel may then find a pipe chase such as in a kitchen or bathroom and spread vertically for the height of the structure.

openings in the area. Again, charged lines must be available. Prior to post-control overhaul, the building should be checked for any structural weaknesses and steps should be taken to illuminate and isolate any unsafe conditions, if found. In addition, operating personnel should be monitored at this stage for signs of fatigue. Utilizing fresh crews for post-control overhaul can help prevent injury caused by fatigue.

Overall (unless anything out of the ordinary is going on at the fire) operations in these private dwellings should be relatively simple. Strategy in the initial stages of the fire should be exclusively offensive. The basic operational action plan should include:

1. Secure a continuous water supply.
2. Conduct a primary search of all areas, especially the upper floors that contain sleeping areas.
3. Ensure secondary egress is secured for personnel operating on upper floors.
4. Stretch an attack and a back-up line of sufficient diameter to locate, confine, and extinguish the fire while protecting the means of egress, the open stairway.
5. Conduct horizontal ventilation opposite to and in coordination with the initial attack.

Many times, all of these operations take place simultaneously. However, if manpower is in short supply, those tactics that do the greatest good for the most people must be accomplished first.

Conclusion

In this chapter, solutions to problems encountered in private dwellings were explored. While the life hazard is usually not as severe as in larger buildings due to the sheer number of occupants endangered, the characteristics of the building that invite rapid firespread continues to make fires in these buildings the deadliest occurrences in regard to civilian fatalities and injuries.

Questions for Discussion

1. Discuss reasons why the home is the location where most fire fatalities occur.

2. Discuss ways that home repair specialists cause unsafe conditions that lead to fire ignition.

3. What are some of the indicators that can be used during size-up to warn firefighters of an unusual hazard in a structure?

4. Describe some of the problems created by combustible exterior walls and how these problems can be remedied.

5. What are some of the ways that the open stairway danger can be reduced from a life safety standpoint?

6. Using the CRAVE acronym, discuss basic firefighting procedures in private dwellings.

7. What are the steps of the problem-solving process and how does it lead to better decision making on then fireground?

8. Discuss the importance of communication and coordination in regard to water supply operations.

9. Discuss the differences between pre-control and post-control overhaul.

10. Discuss the factors that influence the dangers of operating on a peaked roof.

CHAPTER SIX
Multiple Dwellings

This chapter covers larger, residential buildings that house four or more families. This includes apartment buildings and tenements. Technically, smaller buildings that were once one- and two-family residences that have been converted to multiple dwellings (some illegally) are also multiple dwellings. The structural fire problems associated with these smaller types of buildings are more characteristic of private dwellings and will not be addressed in this chapter.

Larger, residential dwellings present problems of a more critical nature because of their size and the amount of occupants housed in them. In addition, like any other structure, the location and extent of the fire impact operations in a significant way. For instance, a fire in the basement will impact very heavily on the life

Multiple dwelling fire incidents present many problems to the incident commander and his troops. Increased life hazard, larger fire loads, and longer hose stretches all contribute to the difficulty of this operation. *(Bob Scollan, NJMFPA)*

The building in the center is being renovated after a fire that spread via the open cockloft to involve the top floors of the adjacent exposures.

hazard problem because the entire building and its occupants may be exposed, whereas a fire in the cockloft will not pose the same life hazard dilemma. However, a cockloft fire will necessitate longer stretches, more pre- and post-control overhaul, and a threat of adjoining building involvement if there are attached buildings to either side of the fire building.

Firefighting Problems in Multiple Dwellings

These buildings are usually constructed of ordinary or wood-frame construction, and while there are differences in the firespread characteristics of each, there are some similarities that impact the firefighting problem regardless of the construction.

Note the lines of demarcation (burn lines) where the fire traveled above the ceiling to the adjacent building via the open cockloft. Note that the partition wall between buildings only extends to the ceiling line.

Combustible cockloft

The cockloft in these buildings can be 3' high or larger and be a dusty, dirty lumberyard of wood and debris. A fire originating in or extending to this space causes tremendous difficulties for the control forces, often burning the roof off of the building. This problem may be greatly multiplied if the building is one in a row of attached structures with an open cockloft space over the entire row. Aggressive ventilation and confinement operations will often be the difference between saving the building and losing the block.

If there is a possibility of cockloft involvement, ensure that enough personnel are on the scene or requested early enough to attempt to stop the horizontal firespread in the cockloft. This is more critical in attached buildings of equal height because the cockloft is often open over the entire row of buildings. Get sufficient manpower to the roof to perform vertical ventilation. Generally, the roof should be cut in an area as directly over the fire as is safely possible. Other factors to consider are:

- *Fire conditions*—If it is not possible to cut directly over the fire due to deteriorating conditions, cut as close as possible to the seat and on the side of the fire's potential travel path, generally, the uphill, upwind side. Always be aware of the fire condition and path of travel. The fire should never get between the escape route and the roof team.

The safest cut is with the wind at your back, maximizing visibility and minimizing exposure to the products of combustion. Remember that the wind direction can change at an instant, so always be prepared to operate in less than ideal conditions. *(Bob Scollan, NJMFPA)*

- *Wind direction*—Make every attempt to cut with the wind at your back, however, if the best place to cut is near the edge of the roof, it may be necessary to work on the safer side of the cut, even if it means facing the wind. A firefighter fell to his death and another was severely injured when venting a fire caused them to lose their balance and fall off the roof. The windward side of the operation was between the cut and the roof's edge.

- *Building orientation*—Know where you are at all times, especially if you are making the cut, which usually necessitates that the other member back up as the cut is made. One method that works well is for an officer or a fellow firefighter to act as a guide for the cutting firefighter. The guide firefighter places his hand on the shoulder of the firefighter making the cut. As long as the cutting firefighter feels the hand

The guide firefighter must operate as the eyes of the firefighter making the roof cut. As long as the hands of the guide are in contact with the cutting firefighter, the cut continues. As soon as the guide firefighter breaks contact, the cutter should stop. *(Mike Borrelli, FDJC)*

High points may be used to forecast fire travel under the roof. Notice how the roof slopes upward from the center to the front and the rear on this "butterfly" roof. Fire will seek the highest point, taking the path of least resistance to get there.

on his back, he continues to cut. When the hand is removed, he stops. If the guide firefighter falls off of the roof's edge, at least the cutting firefighter will remain on the roof. For this reason, the guide firefighter must be cognizant of the roof's edge, keeping his attention on the path of travel. Let the firefighter that is making the cut worry about the vent hole.

• *Roof construction*–Built-up roofs can cause the saw to bind-up. In addition, be sure that the roof you are cutting has not been renovated to house a truss system. If you open the roof and find any structural members whose orientation is other than a 90° angle with the roof boards, raise the flag of caution and take possible steps to withdraw, especially if fire has entered the roof space. Roof structural members, at other than right angles, indicate the presence of a truss.

• *Positioning of natural openings*–All natural openings on the roof should be examined for excessive heat. The larger openings (such as skylights, scuttles, dumbwaiter shaft penthouses, and bulkhead stairway termination) must be opened as soon as possible to alleviate any fire traveling via paths of least resistance (*i.e.*, open vertical arteries). If the natural openings are in close proximity or directly above the seat of the fire, venting them may be all the ventilation required. This need must be coordinated with the reports regarding the comfort levels of the interior crews. More than likely the roof will have to be cut for a fire on the top floor or in the cockloft.

- *Manpower available*–On attached buildings, the speed at which the roof cut is accomplished may very well determine if the fire is localized or it spreads to adjoining buildings via the cocklofts. Opening the roof is a very work-intensive operation. Using one, two, or even three men to carry out this mission will put the confinement operation at a great disadvantage. Thus, it is imperative to reinforce the manpower operating on the roof of all top floor and cockloft fires as soon as possible.

- *Slope of the roof*–Even so-called flat roofs are not really flat. Some have a very pronounced pitch from front to rear that acts as a drainage. Others employ what is termed a "butterfly" roof, which slopes toward the center. Still others, like new-law apartment buildings, pitch ever so slightly toward the middle of the roof. If no other indication about where to cut the roof is present (such as bubbling tar, melted snow, or dry spots on rainy days) make the cut as close as possible to where you think the fire might be at the highest point on the roof. Fire travels from low points to high points, taking the path of least resistance. It eventually finds the high point, if it isn't already there.

Combustible cellar

The ceiling in the cellars are constructed of wood joists, which may be exposed. The wood planking above these joists actually are the underside of the floorboards of the 1st floor. If there was an attempt at some time to compartmentalize the cellar with fire-rated sheet-rock or in some cases, sheet metal, be aware that if the building has

Pokethroughs such as these pipe chases represent a path of least resistance for vertical fire extension from the cellar to the upper floors.

not been properly maintained, this fire barrier may be broken or missing in some places. This negates the intended fire stopping qualities and allows fire to spread up to the 1st floor without interference.

Poke-throughs, created by utilities, are other areas that allow fire to spread from floor to floor. The largest pipe chases are found in the bathrooms and kitchens. In multiple dwellings, these rooms are often stacked over each other and located back-to-back, creating a vertical flue for fire travel. Cellar pipe chases should be properly fire-stopped.

This curtain covers an illegal basement apartment. The fire department responded on "smoke in the basement" and found an illegal tenant had burned food while using an open fire to prepare his meal. Note the pokethroughs above the girder used to run piping and wiring.

Cellars in multiple dwellings are not only large but also present the problem of maze-like passageways, ventilation difficulties, and excessive debris and storage. In addition, there are many of these buildings that have illegal apartments located in the cellar. I have seen what appeared to be a row of storage closets with padlocks on them that were actually one-room apartments for as many as four people. Cellars are also the place where the garbage may accumulate as well as newspapers and magazines waiting to be collected by the recycling center. These are easily ignited and spread fire rapidly. Firefighters entering these areas should also be aware of the presence of dogs that are too big to be kept in an upstairs apartment, so they are chained in the cellar just waiting to take a bite out of a firefighter.

Light and airshafts

These features, also called blind shafts, were designed to bring ventilation to the inner rooms of an apartment before the availability of air condi-

This roof has a parapet at the front, connected end (to the right of the chimney), but no parapet at the rear where the shaft is located. Firefighters have been killed walking off unprotected building edges.

tioning. They can be found almost anywhere in a building. They may be found at the rear, but will not be visible at the front, or vice-versa. They may also be located in the center of a building, which will only be visible from either the roof, the hallway windows, or an apartment served by the shaft. Firefighters operating on smoky roofs should be particularly cautious when working on these buildings as the location of the shafts can change from building to building. You may be able to cross over a parapet at the front of the building on one roof, only to find that the when you step over the parapet on the next building, it is a 5-story trip to the ground. Always be sure of where you are stepping before you put your foot down.

This center shaft is completely enclosed. It is only accessible from a door off the first floor hall or via the fire escape. Ladder company personnel must ensure that the incident commander is made aware of the presence of this shaft and the conditions therein.

Some large enclosed shafts in the center of a building may be accessible via a fire escape or an access door and may even serve the building as a courtyard. However, smaller, narrower shafts that may only serve to bring air to adjoining buildings will only be accessible via a window on the shaft and a fire service ground ladder. A firefighter falling into the shaft from the roof will present an extremely difficult rescue

This tiny shaft can spread fire to the adjacent exposure. It is also an unprotected fall hazard. Rescue of a fallen firefighter from this shaft will be extremely difficult.

problem. In addition, ignition of debris accumulation in these narrow, inaccessible shafts creates an extremely difficult fire to access and fight.

I have also seen very small lightshafts that were diamond-shaped and served only two pairs of windows between adjoining buildings. I have also seen this shaft and window arrangement on old-style, 2-story taxpayers. It is not uncommon for these shafts to be covered by a skylight in order to keep heat inside the building. Renovations to the building may have completely covered the windows up that used to serve on the shaft and the shaft itself. The only remnant of the shaft may be a small patch of plywood, or worse yet, tarpaper over what was once the shaft opening. Always test your footing whenever operating on a roof, especially when it looks like the roof has

This open shaft has been covered over at the roof level. The difference in roof covering as well as the presence of a railing around the area should be a tip-off. These indicators will not always be present.

The narrow shafts at the rear of these attached wood frame multiple dwellings are unprotected and are covered by extremely combustible asphalt siding. A well-advanced fire may be next to impossible to stop from extending to the exposure via the shaft.

been repaired or modified. Different shades or patterns of roof coverings are indications that something may be amiss under the roof.

An important factor regarding firespread in shafts is the construction of the building. While all windows on the shaft are exposed, a wood-frame building will be more conducive to rapid firespread in the shaft via its combustible exterior walls than will a building of ordinary construction because there are more points of ignition in the shaft.

Another important factor is size of the shaft. Larger enclosed shafts and open-ended shafts will not become as large of a heat transmission problem to the exposed building surfaces as a narrower shaft. This is because the larger area of the shaft allows the products of combustion to dissipate more rapidly. A fire venting into a narrow shaft quickly heats the surrounding areas of the shaft and exhibits a greater threat of fire extension via the shaft. Even a large shaft with many windows does not present as much of a fire extension problem as a narrow shaft with only a few windows. The concept here is a simple one of heat transfer.

A smaller area heats up faster than a larger one does and if the rate of heat generation exceeds the rate of dissipation, ignition is the likely result as long as the parent body of fire is unchecked.

The most important factor regarding the shaft is its proximity to the main body of fire. A fire in close proximity to the shaft will most likely vent into the shaft rather than through windows on the exterior walls. If you see flame in an apartment and little or no smoke on the exterior as you arrive, expect it to be venting in an enclosed shaft. This may also be the case for a fire reported on a lower floor where there is a large amount of smoke coming from the roof. Have a ladder company check the roof and the rear of the building immediately. Sometimes the fire may be venting from a shaft that is open to the rear of the building.

Convection heat from the fire will swiftly move upward into the shaft, heating combustibles as it rises. The hotter the fire on the interior and the closer to the shaft it is located, the more exposed the areas bordering on the shaft will be. Furthermore, the hotter the fire, the greater the velocity of the thermal updraft there is in the shaft. This causes the convection currents to ignite wooden window frames, roofing materials, curtains, and other combustibles in the proximity of the shaft. For this reason, it is imperative to conduct extensive reconnaissance operations in all apartments bordering on a shaft. Be prepared to stretch lines to cover these areas in both the fire building and the exposure bordering on the shaft. The strategy most often employed will be one of an offensive/defensive nature with the major portion of the crews operating in the fire building, attacking the main body of fire, and checking the fire building's upper floors for fire extension. Additional crews will be required to operate in the exposed building, taking whatever measures are necessary to prevent ignition of the exposure. Ensuring that adequate manpower is on the scene to accomplish this is a critical aspect of the incident commander's duties. Additional alarms will most likely be required to accomplish all the essential tasks.

The only suitable strategy at this type of fire is to launch an aggressive-interior attack on the main body of the fire, while equally aggressive reconnaissance and confinement operations are being extended simultaneously. Knocking down the main body of fire will cool the hot gases, reducing the velocity of these gases into the shaft and subsequently the convection heat. It will also be advantageous, once the main body of fire is controlled, to direct the stream up the shaft to cool the area and extinguish any fire that can be seen. If the line cannot hit the fire, a report to the incident commander as to its location is critical so that the proper steps can be taken to control it.

I recently responded at about 2:00a.m. to a fire on the ground floor of a 4-story, Class 3 building. In the room adjacent to the room closest to the front,

This is the view of the narrow shaft from the fire apartment. The window across the shaft is the rear apartment of the fire building, an interior exposure.

there appeared to be what looked like a large bonfire. There was, however, a complete absence of smoke at the front of the building. The initial attack line was stretched to attack the fire, the roof was made, and ladder companies conducting a primary search for life and fire extension covered the remainder of the building. The ladder company on the roof confirmed that fire was venting into the shaft because the steam and smoke from the attack line could now be seen emitting from the area near the center of the roof. Fortunately, the adjoining building was also of ordinary construction. The only exposure fire was some ignited windowsills and some of the wood under the roof flashing on the exposed building. The point to keep in mind here is that if there is visible fire, but no smoke evident at the front, sides, or rear, expect the fire to be venting into a shaft. This may be further compounded at a night fire, as was the case at this fire upon arrival. Because it was dark out, the smoke venting out of the shaft was not initially visible. Only after water was applied to the fire, did the smoke turn lighter and mixed with steam, was more visible from the exterior. As always, preplanning, comprehensive reconnaissance once on the scene, and subsequent effective communication will make the difference at these type fires.

Attached buildings

The problem of attached buildings is more critical in frame tenements than Class 3 apartment buildings because fires can spread via the exterior walls, which are many times covered with a combustible covering, such as asphalt shingles. However, from the combustible cockloft standpoint, both of these building construction types are extremely susceptible to rapid horizontal firespread from building to building via the cockloft. This is even more critical in the presence of a stiff wind blowing over the cross-section of the structures. Manpower has to be committed early in order to head off the fire and confine it to the building of origin. This is essentially an offensive/defensive operation.

Attached building problems are not only reserved to the cockloft. Attached cellars are also a route for fire to extend to adjoining buildings. Many times, these buildings are owned by the same person and share common services, such as cable television and telephone systems. These may run via an unprotected conduit between buildings. Especially in a cellar fire, be prepared to fight an exposure fire in an adjoining cellar.

A benefit of attached buildings is that they offer options in regard to roof access. Attached buildings of equal height allow firefighters to easily access the roof of the

Fire and smoke venting into this narrow shaft negated many of the outward signs of a working fire at the front of the building. An aggressive interior attack on the main body of fire and early line placement in the exposed structure controlled the fire to the apartment of origin and prevented extension via the shaft.

Fire travel from building to building via the open cockloft has been the cause of many large fire losses. Early roof ventilation over the main body of fire will slow horizontal spread by drawing the products out of the building via the path of least resistance.
(Louis "Gino" Esposito)

Attached cellars will present some of the same fire spread problems as attached cocklofts. Ventilation, however, will be more difficult. Additionally, the whole building is exposed to fire spread. Get lines into adjoining cellars as quickly as possible to control horizontal fire extension.
(Louis "Gino" Esposito)

fire building by way of the adjoining building's interior stairs and bulkhead door. Firefighters using this route must be cognizant of shaft locations between attached buildings. Many times, shaft locations will not be consistent from one building to the next.

Attached buildings of unequal height must be treated the same way as an unattached building in regard to roof access. The best way to access the roof is via aerial or ground ladder. The next best way is via the rear fire escape. Since most fire escapes are narrow and difficult to climb while carrying tools it may be easiest to use the building's interior stairs to access an apartment on a floor below the fire, access the fire escape via a window, and then make the roof.

Roof access priority at multiple dwellings

Attached

1. Interior stairs of adjoining building of same height
2. Aerial
3. Lower floor (below fire floor) of fire building to rear fire escape
4. Rear fire escape via rear yard

Unattached (or attached, but different height)

1. Aerial
2. Lower floor (below fire floor) of fire building to rear fire escape
3. Rear fire escape via rear yard

Concealed spaces

Other than the open stairwell, concealed spaces are the biggest vertical firespread problem found in these buildings. It is more of a prevalent condition in buildings of ordinary construction, but can still also be found to a great extent in frame tenements.

These spaces, already stated in depth in the building construction chapter, include pipe chases and channel rails. Channel rails are only found in new-law apartment buildings of ordinary construction that are over 25' wide. They are the boxed-out spaces concealing the steel I-beams used to aid in the support of the floors. These concealed shafts run the vertical height of the building, often terminating at the underside of

Channel rails are often hidden from sight and may be found anywhere. These channel rails are only visible because the building is under demolition

the cockloft roof boards. A small vent or soil pipe usually extends through the roof where a pipe chase is located. In contrast, there will be no indication of the channel rail from the roof or probably from anywhere in the structure. Walls will have to be opened to ascertain the location of channel rails in an apartment. There may also be other vertical arteries that extend for the full height of the building, including dumbwaiter shafts, which may be enclosed and no longer used. These, however, will still spread fire throughout the building.

There will be a potential problem concerning concealed spaces and fire travel behind every wall and above every ceiling. Older buildings of ordinary construction will have a plaster on lathe framework or plaster on wire mesh. These are very difficult to open, may fail in huge pieces, and conceal a great deal of fire behind them. Frame tenements also have spaces between the joists, but fire traveling in a joist space usually is confined to one bay when spreading vertically and may not be so susceptible to horizontal spread.

Stacked kitchens and bathrooms

Buildings of this type often stack kitchens and bathrooms one on top of the other. They may also be located back-to-back so that adjacent apartments as well as stacked apartments can share a common pipe chase. This is a convenient way for fire to travel to upper floors. A fire in an apartment flowing across the ceiling often finds these areas, thus allowing the fire to extend to the floor above. Fires that are not in close proximity to a door or window still seek to rise to the highest level in the building possible. Don't be surprised to find fire from the floor below in either the bathroom or kitchen; it may be the least resistant path given the layout of the fire area and the location of the fire in relation to a door or window. Having this knowledge in advance allows firefighters assigned to fire extension prevention to focus on the most likely places for fire to extend into.

Tin ceilings are affixed perpendicular to the ceiling joists and will come down in sheets. It may also be necessary to pull down wood lathe as well as other sheathing above the ceiling before the joists above are exposed. A fire burrowing in this area may suddenly ignite when oxygen is introduced to the area. Have a charged line ready.

Tin ceilings

Tin ceilings are often found in very old buildings. The tin is nailed or tacked to the ceiling joists. This feature in a building can have two different influences on the fire operations, dependent on where the fire is in relation to the tin ceiling. First, if the fire is in an apartment below a tin ceiling, expect higher than normal heat conditions because the tin will not only hold the heat in, but also radiate it down into the apartment. This tin ceiling might also, on the positive side, keep the fire from extending out of the apartment of origin into the floor above. Be sure to examine areas where pipes and fixtures pass through the ceiling.

The other possibility is a fire that has extended into the area above the tin ceiling. These are extremely difficult ceilings to open and usually come down in large sheets. The trick is to find the seam that may be located at the decorative edge near the wall or use a seam created where a fixture or pipe passes through the ceiling. If conditions are not severe, I found it easiest (after the tack welds have been pried loose) to grab the edge of the sheet in gloved hands and tear the whole length of the sheet from the ceiling. It works much better than attempting to make continual purchases with a Halligan hook or pike pole. Before pulling the sheet down, make sure your eye shield is down because sparks are often released when the ceiling is pulled. In addition, make sure unnecessary personnel are kept clear and are prepared for immediate stream application into the now-open ceiling area. Any way you decide to accomplish it, the entire ceiling must be pulled in the presence of charged lines. There can be a massive amount of fire above the ceiling, so be prepared. It must also be stated that there have been backdrafts in these areas caused by improperly opening them from below before the roof has been opened. If there are any indications that a backdraft condition may exist, take all necessary precautions to prevent this from happening by first ventilating at the highest point before the ceiling is pulled.

Also, remember that many of these buildings have been renovated to include drop ceilings, which will lower the height of the ceiling, creating a small cockloft in each apartment. If the building was renovated more than once, there may be two or more cocklofts with a drop ceiling being added each time that the apartment was renovated. The uppermost ceiling above the drop ceiling is often tin. Fire can hide behind it.

Fires where tin ceilings are present are very manpower-intensive operations. A line is required in each involved apartment, as well as a crew of at least two men to pull the ceilings. If the ceilings are not pulled in time and water not applied from below, expect full involvement of the area above the ceiling, whether it is the cockloft below the roof or a space between the fire floor and the floor above.

Unsafe fire escapes

Many of the fire escapes in these structures are in a terrible state of disrepair due to lack of maintenance and lenient building code enforcement. Unfortunately, this only becomes evident when fire department members attempt to climb them at an incident. Firefighting personnel should use extreme caution when climbing, operating, or descending fire escapes.

Starting from the bottom up, drop ladders and counterbalance stairways are the most common structures between the ground and the 1st floor balcony. Counterbalance stairways are usually found on commercial occupancies.

Drop ladders are usually found on residential occupancies such as multiple dwellings. The drop ladder should be tested before use by pulling down on the drop ladder and ensuring it is secured in its tracks before ascending it. If at all possible, a fire department ground ladder should be used. Occupants on a fire escape should be viewed as a life hazard, especially if on a drop ladder. When encountering the elderly and children on a fire escape, it would be better to bring them into a window on a lower floor below the fire and escort them down the interior stairs than down a dangerous drop ladder. An exception to this rule is if the fire and firefighting operations are on the ground floor or in the immediate area of the building's main egress point. In that case, it would be better to take occupants down a fire service ground ladder placed on the wall just next to the lowest fire escape balcony.

Climbing or descending the steps on the balconies should also be done with extreme caution. Many times the steps are either broken or weak enough so that a firefighter stepping on them causes them to collapse under the weight of his foot. Let's face it; firefighters wearing full gear and SCBA are by no means light. Always test the step before placing your full weight on it. Also, step closer to the sides of each step rather than in the middle of it so there will be less stress on the step. If you find one step that is weak or already broken, chances are you will find another. If it is unavoidable that the fire escape must be used, use extreme caution, or safer and better yet, use another method of gaining access to an upper floor.

Gooseneck ladders are most often located at the rear of the building and are used primarily by roof firefighters to either access the roof in an unattached building or, more

One or more broken steps (shown here between the first and second balcony) should serve as a warning that other steps may not be broken yet, but may collapse under the weight of a firefighter.

often, to access the top floor to conduct ventilation, entrance, and search operations at lower floor fires. To ascertain whether the gooseneck ladder is safe to use, the firefighter should actually try to pull it out of its anchoring point. If it holds fast, it is reasonably safe to use.

Any fire escape that is suspect of, or found to be unsafe should be brought to the attention of the incident commander so all personnel can be alerted as to its location and hazard.

Another problem with fire escapes is the amount of debris that may be stored on them. Tenants use fire escapes for everything from flower gardens to bicycle storage areas. I have seen fire escapes that have been sealed with plastic and a bed and dresser placed on them. This was at the rear of a building and out of view from the street. At an alarm activation, the ladder man assigned to recon the rear of the building discovered it.

Another, even more dangerous condition because it is usually discovered at the worst time, is when occupants are trying to flee via the fire escape, is the tying up or

The gooseneck ladder is usually found at the rear of the building. Firefighters should try to pull this ladder right out of its supports before trying to ascend or descend this ladder.

These fire escapes are cluttered with debris and present a major hazard to both firefighters and occupants. Note the bicycle stored on the lower, rear fire escape balcony.

locking of the drop ladder. This is often done to keep burglars out of the upper floor apartments, and is usually found at the rear, in shafts, and other places out of plain site. This can cause an extreme panic condition on the part of the occupants on the fire escape and can overload the balconies as occupants get log jammed on the lowest balcony. It is best, if the area can be accessed, to use a fire department ground ladder to access the first balcony than to attempt to unlock the drop ladder. Proper and regular inspections along with prompt notification if an unacceptable condition is found are the best defenses against this potential problem.

Open interior stairways

This is the most common and dangerous method for fire to extend vertically in these types of buildings (the most effective path of least resistance). The stairway is also the main thoroughfare for building egress by occupants and access by firefighters. If the door to the fire apartment is left open, the escaping products of combustion will quickly fill the stairwell and trap occupants on upper floors. In addition, if the stairs and banister are made of wood, they may ignite, compounding the problem and possibly causing stair collapse. This is why the lower the fire is in the building, the greater the life hazard. In this instance, the firefighters who access the fire

apartment first should attempt to close the door if they are not carrying a charged hoseline. Once all occupants have safely evacuated the building and the roof door has been opened to clear the stairway of the products of combustion, the door may be opened and the fire attacked.

Open, wooden interior stairs will not only allow the products of combustion to spread to the upper floors, cutting off escape, but also add to the fire load of the building. Control and protection of the stairwell is critical at fires in multiple dwellings.

Renovations

Renovations to both multiple and private dwellings are major firefighter killers. With the vast array of multiple dwellings in the response area, especially urban areas, it is impossible to become familiar with every building. Unusual room sizes, access ways, blind stairways, dead-end halls, and adjacent buildings that are now connected in the cellar, on the upper floors, or both have all caused firefighters to become lost, trapped, and subsequently injured and killed. What's more, many maze-like cellars are now home to a dozen families with only one exit. The only indication that people live in these areas may be the sight of people coming out of them in masses during a fire. Also, buildings that had two or four apartments per floor now have four, eight, or even more "living areas". These areas are often separated by a flimsy piece of particleboard or even lightweight paneling. Most have no fire escape access. These buildings become death-traps and complicate the search effort and hose stretch problem exponentially. In North Hudson, it is

These Class 3 multiple dwellings have been renovated at the rear to include an addition constructed of wood frame. The exterior is no longer non-combustible and can cause fire to spread via the combustible wall.

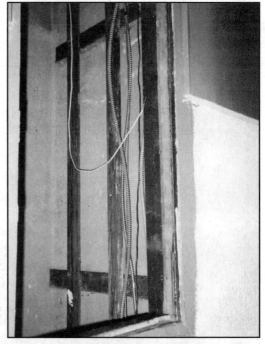

Rarely does a renovation make a building safer against fire. This dumbwaiter shaft is being used to run electrical wiring. This wire will then run horizontally above new drop ceilings via pokethroughs creating both a vertical and horizontal path for fire travel.

probably a good guess that more than half of all multiple dwellings in the jurisdiction have been renovated in some manner.

In addition, those buildings that were once private dwellings and that are now multiple dwellings can be a puzzle. There may be no telling how many people live in the building. This includes both the basement and the attic. Don't be surprised to see an unexpected amount of people on porch roofs, waiting at windows, or worse yet, jumping from these illegal multiple dwellings.

Companies should take every opportunity to make note of unusual features of buildings. In Weehawken, we used to have a report called a "Building Exception Report" (see chapter 12). It became known as the "Weird Building Report". Its purpose was to alert all members of the department to both dangerous and unusual structures in the jurisdiction. When an unusual building feature was encountered, the report was filled out, forwarded to fire prevention, and also placed on a department bulletin board for all members to review. Awareness is the first and most valuable defense in the recognition of factors that could cause firefighter casualties.

Especially in urban areas, cellular phone towers made of unprotected steel are often mounted on top of buildings. These have been bolted to the freestanding parapet wall. These building additions can fail early if exposed to fire venting from the top floor.

Gas-supplied utility fixtures

Before electricity, buildings were usually illuminated by light fixtures that were gas supplied. This condition applies not only to older multiple dwellings, but to most buildings built before electrical lighting was used. Piping was run in walls and above ceilings to supply these fixtures. As electricity became popular and buildings were renovated, the gas-supplied lighting was no longer used. The fixtures were removed and the piping was usually capped off. The piping, however, remained in the walls, concealed behind numerous renovations. Often, the gas supply was still present right up to the cap.

Older structures may still have piping in the ceilings and the walls that were once used to supply gas to lighting fixtures. There may still be gas in the pipe that will cause a "Class A" fire to behave abnormally.

This condition creates some unexpected problems for the fire service. Further renovation may damage the pipes, releasing the gas, which ignites when in contact with an ignition source. Other times, a fire in the building may cause a release of the gas. If the piping was capped by soldering, heat from a fire can melt the solder and release the gas. Other times, the piping may be damaged by collapse or by firefighting operations such as overhaul, which may lead to an explosion.

Gas-supplied fires may be recognized by a brighter flame that does not behave as expected when a stream is applied. In addition, the flame may appear to be emanating from a central point with a noted "velocity" and an absence of smoke. When operating, be alert to a hissing sound that may be present just before the gas ignites.

In all cases, treat the gas-supplied fire as you would a meter fire. Notify incident command, have the utility respond, protect exposed areas, and attempt to shut the gas off at its source. It is best to investigate for these conditions when buildings are being renovated and urge the owner to remove the piping before it has a chance to cause a problem.

Life Hazard Problems in Multiple Dwellings

Some of the same problems found in private dwellings are also found in multiple dwellings, although they will be compounded exponentially due to the greater size and occupant load in these buildings. A substantial life hazard will be present to some degree at all hours of the day. Panic is also more of a possibility at these structures, especially if the products of combustion permeate the open stairway.

Limited egress

Private dwellings, even those converted into multiple dwellings, often offer several means of egress. Multiple dwellings do not offer this same benefit to its many occupants. The egress opportunities are relatively limited in comparison with the number of people attempting to escape. Other than the fire escape, there is usually only one way out of each apartment, the front door. Most occupants will attempt to use the building's front door as a means of egress from the building. This can cause a salmon-like evacuation. If the fire is blocking the main path of egress, such as an arson fire in the front vestibule or a fire that has vented from an apartment to the interior stairs, it can cause panic, injuries, and death to occupants. For this reason, it is critical to gain control of the interior stairs as quickly as possible.

The further the occupants are from the street, the greater the danger. As stated above, this danger will be multiplied tremendously if the occupants of the fire apartment leave the door open when they flee the apartment. This makes the job of the fire department much more difficult as the stairwell on the fire floor and all the floors above will turn into a chimney, venting fire, smoke, and heat up the stairs. Opening of the bulkhead door or stairwell skylight to clear and alleviate heat conditions in the stairwell is critical in these buildings and situations.

Other exits from the building may be varied and include the fire escape, any doors leading to a courtyard, which will be found in some multiple dwellings, but not all, and the bulkhead door at the roof.

If the fire has extended into the hallway in any fashion, using the bulkhead stairs to the roof as a means of escape is a great risk. Some owners, to keep burglars and vagrants out of the building, lock the bulkhead door with heavy-duty chains and locks. If the door is locked and chained shut, the occupants will quickly succumb to the heat and smoke that will rapidly accumulate at these points.

Firefighters should never, under any circumstances in a working fire, use the interior stairs of the fire building to reach the roof unless the stairs are remote from the fire area, such as in another wing of the building. The stairway may be tenable one minute, but if the fire blows out of the apartment into the stairwell, these members will be roasted by the blast of heat traveling vertically up the stairwell. It is best to use an attached building of equal height and cross over the roof to vent the bulkhead. The next best method is the aerial and, lastly, the rear gooseneck ladder. If using the rear gooseneck ladder, it is often easier to ascend the interior stairs to a safe floor, enter a rear apartment, and go out the window to access the rear fire escape.

Speaking of fire escapes, it is vitally important that you know your buildings. Know which fire escapes are safe to use and which are too dangerous. Most occupants of a building attempting to escape a fire will do so via the interior stairs, while the fire escape is usually an afterthought. However, if the stairwell is filled with smoke, expect to find a multitude of people on the fire escape. This becomes a life hazard problem, especially if the fire escape is suspect and there are numerous fire escapes on the building. It is safer to use fire department ground ladders to bring people to the street from the lowest fire escape balcony than to let them descend the shaky, metal drop ladder to the ground.

In one multiple-dwelling fire I responded to, an elderly woman froze in panic on the fire escape drop ladder and stalled nearly two dozen other people above her. She held a death grip on the ladder while people actually were climbing over her to get down. It took firefighters several minutes trying to get her to loosen her grip on the drop ladder so that they could get her to safely descend.

Large multi-winged apartment buildings will often "empty" into a main lobby, where there is one egress point. This could create a logjam that can cause panic and injuries as well as a delay in fire suppression operations.

Large Number of Occupants and Apartments

It will be nearly impossible to account for all occupants of large multiple dwellings. "Shift" living is not uncommon with some tenants occupying the apartment during the day while others occupy it at night. In addition to the large number of living areas in the building, there may also be many illegal living spaces in the basement. Time of day, location and extent of fire upon arrival, and apparatus and manpower responding on the initial alarm will have the largest impact on whether the incident commander must request additional alarms.

It is anybody's guess how many people may be present in this building. Thus, life hazard is a major problem at any hour of the day in multiple dwellings. Recognize this problem and call for help early.

In a serious fire, all areas of the building will have to be searched. This requires a significant amount of manpower in a large building with many apartments. Is there a way that the incident commander can estimate how many apartments are in these buildings that will assist him in forecasting manpower requirements?

The fire escape rule of thumb

In buildings that were originally built as multiple dwellings and not converted from private dwellings, there is a rule of thumb which will allow the incident commander and firefighters involved in reconnaissance and primary search missions to estimate, at a glance or upon recon, how many apartments exist on each floor. It involves counting the number of fire escapes that serve the building and multiplying that number by two. For example, a tenement of wood-frame or ordinary construction that has no fire escape at the front of the building, but one at the rear will most likely have two apartments per floor, with one on one side and one on the other. These apartments will be laid out from the front to the rear and essentially split each floor in half. These are called "railroad flats" because the rooms are laid out in a row, one attached to the other, like a train.

Buildings with a fire escape on the front and the rear of the building, according to the rule of thumb will have four apartments per floor, two located in the rear and two at the front. The rear fire escape, in almost all cases will lead to the roof by way of a gooseneck ladder.

This rule also works for large apartment buildings of the H-type and other similar layouts. Find out the number of fire escapes and multiply by two. The total number of fire escapes in large buildings may not be determinable from a recon of just the front, rear, and sides of the building. There are usually additional fire escapes located in light shafts, which serve apartments on the inner portions of the building. The number and location of these must be ascertained from the roof level.

I used to live in an E-shaped apartment building with six apartments on each floor. There was no fire escape at the front of the building, however there was a rear fire escape with a gooseneck ladder and one at each side located in recessed light shafts; three multiplied by two equals six.

There are fire escapes on both the front and the rear of each of these attached buildings. There are four apartments per floor, two at the front and two at the rear

When sizing up for fire escapes, be sure to check the sides and rear of the building as well as center for the presence of fire escapes in shafts. Often, the only alternate access into the fire apartment is via fire escapes located in these shafts.

I have seen as many as eight fire escapes on very large apartment build-ings that had sixteen apartments per floor. There may be none in the front, two at the rear, one on each side, and four located in a center shaft of the building. The configuration possibilities are many and may vary considerably.

While it is impossible to determine how many people live in each apart-ment, and there may be as many as twenty, occupying the apartment in "shifts" due to work schedules, the Fire Escape Rule of Thumb can be a useful guide at a fire scene in estimating occupant load. Being able to ascertain this occu-pant load information quickly will assist the incident commander in balancing resources with requirements

Dogs

It seems that the only people dogs hate worse than firefighters are post-men and cops.

It is not uncommon to find dogs chained in the cellar. Even worse, some-times the dogs are given the run of the cellar. Firefighters attempting to search or locate utility shutoffs have been attacked and seriously injured by these animals. This problem is not only reserved to cellars. Many apartments also house these firefighter-unfriendly dogs. I was recently at an incident in a "project" building where there was a small fire in the cellar. As the upper floors were heavily charged with smoke, some people had evacuated the building. After the incident, as the occupants were re-entering the building, I counted nine pit bulls. Who knows how many were still in the building. The jaws on these pit bulls are like bear traps. Once they clamp down, it takes great effort to make them let go.

While it is fairly certain that any dog found in a commercial dwelling, espe-cially at night, will be unfriendly, dogs in residences may run the gamut from friendly to absolutely vicious. This is usually not evident until it is too late. Don't underestimate any animal that is not chained up. Treat all dogs as hostile. Take steps to protect yourself.

A brother battalion chief recently related one particular dog story to me. The companies had responded to a stove fire in a multiple dwelling. As they entered the main hallway, a pit bull ran past them and out of the building. The fire was in an apartment on the second floor. As the interior team neared the apartment door, they heard another dog behind the door. The dog was agitat-ed to say the least. The door was forced and as one of the ladder crew attempt-ed to re-close the door, a second pit bull ran out of the apartment. The dog went past three members at the door and latched onto one of the firefighters in the hall in frighteningly close proximity to the most private of areas. Each time, the dog was hit with a Halligan tool, he let go and reattached himself to another part of the now-panicking firefighter.

Meanwhile, the initial firefighter had entered the apartment and found that the fire in the stove was extending to nearby combustibles. He called for a line to be stretched and then went out to the hallway as he heard the commotion from below. Seeing that the dog was still latched onto the firefighter in the hallway, he whistled to get the dog's attention. The dog bolted at him, chased him back into the apartment and latched onto his boot. He managed to free himself from the jaws of the dog and locked himself into the bathroom. All this time, the fire was growing and now the firefighter was trapped in the bathroom with a wild pit bull at the door and a growing fire in the kitchen. In addition to the dog, an exception report was issued from the outside ventilation man who could not access the rear of the building, as there was a third pit bull in the rear yard. The line was stretched and the stream was first directed at the dog, which retreated to a bedroom, where the door was quickly closed, trapping him. The fire was extinguished without significant damage to the apartment. The firefighter who had been bitten suffered only minor injuries, thanks to his personal protective gear. The incident, while laughable later at the firehouse, could have had severe consequences.

The incident commander should also not forget that any escaped animals could also cause havoc in the street, tormenting spectators and disrupting the activity at the command post. The incident commander is responsible for the safety of all participants at the scene. Dog attacks are no exception. Be on the lookout for "Beware of Dog" signs as well as other signs (bones, dog droppings, etc.) that an unfriendly animal is on the premises. Take all necessary precautions to safeguard personnel.

Basic Firefighting Procedures

CRAVE

As the structures themselves are larger, the problems associated with them and the tactics required to bring a fire under control are more comprehensive in nature than the smaller private dwelling. Basic operations, as per the CRAVE acronym, should include the following:

Command

1. Develop and maintain a strong operational command presence
2. Demand that reports of conditions be furnished from all unseen areas
3. Decentralize the fireground as conditions dictate
4. React to reports with proper strategy to match conditions

Rescue

1. Recon all sides of the building, including all shafts
2. Conduct an aggressive primary search
3. Access upper floors by as many avenues of approach as possible
4. Ensure a thorough secondary search is conducted once the fire is under control

Incident command cannot begin to support the operation in unseen areas without proper and timely reports. All areas of the building must be checked as soon as possible and a report issued to incident command.
(Ron Jeffers, NJMFPA)

Basic rescue size-up: a chain across an apartment door is a reliable cue that some-one may be inside.

The Classes of Rescue

Rescues performed by the fire service can be divided into three categories based on the victim location and danger to the rescuers and the victim(s). The following are presented in order from simplest to the most complex and dangerous.

Class I Rescue

- The rescuer(s) know the location of the victim(s)
- Making the rescue does not put the rescuer(s) or victim(s) in any great danger

An example of this would be a child locked in a car, a simple door pop on an automobile, or the removal of a person from an uninvolved exposure.

Class II Rescue

- The rescuer(s) knows the location of the victim(s)
- The rescuer(s) must put theirself and the victim(s) in harm's way

Examples of a Class II rescue would be where firefighters must rescue a victim from an elevated location during a fire via a ground ladder, aerial, or rescue rope. Another example would be a high-angle rescue where the location of the victim(s) is known and can be seen.

Class III Rescue

This is the most dangerous, daring, and difficult rescue.

- The rescuer(s) does not know the location of the victim(s)
- The rescuer(s) and the victim(s) are in harm's way, often in extreme danger

An example of a Class III rescue would be an attempt to find a missing victim(s) at a rapidly advancing fire under deteriorating conditions.

Rescue from an elevated position is a dangerous operation and will place both the fire-fighters and the victims in danger. This is a Class II rescue. Personnel must be well trained to perform any type of rescue, however, the less known about the victim's location, the more danger involved. *(Newark, New Jersey Fire)*

Rapidly deteriorating conditions often make Class III rescue attempts difficult, if not impossible. This room is exhibiting signs of flashover as evidenced by the color and volume of the smoke as well as the glow of the fire. A limited search via this ground ladder may be the victim's last chance. *(Bob Scollan, NJMFPA)*

Attack

1. Establish a strong primary and secondary water supply
2. The first line must protect the stairway
   a. Take the safest, most effective path of least resistance to place the line between the fire and the victims while at the same time protecting the stairway.
   b. Reinforce the attack with back-up lines (most multiple-dwelling fires require at least three lines in the fire building).
4. Coordinate attack operations with support operations

Pump pressures for attack lines. Calculating the pump pressure in your head on the fireground is relatively easy. Flowing 150gpm out of a length of 1¾" hose results in about 18psi friction loss per length. Round this number off to 20psi. The extra couple of pounds per square inch per length will help account for friction loss due to elevation. You can estimate that 180psi will flow 150gpm through four lengths of 1¾" hose with a fog nozzle (20psi for each length and 100psi for the fog nozzle). Any additional hose will result in less water and more friction loss. This is not an exact science, but by using this guide the pump operator will be in the right ballpark.

Applying this same formula to a solid bore nozzle will flow the same 150gpm through seven lengths at 190psi (20psi per length, 50psi for the solid bore nozzle). These pressures will take into account the friction loss in the 1¾" hose at this many gallons per minute and the nozzle pressure for the nozzle chosen. Remember that if the line is flowing the proper water, the attack team will encounter less punishment and a quicker knockdown will be the result. Hit the fire hard with sufficient water and the game is over quickly.

This may raise a question from those departments that use fog nozzles as attack lines, and are required to stretch more than four lengths to reach the fire area. Four lengths and fog nozzles are fine for preconnects and will be sufficient for short stretches such as private dwellings and fires on the lower floors of multiple dwellings.

This 200' preconnect is about the limit you can effectively stretch 1¾" hose with a fog nozzle. Stretching any more lengths with this nozzle or line diameter is counterproductive.

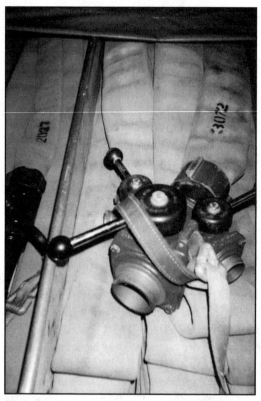

For longer stretches, use of a gated wye (in this case, a water thief) is necessary to keep friction loss and pump pressure to a manageable number. It also allows the second line to be operational in a shorter time compared with stretching a second line off the apparatus.

If the fire is on an upper floor and the preconnect will not reach, the solution will be to use larger diameter hose as a feeder line. Using a 2½" or 3" line as an attack supply line from the engine to a gated wye, where 150' of 1¾" attack line is attached, will allow a small, mobile line to attack the fire and solve the problem of long stretches in these types of buildings. The gated wye can be placed either at the door of the building or even several floors above grade on a stair landing inside the building (if there is a longer than usual stretch). It is also easier to stretch a back-up line from the gated wye. It is in-service quicker and results in less hose in the street and hallway. If you must stretch ten lengths of hose as your attack line from the engine, you have to stretch ten more lengths as your back-up line. This will result in excess hose congesting the main access (and egress) point, the stairway.

Again, estimating the pump pressure is easy. Use 100psi for the fog nozzle or 50psi for the solid bore, 20psi for each length of 1¾" attack line, and 5psi for each length of 2½" supply. Don't worry about the friction loss in the appliance because the formula adds a little extra pressure to each length. In fact, in most buildings, you can probably get away with not adding in elevation. To add a second length to the gated wye, simply change the figure for the 2½" hose to 10psi per length because now you are flowing 300gpm to the wye, where it then splits to 150gpm to the two 1¾" lines.

For this sixth floor fire, this type of hose lay with a fog nozzle flowing 150gpm would require 100psi for the fog nozzle, 60psi for the attack line (3 lengths of 1¾" at 20psi), and an estimate of four lengths of 2½" (20psi) hose. The pump pressure to flow 150gpm on the fire floor would be 180psi. This same stretch using strictly 1¾" hose (seven lengths) with a fog nozzle would result in a pump pressure of at least 240psi to flow 150gpm on the fire floor.

If a solid bore nozzle were being used with the 2½" supplying the 1¾" attack lines, the required pump pressure would be 130psi. Seven lengths of 1¾" hose with a solid bore nozzle would result in 190psi to flow proper water at the fire floor. This is still acceptable, but remember any back-up line must also be seven lengths, resulting in time loss and potential line congestion. Solid bore nozzles may be the answer to friction loss problems with long stretches. If you must insist on fog nozzles, at least try to use one of the newer low-pressure fog nozzles. Some have nozzle pressures as low as 65psi.

This tangent on pump pressures was not intended to teach hydraulics, but to show how an easy rule of thumb can assist the pump operator in "guesstimating" pump pressures to ensure proper water flow. There are no excuses for firefighters injured and buildings lost on account of inadequate water flow caused by ignorance.

Ventilation

1. Aggressive fire confinement tactics must be employed to prevent spread, especially in attached buildings.
   a. Vent at the top of the vertical artery to clear the stairwell
   b. Vent appropriately in relation to the location of the fire
   c. Horizontally ventilate opposite the attack line
   d. Cut the roof at top floor and cockloft fires
2. Coordinate ventilation with attack operations

Extension Prevention

1. Consider the paths of least resistance for fire travel
2. Check all concealed spaces for extension, both vertically and horizontally
3. Have charged lines in areas of vulnerability

In the foreground are scuttle hatches and skylights. In the background is a bulkhead door with a skylight on the sloping bulkhead area. These natural ventilation openings must be opened early to channel smoke and heat into the path of least resistance.

Old-Law Vs. New-Law Buildings

Multiple dwellings can be classified according to how they were built and as a result, how a fire will extend in them. The most critical difference in the delineation between old-law and new-law buildings is the relation of the below-grade area to the rest of the building. Old-law structures have cellars, which are almost entirely below grade, while new-law buildings have basements that are usually located at grade or near it. Old-law cellars are more susceptible to firespread and ignition than new-law basements. This significantly affects tactics and strategies in fighting fires in these buildings, especially the cellar.

Old-law buildings refer to old wood-frame tenements and Class 3 tenements that were built before the turn of the century. The stairwells are constructed of wood and are open for the entire vertical artery, terminating at a bulkhead door or perhaps a scuttle or skylight. The stairs, in addition to being combustible, will usually collapse with the adjoining walls and many times will burn through and aid in spreading fire to the upper floors.

Directly below the interior stairs is the access to the cellar. The cellar is usually separated from the rest of the dwelling by a flimsy wooden door at the top of the stairs, and, if you're lucky, one at the bottom. Recently, codes have been adopted by some jurisdictions that require this door at the top of the cellar stairs to be solid core construction and self-closing. As an equivalency, a panel of sheet metal can be added to the cellar's side of the door to act as a fire barrier. While this is a step in the right direction, the fact remains that a fire in the cellar will still have access to the 1st floor and the rest of the building via the open vertical stairwell. In addition, the ceiling above the cellar is often unprotected, exposing the wood planking that supports the 1st floor. Fires originating in the cellar can spread upward to the floors above via pipe chases before burning through the door at the top of the cellar stairs. Furthermore, as old-law cellars are usually totally below grade, there are very few ventilation opportunities,

These combustible wooden stairs are both the main egress route and the path of least resistance for products of combustion to the upper floors. Protection of this artery is the main strategic focus at an old-law building. The doorway to the cellar is directly beneath these stairs.

which results in a punishing fire advancement operation for attack teams. From a life safety standpoint, these old-law buildings are deathtraps.

New-law buildings were designed as a result of the problem of fires extending out of the cellar and into the building's vertical arteries. In new-law construction, the basement is no longer directly accessible to the building's interior and will be at grade level or only partially below grade. To access the basement, one has to either go through a courtyard accessed via a small tunnel-like corridor under the building or through a door accessed via a half-landing adjacent to the main interior stairwell inside the main hallway. There is no direct interior access from the main building to the basement. The basement door may be located in some type of light or air shaft, either on one of the unenclosed sides of the building or in an enclosed shaft or courtyard.

The concept behind new-law construction is that the basement is only accessible from the exterior. In addition, stairs in the building are no longer made of wood, but are usually of marble or masonry set in steel pans that typically will not collapse with the adjoining wall. Although the stairwells will still

In a new-law building, stretching the attack line via this exterior opening will provide crews with the safest, most effective path of least resistance into the basement. Stretching via the front door will usually result in nothing more than congestion of the main egress route.

This door beneath the main stair leads to the courtyard where the basement door is located. This will require a longer hose stretch when compared to the street-level opening.

be open, they will not be combustible and will terminate on the roof at a bulkhead door. The most significant feature of new-law construction is that the basement is isolated from the remainder of the building. The basement ceiling is required to be covered with a non-combustible material, usually fire-rated sheetrock or plaster. While fire and smoke spread via pipe chases, channel rails, and other vertical shafts is possible, the interior stairs will remain relatively clear for occupant escape while the fire is attacked in the basement.

New-law buildings are relatively easy to identify. They will usually be at least 25' wide and are often called H, E, U, O, or double-H buildings. This is due to their shape. To assist in supporting the floors of these wide-front buildings, steel I-beams are placed at strategic locations in the building's structure. To hide these beams, builders box them out, placing them in closets or other out-of-the-way places. These "channel rails" are a ready path for vertical and horizontal fire travel in these buildings. In a serious fire, channel rails must be located and checked for the presence of traveling fire. If you come across a closet that is not square, but has a box-like protrusion in one of the corners, it should be opened and examined as it may be a channel rail. In addition, check the same location on the floors above, as firespread may be extremely rapid in these areas.

Finally, as new-law basements are either at grade level or partially below grade, there will usually be better ventilation opportunities available to fire-fighters than their old-law counterparts. While usually much larger (such as apartment buildings) new-law buildings are still infinitely more fire-safe than old-law buildings.

Old-Law Construction Vs. New-Law Construction

Old-Law
- Cellar accessible from interior
- Wooden Stairs
- Cellar ceiling exposed
- Ventilation difficulty

New-Law
- Basement access only from exterior
- Non-combustible stairs
- Basement ceiling protected
- More opportunities for ventilation
- Presence of channel rails

Fire-Resistive Multiple Dwellings

Fire-resistive or "fireproof" multiple dwellings are constructed of reinforced concrete and steel. This type of construction is used not only in residential dwellings, but also in schools, hospitals, and dormitories, which are all of low-rise height. There are characteristics in fire-resistive multiple dwellings that act as our ally and others that hinder our efforts to control a fire. Fire-resistive multiple dwellings are often 6 stories in height or more, many times used as low- or moderate-income housing (sometimes known as "projects") and may or may not be equipped with auxiliary appliances.

Regarding firefighting, fire-resistive multiple dwellings will exhibit many of the same characteristics inherent in residential high-rise firefighting due to the fire-containment qualities of the structure. This includes both high heat retention and the ability of the building to limit the fire to the apartment of origin. Some of the characteristics of fire-resistive multiple dwellings are discussed next.

Fire-resistive multiple dwellings are common to "project" areas and are often set back from the street. Setbacks cause increased reflex time in reaching the fire, longer hose stretches, and increased fatigue.

Characteristics

Lack of fire escapes. The building is designed to hold the fire to the apartment of origin, much like a fire-resistive high-rise. In addition, enclosed stairs act to provide refuge and escape from the building as a fire escape does in an ordinary, combustible structure.

Lack of a cockloft. The combustible cockloft is the largest horizontal void space in a combustible multiple dwelling, and as a result, poses the threat of extensive firespread above the top floor, many times involving the roof. The combustible cockloft also causes the fire to spread via this space beyond the building of origin. In a fire-resistive multiple dwelling, this is not possible due to the lack of a combustible cockloft.

Setbacks. These buildings are often built in clusters as part of a "project" property and are often set far back from the street. This causes longer than usual hose stretches with accompanying friction loss problems. It also causes difficulty in reaching the building with the aerial, both to windows for rescue and the roof for ventilation. Fortunately, the building is usually equipped with several enclosed stairwells that are remote from each other, allowing the roof team to access the roof in a relatively safe area provided the proper stairway is chosen.

Difficult forcible entry. The doors are often made of solid core design, using metal doors encased in metal frames that are set in block walls or double fire-rated sheetrock walls. Standard forcible-entry methods are not effective against these doors. The use of rabbit tools or irons substituting a maul for a flat head axe is warranted for gaining entry to these doors. In addition, these doors hold heat in, which is both good and bad. The compartmentalization will result in an earlier flashover, due to the fact that the heat will be confined. However, the door will act as a liaison in keeping the fire contained to one apartment. If the walls are of double fire-rated sheetrock instead of concrete block, it may be easier to breach the adjoining wall than it is to force the door.

Enclosed Stairways. The presence of an enclosed stairway is an ally to the fire attack team. It allows the hose team to prepare the attack in a relatively tenable position, while allowing occupants who have chosen to flee the fire to do so in a relatively smoke-free environment. Companies attacking the fire must ensure that the stairwell is clear of any occupants before opening the door to the hallway. Once the attack has begun, the stairway will be filled with smoke and the products of combustion, thus making it impossible for occupants to use it as an evacuation stairway. This operation must be well coordinated between the attack team and the recon teams. Assigning a chief officer to this position is a wise action.

Scissor Stairs. Scissor stairs are often found in these buildings. If you are unfamiliar with their presence and characteristics, it can lead to a logistical nightmare. Scissor stairs are utilized when two stairwells are run in the

For metal doors, use of a rabbit tool or a sledge in place of the axe will make entry easier. Ensure control of the door is maintained as there may be a great deal of fire behind the door.

same shaft. As such, they cross each other at alternating floors so that a stair serves the north side of the building on even floors, for example, while the other stair shaft serves the north side of the building on odd floors. Scissor stairs require a great deal of pre-fire knowledge as to their presence and which stairwell serves which area of the building. If there is a standpipe in the scissor stairs, it can be even more of an ordeal and cause attack teams to run short of hose if the wrong stairwell is chosen. For instance, the standpipe hookup may be in the north end on the 5th floor and in the south end on the 6th floor. If the fire is on the south end of the building on the 6th floor, hooking up to the standpipe on the 5th floor and advancing up the stairs will put the attack team on the opposite side of the floor and make for a longer stretch, which is often in less than ideal conditions.

Sometimes a return stair is run adjacent to a scissor stairwell. A return stairwell is one that exits in the same location on each floor. If the standpipe connection is located in the return stairwell, this will simplify the operation as the return stair can be used as the attack stair and the scissor stairs as the evacuation and ventilation stairwells. Similar to attack operations in enclosed stairwells, buildings with scissor stairs will also benefit from the assignment of a chief officer to coordinate operations in these areas.

Fire departments must insist that building owners identify these stairways on every floor so that firefighters who are directing people to scissor stairs are able to choose the correct stairway. It is as simple as marking one stair "A" and the other stair "B", but this marking must appear on every door on every floor to be of any use to the fire department.

In addition, stenciling "Roof Access" on the appropriate door on every floor, regardless of building construction or what type of stairway the building has, will aid in making the operation safer and more coordinated.

Long hallways. Long hallways require long stretches that are often in punishing environments, especially if the wind is blowing into the fire apartment from the exterior. Fortunately, codes in most of these buildings require that doors to apartments be self-closing. This makes the advance more tenable. However, if tenants remove the self-closing mechanism, the hallway may

This stairwell should not be used as a vent stair as it does not pierce the roof. It may be used as an evacuation stair. Stairway doors should be marked on every floor of the building.

be a dark, swirling mass of smoke and heat, thus requiring a Herculean effort to advance. Firefighters in this instance should be alert for signs of rollover and periodically test the atmosphere for heat so as not to be caught in a flashover traveling down the hallway.

Dead-end hallways. The presence of a dead-end hallway can cause a firefighter to become trapped if the fire erupts out of a doorway between himself and his egress point. Entry teams should always, when forcing or entering an apartment, position themselves on the egress side of the apartment door. This way, if fire erupts from the opening, escape is possible. This feature must be identified in pre-fire planning visits. Many firefighters have been killed in dead-end hallways.

Conclusion

Multiple dwellings, especially non-fireproof ones, always present a major challenge to the fire service. These are manpower-intensive operations because of the large amounts of exposed occupants and the ability of the fire to rapidly extend throughout the building, especially in the main path of egress, the interior stairway. Be aware of the magnitude of the problems present and ensure the proper resources are summoned to safely mitigate the situation.

Note how the stairwell door is located some distance from the end of the hall. Searching firefighters may miss this opening and wind up looking for an exit at the end of the hall where none exist. If fire blows out of the apartment door on the left opposite the stairwell, firefighters at the end of the hall may become trapped.

Questions for Discussion

1. Discuss fire control in multiple dwellings in regard to the CRAVE acronym.
2. Discuss methods aimed at limiting firespread via shafts.
3. Discuss the differences between new-law and old-law construction.
4. What major differences should be included in the action plan when fighting a below-grade fire in a new-law building as compared to an old-law building?
5. How can the Fire Escape Rule of Thumb be used to assist the incident commander in forecasting manpower needs?
6. Discuss the effect of friction loss in regard to long hose stretches in multiple dwellings and the ways that this problem can be remedied.
7. Why is the top floor a pivotal strategic point in multiple-dwelling fires?
8. Discuss some of the considerations involved in deciding where to cut the roof at top floor fires in multiple dwellings.
9. Discuss the three classes of rescue and give examples of each.
10. Discuss the advantages and disadvantages inherent to fire-resistive multiple dwellings.
11. Discuss some of the differences between fire-resistive and non-fire-resistive multiple dwellings.

CHAPTER SEVEN
High-Rise Operations

A fire in a high-rise building is one of the greatest challenges that the chief officer and the fire service in general will face. The design of the building may or may not cooperate with your efforts to extinguish the fire. The degree of this cooperation will depend in a large part on the incident commander's ability to control the building systems and (hopefully) the mass of troops assembled to fight such a fire. An improperly coordinated operation will result in a logistical nightmare.

Most modern high-rise buildings are of fire-resistive construction. A working definition for the categorization of a high-rise building can be described as any building over 75' in height and equipped with a standpipe and/or sprinkler system. The construction may not always be fire-resistive, and could be of ordinary construction depending upon the age of the building.

This definition of a high-rise is not totally accurate from a strategic and logistical point of view. Many departments have limited or non-existent aerial capability. Thus, any building out of the reach of the department's ladders can be considered, as far as that department's aerial capability is concerned, a high-rise.

While pre-fire plans will aid a great deal in familiarizing firefighters with the building, only a structured approach to the incident stabilization problem will result in the safest execution of tactics required to bring the situation under

High-rise buildings create a myriad of strategic problems for incident command. The main lobby in this building is located on the top of the cliffs, and is actually on the 18th floor (arrow). A fire below this floor may lead to confusion and chaos for those unfamiliar with this building. Effective pre-fire planning is a major component of a successful operation.

control. This structured approach should have been planned for prior to the incident through pre-fire logistical visits. These visits and subsequent analysis of the building should lead to the development of effective standard operating procedures for high-rise operations.

Some cities have many high-rises, some do not. Others only have high-rises in certain districts of the city. Those companies that respond to many high-rises will have the opportunity to hone their high-rise operation through experience. It is this "bread and butter" approach that allows the coordination of command and control operations to be sharpened.

In contrast, in those jurisdictions that have only a few high-rise buildings and fewer high-rise incidents, it is even more critical to have a plan in place. For the department whose experience with high-rise fires is limited, disciplined adherence to the adopted standard operating procedures will be the difference between a smooth, controlled operation and a chaotic, helter-skelter affair of free-lancing and uncoordinated activities.

Plans for manpower, deployment, and tactical reserve must be established beforehand via mutual or automatic aid. The command structure and accompanying standard operating procedure must be simple and, if possible, adopted by the entire mutual aid group so that all responding departments are on the same page.

It is essential that the incident commander have strong command and control over the operation. This requires not only discipline on the part of the participants, but a thorough knowledge of how to establish an organization which will support the efforts of those who are assigned the actual tasks of fire location, confinement, extinguishment, and the accompanying support operations.

This old-style high-rise is equipped with many of the features of a Class 3 apartment building. Fire escapes and operable windows are not found in a modern high-rise. Note the concentrated roof load created by the antennae as well as the extended roof cornice.

The most important aspect of a successful high-rise operation is the establishment of an efficient command organization. The sheer magnitude of the incident requires that this strong command organization be expanded to fit the incident. As such, the incident commander must be prepared to reduce his span of control and decentralize the fireground by establishing control points to organize and coordinate operations in specific areas of the building.

In regard to the action plan and the procedures established, any personnel assigned specific incident command responsibilities are essentially control devices to assist in meeting command and control requirements. Incident commanders who fail to delegate responsibility for the various areas of the building will be quickly overwhelmed. In fact, most unsuccessful high-rise operations are the result of uncoordinated, unorganized operations and violations of the principle of span of control.

Command and Control Operations

Establishment of incident command

The establishment of incident command is the first and most important control operation to take place at a high-rise (or any) fire. The first-arriving chief officer will establish a command post as per established department procedures. If the first-arriving company commander has already established command operations, the first chief will assume command of the operation. His next responsibility is to begin the establishment of control points to serve as "gates" or access points to operational areas. Strategically placed and well-managed command and support positions in these operational areas will give the fire forces the best chance for success at this difficult fire situation.

High-rise control points

Command post. The incident commander must establish a command post in the building lobby or other designated area. A good place for the command post is at the building's communications desk, which is usually located in the building lobby. This allows access to building communication systems, alarm annunciation equipment, and building personnel. This command post location should be arranged beforehand by pre-fire planning. A liaison with building maintenance personnel should be immediately established. A set of building floor plans should be made available for the command post and, if available, a copy should be sent up to the operations post.

This high-rise has a communications station complete with public address capability to all areas of the building as well as hard-wired telephones for designated control point to command post communications.

Fire department personnel should get control of, and ensure the proper operation of the building's systems. These will include ensuring operation of fire pumps, shutting down of HVAC systems, controlling elevators, and initiating the use of any specialized communications systems such as standpipe telephones and building public address systems. The early control of building communications will allow the command structure to expand in an organized manner and is critical to the safety of all operating personnel.

From the command post, the incident commander can formulate a strategy to mitigate the incident based on reports of conditions generated in the fire area. Additional alarms should be transmitted as required, with responding companies reporting to an established apparatus staging area. This staging area should be close enough to the building so the reflex time is not so long, but far enough away so as not to endanger personnel or apparatus from the dangers of falling glass and other hazards from the building.

An important, often overlooked matter that the incident commander should order is the silencing of the alarms. It is tough enough for firefighters to operate on the fire floor and adjacent areas without the incessant blaring of an activated fire alarm.

Command company. In the initial stages of an incident, especially at high-rise operations, the incident commander must take actions to prevent being overwhelmed while trying to sort out the mountain of preliminary information being directed to the command post. In addition, there are other tasks that directly support the command post and fire operation that need to be addressed as quickly as possible.

One method of breaking down the initial organizational demands to a manageable degree is to request a command company, also known as a support company. This can be an engine company that is assigned as part of the initial alarm response, an additional alarm company, or it may be a special call. Having an extra few sets of hands in this early juncture of the fire operation will be critical to the quick establishment of functions and positions that must be put in place to support both incident and firefighter safety.

The command company operates to support command post activities in the initial stages of the operation. The company may operate as a team or be split up. Some of the functions they may perform are:

- The officer may be assigned as fireground safety officer.
- The officer may be assigned to the command post to assist in command functions.
- The officer may be assigned as required to staff the position of water supply officer, rehab post officer, communications officer, or other command positions such as planning or liaison officer.

- The officer may be assigned a division to effectively decentralize the operation into manageable parts.
- The firefighters may stay with their officer to assist him with assigned duties or they may be assigned to other duties or officers.
- Firefighters may be assigned the task of initiating accountability procedures, such as setting up a command board and coordinating the personnel accountability tags (PATs) or riding cards of companies as they report to the command post.
- Firefighters may be assigned such necessary, but peripheral duties as silencing alarms, checking on the status of fire pumps, shutting down utilities and HVAC systems, checking areas of importance that may not as yet have companies assigned, and coordinating SCBA filling operations.
- Firefighters may be assigned to supplement the manpower contingent of FAST or RIC teams.
- Firefighters may be charged with assisting the apparatus staging officer or resource post officer in demobilization activities such as placing companies back in service and re-staffing relocated firehouses.

In instances where more than three engine companies are assigned to the response, a unique option opens up to the incident commander if he chooses to take advantage of it. The incident commander can choose to use the entire first-arriving engine company as a command company. In a three man engine company, the officer becomes the initial operations officer, coordinating the attack on the fire floor in the crucial beginning moments of the operation. The chauffeur becomes the elevator control. The third firefighter becomes a command aide at the command post or an aide to the operations officer. If a fourth firefighter is assigned, he can operate as the aide to his officer, the initial operations officer. This option is acceptable in a high-rise fire, where the fire is usually compartmentalized and possibly being controlled by sprinkler activation. At this juncture, command and control (essentially firefighter safety) are more important than the attack operation in the initial stages of the fire. In addition, using the first-arriving company in this manner takes the burden off the other members from the second and third arriving companies who will now be responsible for the initial attack line. The result is increased coordination and safety due to the presence of proper supervision in the fire area during what is usually the most chaotic portion of the incident.

The assignment of a command company is not exclusively reserved to a high-rise operation. It should be utilized whenever the incident commander foresees a need to delegate specific functions on the fireground. Likewise, the number of command companies is not limited to one. At a very large incident, there may be a requirement for additional command companies to meet the needs of the incident. It is always better, from an organizational point of view, to break the fireground down into manageable parts in the initial stages than to be overwhelmed at a time that will impact a great deal of the outcome of the operation. It has been said the first five minutes can dictate the next five hours. The use of additional resources to help set up the organization will make the experience more palatable and easier to manage.

Lobby control post. The lobby control post should be the second job assigned after the command post is established as the control organization literally builds from the bottom up. Initial lobby control may be handled by the member that is designated elevator control, but this responsibility should be assigned to another officer as quickly as possible. Headed by a lobby control officer and possibly a staff, lobby or access control is an extremely crucial item to establish, especially in the initial stages of the operation. The lobby control post is the access or jump-off point to upper floor operations. The lobby control officer is responsible for the funneling of crews from the command post to the proper elevator or stairwell to ensure they arrive safely at the resource post on one of the upper floors. The lobby control post is the foundation of the personnel accountability system at an incident. No personnel are to be admitted to an upper operational area without first passing through the lobby control post.

More often than not, there will be several stairways to choose from. Even if the elevators are used, accessing the proper stairwell from either the lobby or from an elevator bank on an upper floor is one of the most important aspects of controlling personnel at the scene and ensuring their safety, especially in the area of the fire. Personnel taking the wrong stairwell can get themselves into serious trouble.

The lobby control post, often staffed by a subordinate officer or the chief's aide acts as the gatekeeper of the access to the operational area. This point should also initiate accountability functions at the scene by keeping a secondary or separate status board that will account for all companies passing through lobby control, and eventually passing back through as they are released from the scene. Whatever personnel accountability system the department uses, whether they be tags, riding cards, or what North Hudson calls MACs (Magnetic Accountability Cards), it should be dropped off at the lobby control post if it isn't already being handled by a designated command company.

Elevator control. This post is usually staffed by one or two (if you're lucky) of the members of the first-arriving engine or ladder company, or may be assigned to a chief's aide, depending on department SOPs. The duty of elevator control is to be a manpower and equipment shuttle between the lobby and the resource post two floors below the fire. The elevator must be placed in the "Fire Service" mode. As soon as this is accomplished, a report must be made to the command post, informing incident command that the elevators are under fire department control.

Forcible-entry tools should be kept in the elevator at all times in case there is a need to force open the elevator doors during a malfunction or unintended electrical shutdown. The member or team placed in charge of elevator control must remain in control of the elevator and not relinquish this position until incident termination or properly relieved by fire department personnel at the orders of a chief officer.

A key responsibility of the member assigned elevator control is to limit the amount of manpower and equipment that enters the elevator at any given time. Some elevators, especially in older high- and low-rises were just not designed to handle the load placed on it by firefighters and their equipment. The result may be failure at the most inopportune time. Worse yet, the elevator may get stuck between floors and expose the members inside the elevator to serious danger. Depending on the age and weight limit of the elevator, it is usually not a good idea to place more than four of five firefighters and their equipment inside the elevator. This number includes the elevator control man. Always read the occupant and weight load placard inside the car and adjust accordingly.

At incidents where elevator use is not warranted, the elevators should still be placed under the control of the fire department to prevent any occupants from using the elevators. This does not mean that a member must be assigned to elevator control, but by placing the elevators under fire service control, this helps ensure that they are unavailable for occupant use or unauthorized firefighter use for the duration of the incident. Placing barrier tape across the elevator area also reinforces and prevents any unauthorized use of the elevator. The best action to take may be to shut the elevator down and lock out the operating controls to ensure that unauthorized use is prevented.

Operations post. The second-arriving chief officer, usually a battalion chief, will be designated as the operations chief and will set up the operations post one floor below the fire. The problem here is that inevitably, the initial attack teams will usually arrive at the fire area and begin to operate before an official operations post can be established. Command and control is critical in this initial operational period and must be maintained in this area. Information about conditions are at their most nebulous at this period. This is the time when the incident commander at the command post in the lobby needs the most information and usually gets the least because companies on the fire floor are operating and a chief officer has not yet arrived at the fire area to establish a control point. It is a very helpless feeling to be standing twenty floors below where companies have been sent and not know what is going on up there. If things are going to go wrong, they will usually do so during this initial phase of operation, most often due to uncoordinated actions. It is critical that someone take command of the operation on the fire floor prior to the operations post being formally established. The best person to perform this function is the first-arriving officer in the fire area, possibly from a designated command company. He should stay out of the hands-on operation and become both the fire attack coordinator and the communications link between the fire area and command, a sort of functional operations officer, until relieved by the designated operations chief. This will allow for a safer operation because the command function will be maintained in the fire area as soon as the first personnel get there.

Once the designated chief officer arrives, a formal operations post must be established. The location of the operations post should be in close proximity to the designated attack stairwell, which should have been established by the initial attack units. Usually, the operations post is set up after the placement of the initial attack teams, but if it is not, one of the responsibilities of the operations chief will be to establish an attack stairwell, and if possible, an evacuation and ventilation stairwell. This information should be relayed as soon as possible to the command post. This information will enable the lobby control officer to send additional companies to the operations post via the safest route.

The operations chief will be responsible for bringing to life the strategy being developed at the command post. He is responsible for operations on the fire floor and floor above. For this reason, he may have to be positioned in the attack stairwell part of the time, on the fire floor or floor above part of the time, and at the operations post part of the time. It is crucial to have an aide assigned to the operations chief who will remain at the operations post when the chief is operating elsewhere.

The operations post must furnish progress reports to the command post at regular intervals, informing the incident commander of current conditions, manpower, and equipment requirements. It is best to operate on at least two frequencies, one as the incident command channel and one as the operations channel. This separation of strategic and tactical frequencies will not unnecessarily tie up the radio. If standpipe telephones are available, it will be even more efficient to communicate to the command post in this manner, as portable radios are often inefficient at these buildings. Even cellular phones may be used in a pinch.

Communication between the operations post and the command post is essential. To allow for a more efficient communications link to incident command, the operations post should be established, if possible, in the vicinity of a hardwired standpipe telephone.

The operations chief is also responsible to keep an account of companies operating in his assigned areas. One simple way to keep track of companies is by using a black magic marker and a wall to account for personnel. This requires no other equipment than the marker. It doesn't require operations boards and tactical worksheets and can be as large as required. The walls will most likely be painted after the fire anyway.

The initial manpower compliment assigned to the operations post should be at least two engine companies and one ladder company on the fire floor and the same on the floor above. In large buildings, if the manpower is available, twice as many men may be required. This may require additional alarms and further incident command decentralization to maintain the proper span of control. The FAST team will also be positioned at the operations post, properly equipped and at the ready. At these operations, there may be a need for two or three FAST teams. If there is an abundance of men operating, provisions must be made to arrange for their safety in the most efficient manner possible.

Resource post. The next control point to be established will be the resource post. The resource post is to be located one floor below the operations post and is the jump-off point for companies reporting to the operations post for assignment. It is essentially the personnel and equipment staging area for the operations post. The resource post will also be staffed by a chief officer and possibly an aide. The reason it is set up after the operations post is that just about all of the companies on the initial alarm will be working in the operations area and there will be no need for the resources post in the very early stages of operation. As soon as additional alarm companies begin to arrive, they report via the lobby control post to the

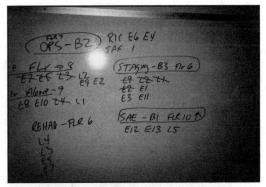

To keep track of operating personnel, find a good wall and a good marker and you are in business. This blackboard approach will help organize the ever-changing participants in the fire area. This crude tracking system will also work in the resource and SAE posts.

resources post to await assignment. In this way, unassigned companies will not be streaming into the operations post, which could lead to freelancing and a loss of operational control in the fire area.

The resource chief is responsible for accounting and staging all companies reporting to the resource post. Again, a black magic marker used on a wall will provide an easy way to account for personnel moving in and out of the resource post. In addition to personnel staging, the resource post is also the place where the equipment pool is established which will be fed up to the operations post as required. The resource chief must coordinate manpower and equipment requirements with the command post, the operations post, and SAE (Search And Evacuation) post and request additional companies as needed. He should keep the resource post staffed with at least two engine companies and one ladder or rescue company at all times. Upon request, companies are sent to the operations post or SAE post for assignment.

A further duty for the resource chief is to set up a rehab post three floors below the fire (one floor below the resource post). Medical personnel and equipment, water, and possibly food should be made available at the rehab post. It is the responsibility of the resource chief to send companies to the rehab post for relief, while still maintaining proper manning levels at the resource post. If responsibility for the rehab post can be delegated to another officer or maybe to the EMS chief, it would allow the resource chief to concentrate on the manning requirements of the incident. This delegation and further decentralization of the fireground provides for increased safety for operating personnel.

Search and evacuation post. The search and evacuation post should also be supervised by a chief officer. It is one of the more dangerous operating areas in the building as the entire operation occurs above the fire. The SAE post must be kept informed of conditions on the fire floor and floor above. Therefore, it is critical that progress reports regarding conditions on the fire floor be furnished by the operations post to the SAE post at regular intervals. The SAE post should be established in a safe area. A position in close proximity to the elevators and/or stairwells where there is a standpipe telephone jack and/or a telephone is an ideal place for the SAE post in case radio communication is not working properly.

The responsibility of the SAE chief is to reconnoiter all floors from two floors above the fire to the roof. Operations include a search for fire extension, especially in shaft areas and utility closets. However, the main function of the SAE post is to determine whether occupants should be evacuated or protected-in-place. This decision will be based primarily on information received from the operations post. If evacuation is not necessary, any residents already out in the hallways and especially in the stairwells can be placed in nearby apartments with neighbors until the emergency is over. No one should be wandering the hallways. Any occupants who must be evacuated should be brought down designated evacuation stairwells. It is critical that this operation be closely coordinated with the operations post. In addition, any upper floor ventilation operations must also be coordinated with the operations post.

Manpower and equipment requirements at the SAE post as well as rehab requirements are to be coordinated with the resource post. The SAE post should be manned with at least four companies at all times. In large buildings, it may be best to split companies into two teams, assigning one team of two companies to search from the roof down while the other team of two companies search from two floors above the fire upward. Progress reports of search results and conditions should be made to the SAE post every five floors or as established by SOP. It is the responsibility of the SAE chief to keep the operations post informed of the status of SAE post operations and conditions.

Firefighting Problems in High-Rise Structures

Extreme heat

Most high-rises are constructed of fire-resistive materials. By their very nature, they are designed to contain a fire within the area of origin. To accomplish this, some very heavy building material must be used. In the case of fire-resistive high-rises, this is usually reinforced concrete, on the ceiling, the floor, and possibly the walls. Double-sheetrock attached to aluminum studs between apartments may be used, but will provide less fire-resistance than concrete.

Concrete walls, ceilings, and floors will tend to compartmentalize the fire, but will hold heat in like an oven. If the wind is blowing into the apartment, attack crews may have to resort to alternative methods for accessing the seat of the fire. *(Bob Scollan, NJMFPA)*

Reinforced concrete tends to completely compartmentalize the occupied spaces, whether they be offices or apartments. A fire within the confines of a compartmentalized area is akin to entering an oven. Since the fire's heat has basically nowhere to go, all the heat is confined to the compartment of origin. While this is an ally to the cause of fire containment, the heat produced may not allow a close approach to the fire area. For this reason, 2½" lines for commercial high-rises or 1¾" or 2" lines for residential high-rises, both equipped with solid bore nozzles for penetration and reach, are preferred. If the fire floor or apartment is still untenable, then some creative solutions may be in order. These include attack lines advanced in tandem, master streams from the exterior if they reach the fire floor, or interior deluge sets from the hallway. Other alternatives to a close approach may also be to breach the wall of the apartment next door and apply a large diameter handline into the fire area to cool it down so the attack may be resumed and the line advanced. Still another alternative that is gaining acceptance is the draft curtain. It is designed to defeat the blast-furnace hallway effect of a wind-driven fire. This curtain, made of material similar to a welder's tarp, is draped in front of the fire apartment windows from the floors above to cut off the effect of the wind, allowing the attack team to advance. The prevailing wind blows the tarp against the building, sealing off the window, easing conditions in the hallway.

The best protection in any building would be an automatic wet pipe sprinkler system, but these are usually only found in commercial high-rises or only in the common areas such as the hallways and lobbies in residential high-rises. Rarely are sprinklers located in the individual apartments.

Spalling concrete

The extreme heat condition due to the compartmentalization characteristics of fire-resistive construction may cause a spalling of concrete in the fire area. When concrete is poured, it is nearly liquid. As it hardens or cures, much

of the moisture evaporates out of it. However, some of the moisture always remains trapped in the concrete. The heat of a fire causes this moisture to expand, causing spalling, the localized collapse of an area due to this expanding moisture. Caused strictly by direct flame contact at temperatures around 2000°F, spalling will result in small pieces or even relatively large chunks of concrete to

The heat of this fire caused extreme spalling of the concrete ceiling above. Note the exposed steel reinforcement rods. *(North Bergen Fire)*

be jettisoned from the structural mass of concrete. In fact, if a fire from below heats a steel roof or floor deck upon which concrete has been poured, it will actually cause the concrete to collapse or spall upward. This is because the steel deck keeps the concrete from dropping down toward the heat source.

The key to the spalling threat is directly related to the amount of moisture present in the concrete. In respect to moisture, let's compare concrete to wood. Just like wood, the longer the concrete is in place, the less moisture will be present. However, this factor has an inverse impact on concrete. Wood that has been in place for long periods of time is drier, and therefore easier to ignite. Concrete that is in place for a long period of time will also have less moisture, and this will make it more resistant to spalling, which is a benefit for the fire personnel working in the area. Streams applied ahead of the advance to cool the concrete overhead will alleviate some of the spalling problem of concrete, but may create steam, generating another problem. If this tactic must be used, ensure adequate ventilation is provided opposite the hose line.

Manpower requirements

A substantial fire in any large building will require a considerable amount of manpower. High-rises are no different. Any fire is these buildings will require a larger than usual contingent of men to control. In fact, it has been stated that the manpower requirements will be four to six times greater than that for a normal fire operation. No matter how you slice it, that is a large commitment of manpower for just about any department.

As we saw at the beginning of this chapter, just setting up the organizational structure to initiate any control tactics requires about a dozen men. This

Attack equipment must be transported to the fire area in the easiest way possible. This high-rise attack kit, known as the "SPU" (Standpipe unit), was developed by three North Hudson members. It is lightweight, detaches easily from the wheel assembly in tight spaces, and in addition to an attack line, has a compartment for spare cylinders and standpipe tools.

manpower requirement should be calculated into the initial response any time companies are dispatched for a reported fire. Pre-incident planning will play a large role in the estimate of manpower requirements. This is where, if you don't have the resources available, you must come up with a way to make them available. A solid mutual aid agreement between several municipalities is a large step toward ensuring the safety of the men through numbers, for manpower is the name of the game at high-rise fires. You must literally throw men at the fire. This will require determining, in addition to command and control organizational requirements, the manpower needs to simultaneously have men operating to control the fire, men recuperating at the rehab post, and men staged at the resource post waiting to be sent into the battle. If you don't have them, you better know where to get them, and the sooner they are summoned, the safer and usually more successful the operation will be.

Reflex time required to mount an attack in an elevated position will allow fire to grow and spread. The prudent incident commander takes reflex time into account and ensures sufficient resources are summoned early. *(Mike Johnston)*

Reflex time

Reflex time is the time that lapses from when you first think you need a resource to the time that is operating at the scene or at least standing by in the ready mode in staging. Chances are if you summoned help early or with the initial dispatch, you will already have companies in staging or at the Resource Post and be a step ahead of the game. If, however, you have exhausted men and have an empty Resource Post, you are in deep trouble, my friend. No matter how good your action plan is, when you run out of people to put it into operation, you run out of options and, essentially, forfeit your plan and usually the game.

Resource needs are about forecasting, and forecasting is about foresight. This requires proper supervision from the start of and even before the incident. The very word "supervision" means "extra vision". This is exactly what is required at these incidents, the insight to plan in a proactive manner in regard to manpower and have it on the scene before the momentum of the game shifts in the fire's favor.

Control of building systems

This is another critical factor in the incident commander's attempt to initiate proper command and control over the incident. These critical building systems include:

Communications

- Where the communication center for the building is located
- If standpipe telephones are available
- Where radio "dead spots" may be located

Elevators

- Location of keys
- Which elevators serve which areas, including blind shafts and sky lobbies
- Where the elevator control room is located

HVAC Systems

- Where system deactivation controls are located
- How to access the duct shafts
- Location of any emergency power sources

Auxiliary Appliances

- Location of connections
- What each connection serves
- How to shut the system down and drain it
- Where the alarm panel is to silence alarms once the fire is located

Fire Pumps

- Location of pumps
- How to operate pumps
- How to override the system

Building Maintenance Personnel

- Know who the reliable contact people are
- Blueprint availability
- First-hand knowledge of building features and systems

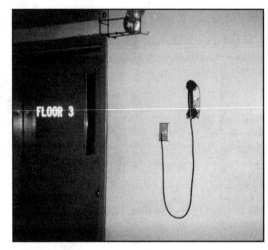

Communications via portable radios may be impossible due to the infrastructure of the building. A standpipe telephone system is one way to overcome this problem. Jacks should be located in strategic spots throughout the building.

In large-area buildings, not all fire department connections supply all areas. Having the system marked reduces confusion.

All of these systems should have been accounted for in pre-fire plan surveys and accompanying SOPs. Again, a proactive approach will save a considerable amount of time. Building familiarization on a regular basis will help reduce game day problems. Knowing where the building systems are located and how they can be used to our advantage will positively impact on the fire situation every time.

Standpipe pressure reducers

Pressure reducers are utilized on standpipe systems where occupants may be expected to operate a hoseline. The reason for this is that building occupants are not trained to handle properly pressurized fire streams. To counter this problem, pressure reducers are added to the system, usually in the area of the hoseline connection or the operating valve. Firefighters must be familiar with the various types of pressure reducers that, if not removed or overridden, will prevent the development of a proper stream on the fire floor. There are several types of pressure reducers. Some are designed to create additional friction loss, thereby reducing flow, while others are intended to limit the control valve operation.

Those designed to limit flow by increased friction loss are usually found attached to the supply side of the standpipe connection where the house line is threaded. It may be a set of overlapping holes that reduces flow or a special 2½" adapter with an inside diameter that has been reduced to 1½". These must be removed before the fire department connects the attack lines.

This flow reducer may be found on the supply side of the standpipe system. It is intended to restrict the flow to the line. It must be removed to ensure the total amount of flow is available to the attack lines.

The other type of reducer operates to reduce flow. Called a "seco pressure reducer," it is a pin similar to a fire extinguisher pin, which does not allow the control wheel to be opened fully, thus limiting flow. Firefighters operating the attack line must remove the pin to acquire maximum flow

Don't be surprised to find both types on the same outlet. To avoid unsafe and

A plate installed inside a reducer is used to restrict flow to the occupant hoseline. This reducer must be removed prior to connecting a fire hose to the standpipe connection.

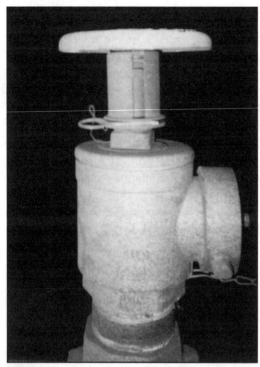

This "Seco" pressure reducer restricts the operation of the control valve, limiting flow. Removal of the pin is similar to that of a fire extinguisher.

inefficient pressures, always probe the inside of the standpipe to check for obstructions before attaching the fire department lines. It is also a good idea to quickly flow the standpipe to not only ensure that there is water available, but also to flush any debris in the pipe that could clog the hose line and/or nozzle.

Another condition, although rare, may occur when a standpipe system similar to a pre-action sprinkler system is used. Water supply to the system is held back by a deluge-type valve. The system requires a fire detection device to initiate a water supply to the standpipe. If the fire and smoke detection system is out of service for repairs or due to tampering, no water may come out of the standpipe when the control wheel is opened. Likewise, if a fire of a minor nature is present or is blowing out a window and issuing little or no smoke on the interior, the detection devices may not activate. If this is the case, the standpipe may again remain dry and deliver no water to attack teams. What will be necessary to activate the system will be to manually pull a fire alarm to activate the system. Departments must preplan the types of auxiliary systems installed in each building and be prepared to address system-specific problems that may arise.

Difficulty in ventilation

Due to the fact that these fires are often out of reach of aerial devices, ventilation opposite the advancing hose line will be difficult, if not impossible. Fortunately, very hot fires will usually blow out windows. However, if the wind is blowing into the apartment and the apartment door is open, it will make the advance down the hallway akin to advancing into hell. For this reason, it is a safer move to check the wind's direction and strength from the floor below the fire before venting the fire apartment windows. When the hallway proves untenable, large diameter lines with solid bore nozzles are mandatory.

Ventilation, whether caused by the heat of the fire or by firefighters will also cause a flying glass hazard around the perimeter of the structure. Scene control is critical to ensure no one is beheaded or sliced to ribbons by falling glass.

Add to this the problem of stairwell coordination. It is generally frowned upon to inadvertently use the evacuation stairwell as a ventilation stair, especially by the occupants who are trying to escape. Careful coordination is critical when venting the fire floor via a stairwell.

Other problems regarding the movement of smoke are stratification and the stack effect.

Stratification of smoke

As heated gases rise in a building due to convection, they will, in time, lose their buoyancy. This is due to the physics of heat transfer, which seeks to balance out all matter. Heat in the smoke will be lost to the surrounding areas as the smoke rises. When the temperature of the smoke balances out with the ambient temperature of the surrounding area, it will no longer rise, but will sit or stratify at wherever that point of temperature balance happens to be. As such, it will be extremely difficult to ventilate utilizing natural methods. It is sometimes referred to as "cold smoke" and is usually present where sprinklers have operated. It is equally as difficult to ventilate properly using natural methods.

Stratifying toxic gases have caused many deaths on floors above the fire. This is especially true in buildings built many years ago that have open stairwells and large atriums that allow smoke to travel upward unimpeded. This smoke will have to be removed mechanically by fans or via close coordination with building engineers in the use of the HVAC system.

The stack effect

The stack effect is caused by temperature differentials inside buildings. The larger the building, the stronger the effect. Stack effect can cause air movement, including smoke movement, to travel in directions it wouldn't normally move. This effect will be even more pronounced when the temperature differential between the outside air and the inside environment are extreme.

Normally, heated air rises to the top of the building. However, when the atmosphere inside is much colder than the outside temperature such as in the summer when the air conditioning is on non-stop, colder, heavier air on the inside will sink to the lower portions of the building. This may pull the smoke with it, causing a smoke condition below the fire. If a stair shaft is being used as a ventilation shaft to allow smoke to move to the upper floors and out via a bulkhead stairwell door, the ventilation company may be in for a surprise.

A similar problem may be encountered when attempting horizontal ventilation. As stack effect will pull air into the shafts, it may cause an effect exactly opposite that which is desired when attempting to take out windows on the fire floor. This is likely when attempting to vent high floors when the temperatures are high or lower floors when the temperature is low. Taking out windows on high floors when the temperature outside is high will, pull the products of combustion toward shafts where the dense cooler air will pull it to the lower floors. On the other hand, venting windows on lower floors when the temperature outside is also low will cause the products of combustion to again be pulled to the shaft, but this time they will rise with the heated air being produced by the building's HVAC system. Again, it is wiser to check the wind's reaction to venting on a floor out of the area of the fire. The point to remember about stack effect is that if there is a great disparity in the temperature between the building's interior and the outside air, be very cautious in your horizontal venting.

Unsafe aluminum balconies

Many high-rises have aluminum balconies allowing the residents a scenic view, which is one of the major lures of high-rise living. A tubular aluminum railing usually encloses these balconies. Sometimes the only way to vent the apartment windows or balcony glass is to use a tool from above or an adjacent balcony to accomplish this task.

Aluminum melts at approximately 1200°F. A fire venting out of an apartment can easily surpass those temperatures, subjecting the aluminum to failure. In fact, it will have lost a great deal of its strength before this point is reached. A firefighter leaning against the balcony railing to attempt an outside vent can cause a collapse of the railing. It is usually a long way down and, unfortunately, gravity never takes a day off. In addition, these railings are often set in concrete. A fire from below can cause spalling of the bottom of the balcony above, furthering weakening the bond between the railing and the balcony.

Fire exposing a balcony may not only spall the concrete above it, but also cause the aluminum railing to become weak and collapse when leaned against. The balcony on the floor directly above the fire will be most susceptible to this weakening. *(North Bergen, New Jersey Fire)*

Firefighters performing overhaul of a fire apartment sometimes seek fresh air on the balcony when taking a breather. Do not, under any circumstances, lean against, put pressure upon, or rely on the railing for any reason. In fact, it should be avoided altogether. To prevent unnecessary tragedy, it would be a good idea to cordon this area off with whatever means are available.

Enclosed underground parking garages

The fires most likely to occur in an underground parking garage will be a vehicle or rubbish fire. While these are usually routine fire responses, the location of the fire creates some unique problems for responders. If the garage is on a slope or a gradual incline from the lowest floor to the entrance of the garage or vice-versa, the fire will, much like a wildfire on sloping terrain, follow the path of least resistance and spread to adjacent vehicles on the uphill side. As vehicles are usually parked in close proximity to each other, several vehicles may be involved by the time the attack line is in place. Reflex time will impact greatly on a fire in these areas.

Most enclosed garages have extremely low clearance. It will not be possible to fit apparatus into the garage. Therefore, unless the fire is in close proximity to the garage entrance, the best way to attack is via access stairs to the garage running from the interior of the building or by standpipe, if available. Remember also, that if the apparatus "just makes it" into the garage, once the booster tank is empty and several tons of weight is lifted off the apparatus springs, the apparatus may not make it back out. Know how high the apparatus is with and without a full booster tank.

The greatest protection against the spread of fire in these areas of the building will be an automatic wet sprinkler system. While it will probably be successful in holding the fire in check and possibly preventing a major extension problem, there will be a significant problem of cold smoke, decreasing visibility. Smoke, whether hot or cold will be extremely difficult to ventilate in these parts of the building.

The manner of operation will depend on the location and extent of the fire and the presence of auxiliary appliances, namely sprinkler systems and standpipes. If a standpipe is present in the area, operating in these structures will be like operating in an upside-down high-rise. If possible, the fire should be attacked via an enclosed stairwell from the floor below the fire just as in a high-rise. If this is not possible, attack from the fire level, using the standpipe in an enclosed stairwell.

If there is no standpipe outlet present in the garage, the fire will be attacked similar to a cellar fire. This may require long stretches. In this case, the use of a 2½" supply line to a gated wye near the fire will allow two lines to be stretched from the same water supply. This will be a very manpower-intensive and time-consuming operation. Fortunately, enclosed stairways will most likely be present so the stretch will be made in a relatively smoke-free area.

Whichever way the line is stretched, care must be taken to ensure that hose stretched to the fire area does not get caught and kinked under the wheels of a vehicle. This will have the same effect as closing a door on the hoseline and cause a properly supplied attack line to be less than efficient. Extra personnel may have to be given the assignment of hoseline management.

Speaking of smoke, if the fire is deep within the garage, the smoke will be hot, dense and black, and have nowhere to go. Visibility in the area of the fire will be zero. Just as in a fire in the building proper, the HVAC system should be shut down to prevent the products of combustion from spreading to other levels of the garage and possibly to the occupied areas of the building. Use of a lifeline and thermal imaging camera to locate the fire is mandatory, as is SCBA. Once the fire is located, an attack line should be stretched to confine and extinguish it.

Ventilation is another problem. In some newer structures, the HVAC system may service the garage area. If so, once the fire is extinguished, the HVAC system can be utilized to clear smoke from the area. This must be carefully coordinated with building engineers or maintenance personnel. In some advanced systems, smoke control measures can be taken by placing the fans in the fire area on exhaust and the fans in adjacent areas on blow. This confinement tactic is used successfully in vehicle fires in tunnels such as the Lincoln and Holland tunnels that connect New Jersey and New York City. These measures, if available and performed properly, can help to confine the fire provided it doesn't spread it into the ductwork. Be careful and preplan comprehensively before attempting this.

To ease the duration and difficulty of the operation, know the location of any service entrances to the garage. Additional openings for these service purposes, such as sanitation or recycling, will allow a shorter stretch from a safer area if conditions and the location and extent of the fire allow. Taking advantage of these structural openings will significantly reduce, in addition to the amount of hose stretched, the access and possibly the ventilation problem. However, it must be stressed that this takes pre-fire planning. If you don't know a building feature exists, it will be of no value to the operation.

Another asset, if they are present, are wall fans that are set in an exterior wall and used to vent the garage area. A carbon monoxide sensor often activates these fans when car exhaust levels reach a specified level. Smoke from a car fire will quickly activate these fans, aiding in the ventilation operation. Although the fans may not be of sufficient size to adequately vent the area, their presence cannot hurt the operation either.

Fans that are activated automatically by excessive carbon monoxide will certainly activate when a fire is present. Wall fans issuing smoke may also be utilized by arriving companies as an exterior cue as to the approximate location of the fire.

If there is no ventilation system, you may have to utilize some creative smoke-removal tactics. One method may be to use one stairway as attack while the other is used as ventilation. This will work well, but only if the stairwell can be exhausted to the outside air. In most of these high-rises, the garage stairwells lead to the lobby and/or continue up into the occupied areas of the building. If the stairs do lead to the exterior, a second fan can be placed at the opening to be used as a fan relay and exhaust the products of combustion more quickly. Another method, which is more questionable and will depend to a great degree on how deep into the garage the fire is located, will be to use fans in a relay to send the smoke out the entrance. It may be that, absent any other means, the smoke will have to dissipate via natural means, which usually takes quite a bit of time.

Sometimes a garage is not underground, but merely enclosed. Others are not enclosed, but their location relative to the attack points will make the operation difficult. If the fire can be attacked more easily from another area of access, it should be done. If, however, there are no other openings, the fire attack

This garage, although it is not enclosed and will not create a difficult ventilation problem, may create a difficult access problem as it is set in the side of a cliff and offers only access via the interior which may expose the building proper to the products of combustion.

tactics will still be the same as stated above, as the only access will probably be through stairs inside the building. In this case, there may be windows on the stair landings. This may allow a properly coordinated positive pressure ventilation operation to efficiently vent the fire area once the fire is knocked down. It may be possible to utilize fans in relay to clear the area. One fan in the fire area and one at the bottom of the stairs can be used to provide positive pressure. An additional fan can be placed in the window to be used as a negative pressure vent or opposite the window to direct the smoke to the proper opening.

A fire in an enclosed garage of a high-rise will require the same command and control demands of a fire on the upper floors. Proper pre-fire planning will go a long way in addressing the problems that may be encountered. Attempting to solve the problems in a proactive manner before the incident will afford the fire forces the best chances of success and the greatest margin of safety.

Utility control

These buildings use up a substantial amount of "juice". If not properly controlled, being in the wrong place at the wrong time or coming in contact with the wrong thing at the wrong time can have fatal consequences. Firefighters engaged in search operations should be aware of the location of such areas as electrical closets (can you say high voltage?), elevator shafts (it's not the fall that will kill you, but the sudden stop), and service and storage areas that could contain hazardous processes. Preplanning a structure to find these hazards before an incident is the best way to gain a thorough knowledge of the building. However, this is not always possible.

Firefighters should be aware that when searching buildings, most doorways that open into the hallway, toward the firefighter, usually lead to trouble. These include closets, high-voltage utility areas, elevator shafts, and mantraps such as storage areas. Those doors that open inward, away from the hallway and the fire-fighter, usually lead to apartments in residential high-rises and offices in commercial high-rises. It is also crucial when searching an area suspected of containing an electrical hazard to search with the back of the hand or a closed fist. The hand's natural reflex is to grab. Using the back of the hand or a closed fist will prevent the grabbing reflex from latching onto an undesirable object.

Curtain walls

Sometimes called panel walls, these non-load bearing walls are used to provide an enclosure in fire-resistive, steel skeleton buildings. Usually limited to one story in height, the walls support no weight other than their own. Often made of glass and a light-gauge metal such as aluminum or a lightweight masonry, the installation of these walls may create a vertical path for fire travel from floor to floor via the structural voids between the framework of the building and the curtain wall itself. These voids must be properly fire-stopped. A fire blowing out of a window in a building with curtain walls is not only

Large buildings create large utility hazards. Beware of doors that open toward you in hallways. These usually contain dangerous building systems. Doors with louvers in them, especially at the bottom, are also indicators of potential danger.

A crane lowers a curtain wall into position. If not properly firestopped, fire may spread vertically in the space between the exterior curtain wall and the steel infrastructure of the building.

a fire spread threat due to autoexpsoure, but also to extension via the space created by the curtain walls. A fire originating in the vicinity of, spreading to, or exposing the exterior walls should prompt the incident commander to order a thorough fire extension investigation to be made inside curtain walls. It is imperative that fire personnel check these perimeter areas for fire extension whenever a serious autoexposure condition is encountered.

Buildings under construction

One of the most dangerous incidents the fire service will face in a fire-resistive high-rise will be in a building that is under construction. It is at this time that the building will be least stable in terms of structural integrity, most vulnerable in regard to fire ignition and spread, and the most inundated with hazards that will be present in and around the building.

Fire-resistive buildings for the most part are constructed of reinforced concrete. This construction is either precast and shipped to the site in already completed sections or cast-in-place, where the structural concrete is actually poured on the site. Concrete alone has great compressive strength. Forces pressing against each other cause compression. Concrete has very little tensile strength, the forces that act to pull things apart. Without strength in both areas, reinforced concrete fire-resistive high-rises would not exist. What gives the concrete tensile strength and justifies the moniker of reinforced concrete is the presence of steel rods. Steel is very strong in tensile strength. The rods are inserted under tension in strategic places in the uncured concrete, which, when dried, will receive the required tensile strength from the rods.

The concrete slabs in this precast building were delivered to the site intact. If fire attacks any temporary connections, which are usually unprotected steel, the building may suffer complete collapse.

As stated earlier, wet concrete is a very elastic material with practically no shape or strength. To ensure that the concrete takes the proper shape desired, wood framing called formwork is erected to hold the wet concrete. This formwork consists of 2" x 4" and 4" x 4" bracing members as well as plywood and other wood scraps. It is all nailed together to hold the concrete in place. A fire attacking this formwork will cause it to fail and drop the concrete load it is supporting onto areas below. This collapse may take successive floors with it, pancaking the structure to the ground.

Cast-in-place buildings involve the pouring of wet concrete at the site. Note the steel reinforcement rods being covered by the concrete. The steel will give the floor tensile strength, while the concrete provides the compressive strength.

Sometimes, instead of the 4" x 4" posts, the formwork will be held up by steel screw jacks. The problems of fire attacking unprotected steel will apply to these jacks. These jacks will fail before the wood posts will.

The formwork on the top floor of this cast-in-place high-rise began to fail due to an excessive wind load. Beware of any building where formwork is present.

Use extreme caution when operating around these weak supports.

It takes concrete approximately 28 days to reach a stage of structural integrity where it is no longer a threat to collapse. However, the construction process moves much quicker than this. After about two or three days, depending on the weather, most of the formwork is removed to be used on other floors above. Some bracing will still remain in place to support the newly poured and still-curing concrete. If you observe this bracing or get a report that it is present, then the threat of a collapse of this and the floors above is a real possibility. If heavy fire is present in or near these wood-bracing members, it may be time to re-evaluate what your companies are doing in the area. Defensive operations outside established collapse zones might be the only safe strategy available.

Another problem relates to our ability to get streams in place. Standpipe installation usually does not keep up with the rest of the building. In addition, the valves on the individual floors may be open, not yet equipped with threads, or missing altogether. This will allow the fire to extend to major proportions precipitating a defensive strategy. Furthermore, our ability to supply the system may be thwarted by the same elements that prevent us from connecting lines on upper floors. Moreover, construction materials and debris may block the fire department connection.

Auxiliary appliance installation may be incomplete and primary water supplies may be inadequate at a building under construction. This building was already fifteen stories tall.

In addition to the ordinary building construction debris and material on the site, there will also be more flammable liquids and gases around than when the building is completed. These may either cause a fire to ignite or intensify one that is already in progress.

Fire officials and fire companies should routinely visit the construction site as the work progresses. This will serve two purposes. First, it will familiarize the members with the building from the ground floor up, so to speak. Second, it will allow the department to correct any violations or operations that may complicate our ability to fight a fire in the structure.

Hazardous materials such as these propane tanks may be scattered amidst the debris. Flammable liquids are also common. These materials will not only create an explosion potential, but also cause a Class A fire to burn hotter and more intensely.

Life Safety Problems Related to High-Rise Structures

Occupant indifference

It is appalling when one considers the extent to which the public is ignorant about survival and escape in a high-rise fire. A story was relayed to me by a captain from another city. A woman told him that she was not worried about a fire in her high-rise. She had a plan. Her plan was to take her mattress to the balcony and jump with it under her knees. When she was about to hit the ground, she would simply get off the mattress. And she was dead serious. In the fire service, you will find that many times, truth is stranger than fiction.

Occupant indifference has already been mentioned to some depth in the discussion of the Westview Towers fire. Whether as a result of laziness or improper attitude, it has caused many people (including firefighters) their lives. It must be impressed upon occupants that the features built into the building are there for just that: fire safety. A solid, aggressive, and proactive fire prevention and community awareness program is the best defense against this problem. When prevention fails, a solid, aggressive fire attack with related support will be the only measure available to safeguard occupants who refuse to listen to fire-related warnings and education.

Control of occupants

The effective control of occupants will either make the incident a manageable operation or a nightmare for the incident commander. This control factor is directly related to the fire safety education of the occupants. While this cannot be counted upon at the fire scene, the incident commander can hopefully count on fire personnel to remain disciplined and work within the confines of the action plan. This is directly related to incident command's ability to manage the fire personnel on the scene. Personnel operating as per standard operating procedures and unified under an established action plan will be positioned in their assigned places, taking care of the business at hand: the protection of life and the confinement and extinguishment of fire. Freelancing will be eliminated or at least minimized substantially.

The incident commander has two options in regard to occupants: evacuation or protection-in-place. Occupants on the fire floor and directly above it must be evacuated when any fire of significance is in progress. It might also be a good idea to evacuate or move to another area, occupants of those floors that will serve in the support roles such as the operations, resource, and rehab posts. To leave occupants in their apartments, offices, or other areas in proximity to the fire and fire operations is an unnecessary risk unless the fire is of a very minor nature and rapid control is certain. If there is any doubt, evacuation of these areas will eliminate headaches later. As always, the evacuation must be planned. The safest routes must be established and communications coordinated to effectively evacuate the necessary areas. Pre-fire planning can help establish the best routes prior to the incident.

For occupants not on the fire floor or the floor above, protection-in-place is the method of choice for safeguarding occupants at a high-rise fire. Pre-fire planning and public fire education are critical if this operation is to be successful.

Once occupants are evacuated, the incident commander must also have a plan as to where to send them. If it is cold outside, the occupants cannot be expected to freeze outside, however, it is also undesirable to have them milling about the command post. Again, preplanning will offer the solution. There may be conference rooms or other areas, including nearby buildings, schools, or other shelters that are large enough to accommodate the displaced occupants.

Protection-in-place must also be planned for in advance. This means that people should not be wandering the halls on the floors not directly involved in the operations. Panicky occupants may be better off if they are placed in an apartment with a calmer neighbor, or if manpower is available, in the company of a firefighter until the emergency is under control. It is usually proper to use protection-in-place measures with any occupants not on the fire floor or the floor above. The SAE post must be constantly aware of conditions, as additional areas may have to be considered for evacuation as the situation dictates. It may also be feasible to relocate some occupants to safer, less-exposed areas than to take them down the stairs past the floors of fire operations. In a large building, more people will usually be protected-in-place than evacuated. The value of the

An exploding oxygen cylinder blew out the hallway wall (note what is left of the wall at bottom left). The occupant directly across the hall survived the explosion and fire by remaining on the balcony in her apartment. *(North Bergen, New Jersey Fire)*

protection-in-place option cannot be overemphasized. In fact, protection-in-place can be so effective that at the Westview Towers fire, a woman directly across the hall from the fire apartment spent the entire incident on her balcony. Recall that at that fire, an oxygen cylinder exploded and blew a large hole in the wall between the fire apartment and the public hallway.

Control and coordination of stairwells

To run a smooth operation, it is imperative that stairwell operations be well coordinated. This particular item in the control inventory is critical to both fire-fighter and occupant safety. All stairwells designated as crucial to the operation must be identified and made known to all fire personnel, preferably beforehand. Obviously, the stairwell being used for the fire attack is out as an evacuation artery as smoke and heat will vent up the stairs once the hallway door is opened and the attack begun. If there are occupants already using this stair, the attack may have to be delayed until they are out of danger or re-routed to a safer stairwell. The stair with the standpipe connection will usually be the attack stairwell unless the fire is on a lower floor and the attack lines can be stretched from the apparatus on the street. Other stairwell designations will

Stairwells with roof access are excellent choices for ventilation stairs. Ideally, these stairs should be on the opposite side of the attack stairwell and remote from the evacuation stairwell.

be established as the situation dictates. If possible, the evacuation stairwell should be as remote from the fire area as possible. Ventilation stairs should be as opposite the attack line movement as possible. This vent stairwell must pierce the roof or at least have good venting capabilities via windows in the stairwell. This is many times not possible, as enclosed stairwells will usually be windowless in high-rise buildings. This, again, should also be known beforehand and figured into the attack plan.

As the reader can see, if the arteries leading to the operational areas are not controlled, it will set the stage for possible death and injury as uncoordinated operations result.

Accounting for personnel

Personnel accountability is a critical concern at any incident. There is almost no incident more accountability-unfriendly than a high-rise fire. At conferences and seminars I've attended as well as feedback from classes I've taught and literature I've read, the methods of ensuring firefighter accountability are widely varied. From simple tag systems and apparatus riding lists to elaborate systems that implant computer chips in the collars of members where they can be tracked by a global positioning satellite (which incidentally do *not* work inside buildings), there is one thing these systems overlook. They do nothing to protect the firefighter prior to the dreaded occurrence of being trapped or lost. They are, in fact, reactive. They only come into play when the firefighter is reported missing. In fact, they do nothing to prevent this from occurring. And, to take this a step further, I have never seen a tag hanging on a board lead the incident commander to where a firefighter is located. Usually, the tag is indicative of where the member should be, but isn't. Most firefighters who get lost or trapped do so alone, most often because they were freelancing and operating outside the action plan in a place they weren't supposed to be.

On average, we still kill about 100 firefighters a year. This was happening well before the implantation of the tag and other systems, and it is still happening. The system is obviously not working, not being followed, or a combination of both.

I was recently at a meeting where the merits of the current accountability system of using tags was being discussed. There were varying levels of chest-thumping, gnashing of teeth, and sermons. However, when asked how the system ultimately protects firefighters, a panel of chief officers could come up with no better statement than "it allows us to identify a body"! A member in the meeting remarked that if that was the rationale, why don't we just tattoo the guy's name across his butt. It didn't go over too well, but the point was made.

Tag systems are essentially a reactive system. Over-reliance on these systems is dangerous. It must be the company officer's responsibility to account for assigned personnel at all times. *(Bob Scollan, NJMFA)*

A Magnetic Accountability Card (MAC) riding list is placed on the door of the first-arriving ladder company at the front of the fire building. The member assigned to accountability and the command board then collects the cards. This system still leaves much to be desired.

In North Hudson we have found that the tag system was not effective for our needs. The rings on the apparatus that held the member's tags were rarely collected from the initial-arriving companies or brought to the command post by later-arriving companies. A simpler system was required. What has worked better, although it still leaves a lot to be desired like any reactive accountability system, is the use of magnetic riding lists kept on the dashboard of the apparatus. It is filled out by the company officer with a grease pencil and updated throughout the shift to reflect any changes in personnel. Called a magnetic accountability card (MAC), it is similar to a riding list. At the onset of a fire operation, the card is placed on the door of the ladder company in front of the building where it is later retrieved by the member charged with setting up the command board. It has proven to be a more effective system. However, it is still a reactive system, used only when a member is in trouble or reported missing or trapped. Hopefully, it will never have to be put into operation.

There is a better way to achieve personnel accountability on the fireground in a more proactive manner. The feeling of the author is that these tag systems just do not work in the urban/structural firefighter setting. While these systems may be effective in the wildland setting where men are operating as task forces and report and operate as a team in an assigned area, this is not usually the case in a structure fire. If a company entering the building leaves their tags at the front door or on a spot on the command board, how is the incident commander to know what room they are operating in or even what floor they are on. Truthfully, he can only guess. This method is very difficult to control, especially as time passes and/or the incident escalates.

One method that may be used to support and cut down on this guesswork is through the use of a personnel accountability report (PAR). This system is much more proactive in verifying the location of personnel than the lackluster tag system.

The PAR is verification, via radio, of the status and location of operating companies. Essentially a roll call, the PAR is conducted by dispatch and should be initiated using a special radio tone, one that differs from the emergency transmission tone adopted by the department. Following the tone should be a general boilerplate statement announcing PAR. The PAR will require that each company on the fireground, upon radio prompt, give their current location and status at the time of the PAR. For example, "Dispatch to all operating companies on the fireground, stand by for PAR…Engine 1, state your PAR." Engine 1 should then announce their PAR status and location, such as, "Engine 1, all members accounted for, operating on second floor…"

Company officer response to the PAR may have to be modified when companies are split up by SOPs. In high-rises, ladder company members usually stay together as a unit. In smaller residential and multiple dwelling fires, the company may be split up to more effectively provide coverage of the building. In this case, it may be necessary to contact each individual ladder company member or team for location and status during the PAR. The same procedure may be necessary for the engine company whose chauffeur is operating the pumps. In this case, the company officer can report the status and location of the company with an additional statement that the chauffeur is operating the pumps. This must be decided at the local level.

Upon completion of the PAR, dispatch reports to command the results of the PAR with the time, for example, "Incident command from dispatch, PAR complete, all companies accounted for; time is 1344." Any company that does not answer the PAR will cause the initiation of firefighter rescue tactics, namely the activation of the FAST team.

A personnel accountability report should be initiated at specified operational time periods on the fireground. Generally, the PAR is conducted every twenty minutes elapsed or about the time it takes to exhaust an SCBA cylinder. This does two things for the incident commander. It keeps his head in the game in regard to fire progress vs. time lapsed. It also allows the tactical worksheets and command boards to be updated and kept current. The PAR system is usually not implemented until the request for a second alarm is made by command. As these additional companies arrive, proper span of control requirements force the need to decentralize the fireground. More companies, and thus more manpower, require a way of keeping current track of personnel. The PAR is one way of attempting to accomplish this.

In addition to the second alarm/twenty-minute rule implementation, a PAR should also be requested during the following conditions:

- Firefighters are being withdrawn from all or part of the fire building (initiation of strategic mode change)
- Explosion, collapse, or unexpected fire extension
- Any report of a missing or injured firefighter
- Fire control benchmark is reached
- Any time incident command or the safety officer requests one

The PAR system does have drawbacks. It ties up the radio. The more companies on the scene, the longer the PAR. This may cause important radio transmissions to be delayed. It also creates the possible problem of deploying

Any collapse of the fire building should prompt incident command to initiate a Personnel Accountability Report (PAR) immediately. All members must be accounted for during all phases of the incident. The PAR helps accomplish this. *(Newark, New Jersey Fire)*

a FAST team to "rescue" companies that have simply not heard the PAR announcement. For the system to work, all members must be strictly disciplined to listen for and be aware of the PAR tones. In the infancy stages of the implementation of the system in Jersey City, the PAR was initiated just prior to termination of command at all multi-company responses. Command operations were not terminated until PAR was complete. To make this an almost Pavlovian response, the tones should be tested each day at least once and more importantly, it should be injected into training sessions. An effective method of reinforcing this system was used by the Jersey City Fire Department. Members were able to get used to the system in a less that hostile environment.

It may take time for a department to grow into the most effective way of implementing the PAR system, to work out the kinks. It still may be the best way to provide simple up-to-the-minute personnel accountability to the incident commander.

It is apparent that accountability systems leave much to be desired. The slack must be taken up in other ways. The most important resource, often overlooked as the panacea to effective accountability, is directly available to the incident commander at all times. This resource is called the company officer, and at larger incidents such as this and other high-rises, accountability also becomes the responsibility of the chief officers working in designated command positions (control points).

Strict adherence to apparatus positioning and scene assignment SOPs as well as disciplined, regular, and consistent progress, exception, and completion reports to incident command will ensure the best accountability on the fireground. To put it more simply, disciplined officers with well-trained and equally disciplined firefighters following effective standard operating procedures are the best method of accountability. SOPs, by their very nature, establish accountability on the fireground. An effective scene assignment SOP, coupled with prior knowledge of the building as a result of proper pre-fire planning, will place members in strategic positions as a matter of standard operation. When companies are operating in the positions they are assigned according to the SOP, then the most effective means of accounting for not only their whereabouts, but their general activities, will be maintained. If companies operate in this manner, the tags, riding lists, and computer chips will function as they were hopefully intended, as a system to be used after the fact and probably not at all.

The primary responsibility of the company officer (and chief officer) is the safety of his assigned personnel. If he places reliance on a tag to ensure that safety, then he displays a misunderstanding of the responsibilities of his rank.

To take this a step further, when a building collapses or is evacuated, and a roll call is ordered, it is the company officer who is contacted to account for and report for his personnel. If he does not know where they are, there is something tragically wrong. He should not need a tag to tell him who his men are or where they should be. Moreover, if a tag is displaying where the man should be and he is not, what is the value of this system? The tag is of no help in locating his whereabouts.

As I mentioned in the introduction to this book about fire service fads such as confined space psychosis and Haz Mat hysteria, this accountability system paranoia will find its rightful place in the fire service as sensible fire service leaders take a hard look at it.

Departments who lean too heavily on the tag or other accountability systems and allow officers to avoid the responsibility of accounting for their assigned personnel are playing a dangerous game. This is essentially command-responsibility avoidance and has no place on today's fireground. Avoidance of responsibility should be construed as a weakness in incident command and should never be a reason for using an accountability system. If the department is operating in a disciplined, coordinated fashion, there will never be a need to put the tag system into action.

Reliance on systems that are reactive and doomed to failure must be placed on a lower priority. The fire service must stop all this silliness with elaborate accountability systems. Keep it simple and demand that the company officer do his or her job and the incidence of firefighters lost and trapped in areas where they don't belong will be minimized.

In 1999 (the current height of the accountability furor), 112 firefighters died in the line of duty. That is more than two firefighter deaths per week! This is the highest death toll since 1988. How can this be if we are so accountability conscious? Obviously, the system, in its present state, is not working. Wake up, people! It is time to put the true responsibility for accountability back on responsible personnel, not on a piece of plastic.

I am not against the development of an effective system that will ensure we never lose another firefighter. I feel the PAR is a step in the right direction, but a still leaves much to be desired. The problem is that funding for these systems seems to be a matter of priority on the part of legislators; low priority, that is. Why is it, that the Police can track a criminal under a house arrest wherever he goes, but we can't find a firefighter inside a building. Something is wrong here. It is obvious that in the minds of the fund providers, fire is not viewed on par with criminals as a menace to society. If it was, the amount of money made available to the fire service for this type of equipment would not be an issue.

All types of funding is made available for law enforcement programs, and rightfully so, but the fire service has had to rely on donations and virtually begging to purchase thermal imaging cameras and other equipment to save our own. I guess the lives of firefighters are not worth that much. It is sad and glaring that if we are not going to be taken care of by the very people we are sworn to protect, including those who create and allocate these funds, we must rely on our own discipline to maintain accountability.

Basic Firefighting Procedures

Firefighting procedures will not be of the "bread and butter" type found in other residential occupancies. These buildings offer problems of a more complex nature due to the inherent firefighting problems mentioned. Even a small fire will require a much greater response compliment than if it were in a non high-rise building. Prior knowledge of the building is imperative. Prepare for a large, resource-intensive incident.

CRAVE

To address the CRAVE acronym, employ the following general guidelines:

Command
- Forecast and provide for an expanded Incident Management System.
- Establish operational control points as soon as possible, preferably using chief officers.
- Establish effective communication as soon as possible; keep these channels open at all times.
- Demand timely reports from all control points in the operational area.
- Control the building systems.
- Provide adequate manpower to outlast the fire.
- Be prepared to supply more than one FAST team.
- Ensure that an adequate tactical reserve is in place.

Rescue
- Reconnoiter all areas of operation.
- Determine the evacuation stair.
- Provide primary and secondary search on fire floor and floor above.
- Decide on evacuation vs. protection-in-place.

Attack
- Ensure a primary and secondary water supply is established.
- Ensure auxiliary systems are supplied.

- Determine the attack stair.
- Provide at least two companies for each line being stretched.
- Stretch lines to the fire floor and the floor above.
- Coordinate attack with support activities.

Ventilation

- Determine ventilation stair, if applicable.
- Seek out opportunities for effective ventilation dependent on:
  a. Location and extent of fire
  b. Wind velocity and direction
  c. Building and fire floor layout
- Be aware of consequences of indiscriminate ventilation operations.
- Coordinate with attack operations.

Extension Prevention

- Ensure that the HVAC system is shut down.
- Reconnoiter all areas for vertical and horizontal fire travel.
- Be prepared to fight fire in remote areas from the main fire.

Conclusion

High-rise operations will be a horror for the department that has not taken adequate steps to plan for it. Pity the incident commander who has this mess dumped on him without adequate departmental planning. If your department does not have a high-rise SOP, take immediate steps to create one, or even better, steal one from a neighboring department and adapt it to fit your needs. There is no need to reinvent the wheel. Just make sure that the source you choose to steal it from is reliable.

It is the wise incident commander who realizes that a strong command organization is the backbone of any high-rise operation. The best (and safest) plan is to amass and coordinate the troops. Remember that fast action is not as critical as safe action. Take the time to find out what is happening. Those few extra seconds may make the difference between a successful outcome and losing both the building and the lives of firefighters. Look at the conditions. Evaluate your resources. Balance your strategy and action plan with the manpower on hand, responding, and still to be requested. And always act on the side of safety and accountability.

Questions for Discussion

1. Discuss the control points that should be established at a high-rise fire to meet the expanded organizational demands of the fire.
2. What are the major responsibilities of the following incident command positions at a high-rise fire?
   a. Command post
   b. Operations post commander
   c. Resource post commander
   d. Search and evacuation post
3. Discuss some of the possible assignments for the command company at a high-rise fire.
4. Discuss the vital building systems that must be controlled as part of the incident commander's action plan.
5. Discuss fire control in high-rise structures in regard to the CRAVE acronym.
6. You respond to a fire in a residential high-rise building. There is a fire on an upper floor. The wind is blowing into the fire apartment, making it extremely difficult to advance against the heat condition. What are some alternate methods of attack that the incident commander should consider?
7. Discuss the types of standpipe pressure-reducing devices that may be found in a high-rise.
8. Describe some of the problems encountered and possible solutions to fires in enclosed underground parking garages.
9. Discuss some of the problems inherent in high-rise buildings under construction.

CHAPTER EIGHT

Contiguous Structures: Row Houses, Garden Apartments, & Townhouses

Contiguous structures are buildings connected to one another without separation, often having non-existent or compromised firewalls. These structures create a fire involvement and spread potential that will tax the resources of even the best-staffed fire departments. Incident commanders must be proactive in their strategic approach to these buildings. An offensive/defensive or defensive/offensive mode of operation will usually be appropriate, dependent on arrival conditions and the initial manpower compliment. A strategy must be developed based on priorities regarding the location and extent and subsequent life hazard in the fire building compared with where the fire is going and what actions must be taken to head it off.

When addressing the potential fire area, these residential complexes should be treated as one building. The incident commander who fails to react to a fire in these structures in this manner sets the stage for the possible loss of the entire row or complex.

Contiguous structures can be broken down into two categories: old-style and new-style. Old-style structures include row houses, and older garden apartments of both wood frame and ordinary construction. New-style structures will include townhouses, condominiums, and newer garden apartment complexes constructed of lightweight materials. Often, these new-style structures are given the appearance of old-style buildings by using a decorative veneer on the exterior. The best way to ascertain this is during the construction phase. Without prior knowledge, what may look like a building of ordinary construction or a row of "brownstones" is actually a wood frame building of lightweight construction utilizing a brick or stone veneer to give the appearance of another type of construction. Do not be misled by these aesthetic cover-ups. There are very few, if any, buildings of ordinary construction being built today. The lightweight material being used to construct these buildings includes both the truss and laminated I-beams used in the construction of the roof and the flooring.

This chapter will address the concerns common to four types of contiguous structures. These are row houses, both of ordinary and frame construction, garden apartments, and townhouses, of which condominiums will be included.

Contiguous structure fires can cause the loss of a whole block or more. Aggressive tactics will be required to confine the fire to the building of origin. This includes ventilation above the main body of fire, proactive recon operations, early line commitment into threatened buildings, and extensive pre-control overhaul. Manpower is the name of the game in contiguous structures. (Ron Jeffers, NJMFPA)

Be very suspicious of new "brick" buildings. This building, which gives the appearance of a Class 3 building, is actually a lightweight wood-frame building with a decorative brick veneer. There are virtually no new buildings of Class 3 construction being built today.

Firefighting Problems in Contiguous Structures

Lightweight construction

Old-style contiguous structures, although combustible in construction, will stand up better to the ravages of fire than their more modern counterparts. The floor joists and the roof rafters will usually be made up of larger structural members such as 2" or 3" x 10" or 12" beams. Row frames of ordinary construction will not be vulnerable to the exterior flame spread inherent in wood-frame structures. Old-style garden apartments built before the lightweight construction frenzy will often be platform construction with a brick or other type of masonry veneer finish on the exterior wood walls. This is still better than lightweight construction.

This roof rafter is at least 3" thick. Although charred, its size and mass allow it to remain intact. This cannot be said of the newer, lightweight construction being built today.

Lightweight construction materials cannot be expected to last long under heavy fire conditions. The chief officer on this balcony is not in a good place. The peaked roof trusses above him are clearly involved in fire as evidenced in the peaked roof window. *(Capt. Mike Oriente, NHRFR)*

Newer contiguous buildings are constructed of lightweight 2" x 4" wood truss members. These members will usually be constructed of the parallel chord type in flooring and of the peaked type in the roof area. Lightweight laminated I-beams are also used, but usually only in floor construction. Both of these structural members, the lightweight truss and the laminated I-beam, cannot be expected to withstand a fire for a long period of time, often failing in as little as five or ten minutes. I have seen these roofs and floors hold up for longer than this time frame, but this is no guarantee and should not be used as a guide. Any roof cutting operation should be accomplished from the safety of an aerial ladder or platform. The roof cut may also take place ahead of the fire (for example on the roof of an attached exposure) in an attempt to stop its spread, but it is too dangerous to put crews on this type of lightweight roof without any support. The structural members are too flimsy and will fail without warning. This failure is mainly due to weak connection methods, namely the sheet metal surface fastener, also called the "gusset plate" or "gang nail". This connection device is prone to failure for six reasons:

1. Insufficient depth of the penetrating surface, often only ¼" to ½" into the wood truss member.
2. The truss may have been damaged prior to installation, usually during transport or storage.
3. The truss may have been insufficiently fastened during the construction process.
4. The truss may have been compromised due to excessive moisture in the roof and floor area. This can cause the fastener to rust and the wood to corrode, further weakening the bond.
5. The fastener can curl and pull away from the wood member as a result of heat exposure.
6. The fastener may pull away from the wood member due to the impact load when a falling floor or roof from above drops on it.

Lightweight wood trusses are usually manufactured off-site and shipped via truck. Note the damage to the top chord of the trusses at the rear of this truck. These may be reinforced with still more gusset plates to "repair" the defect.

Impact of the roof and top floor collapse of this townhouse caused the secondary collapse of all lower floors in a pancake fashion. Note the relatively undamaged trusses standing straight up at the center of the photo.

A room and contents fire in one of these lightweight structures should not cause an early collapse, but once the compartment flashes over, the fire begins to attack the structural members, of which the connection point is the weak link. Recon of the roof area, the floor space, and cockloft is critical to the safety of operating personnel. When in doubt, withdraw and use the reach of the stream or use master streams to control the fire. It may also be possible to withdraw from the fire area and use a "pincer" or pinch-off attack to confine the fire to the area of origin. This interior defensive strategy is the most desirable action to take when it can be safely done. The other alternative, complete withdrawal, allows the fire to consume the entire row.

Combustible open cockloft

Both the new and old-style contiguous structures will be constructed with a combustible cockloft. This cockloft may or may not be fire-stopped between adjacent structures. The incident commander should be of the opinion that firewalls between buildings are non-existent or compromised until proven otherwise. This skepticism will allow him to take the necessary action to cut off the fire early and not be caught off guard by a rapidly spreading fire.

Sometimes, it will be easy to detect the presence of a firewall between two adjacent structures. In the Class 3 row house, the parapet wall will act as a fire wall and may rise several feet between buildings. This indicator is rarely present on a wood frame row house. Many times, if there are walls that act as fire stops, they only rise to the top of the top floor ceiling or just underneath the roof boards. This condition will not prevent fire spread in the cockloft and will not be visible from the roof. Any building that does not have a visible fire stop or firewall showing above roof level should be treated as having an open cockloft. Take steps to fight a serious fire in at least the leeward exposure, but also get lines and personnel into the windward exposure as fire can spread both ways.

The row house will have a much smaller cockloft than the new-style contiguous structures. There are both advantages and disadvantages to this. Being smaller and of larger dimension wood, the chance of early roof collapse is not as great as the newer lightweight truss cocklofts. The roof boards will maintain their integrity for a relatively long period of time. This is a distinct advantage for those members working above the fire on the roof. The disadvantage of the smaller cockloft area is that the heat build-up will be faster, leading to the potential for a backdraft condition. Ventilation of this area should always be conducted from above before the ceilings are pulled from below.

In newer construction, the open roof space is larger and extends for the whole structure without any firestopping. A fire originating or entering this space will be extremely difficult to access and control. Command must be proactive and order additional resources as early as possible.

This U-shaped garden apartment complex is likely to have an open space over the entire roof area. Do not wait until the fire is showing at the roof level to address this problem.

The larger cockloft area of the new-style building will exhibit the same firespread characteristics of the old-style, but will usually burn through and collapse earlier due to the lack of structural mass in the lightweight construction materials. In newer townhomes and condos, rarely is the roof area firestopped between building clusters, giving the fire open access to all the buildings in the row. Windy days will especially cause difficult fire control problems. Due to the larger cockloft space, the conditions that create a backdraft condition may not build up as quickly as in the older, smaller, row houses. Nevertheless, it is still a threat and should prompt all firefighters to size up the roof area for the indications of backdraft.

In the new-style contiguous structure, be prepared for an open cockloft over the entire row. There may be firewalls between every few occupancies or, instead of firewalls, the height of the roof may vary or the buildings may be offset.

In garden apartments, the roof is usually open over the entire area. When I was in college, I lived in a large U-shaped garden apartment complex in Connecticut. You could literally walk upright (the space was about ten feet high) across the ceiling joists from one end of the complex to the other. Fire entering this space would likely spread to the entire row and burn the roof off. If you arrive at one of these buildings and see smoke pushing from the eaves of the roof at locations other than the fire area, especially downwind, expect that the roof is not partitioned and the fire may already be raging throughout the cockloft.

Combustible exterior and roof

As the exterior is combustible, fire will spread rapidly across its surface. If you are faced with this condition, it may be necessary to use an exterior stream to knock down the exterior fire while interior lines are being stretched or operating. This defensive-offensive operation requires both strict control on the part of the incident commander and discipline on the part of the firefighter operating the exterior stream. The stream cannot enter the structure at any time for it may cause an adverse effect on interior operations. As soon as the exterior stream has accomplished the objective of knocking down the outside fire, it should be shut down. This strategy is acceptable on all types of wood-frame structures, but it is most critical on a fast-spreading fire in contiguous structures, especially those covered with asphalt shingles or rough, decorative wood siding.

The same problem may exist on the roof of the new-style structures. The roof is often peaked in a multi-level fashion with peaks varying in pitch and height. Fire extending out of a window can easily ignite this roofing material and spread across the roof to adjoining structures. This is

Varying roof heights will not deter the spread of fire from area to area. All portions of this building are combustible. Water applied to the surface to keep it cool is the best way to prevent ignition in this case. *(Capt. Mike Oriente, NHRFR)*

especially prevalent when a strong wind is present. The fire on the roof may jump firewalls and spread into windows and under eaves into the cockloft. Again, the exterior stream, used judiciously, may be the answer to stopping this spread. In addition, lines must be stretched to the threatened areas and ceilings must be rapidly pulled if any attempt to stop the fire is going to be made.

Old-style contiguous structures (the ordinary and wood frame row houses) will also have a combustible roof surface, but the roof will invariably be flat. There is an excellent chance that it is also built-up with many layers of tar and tarpaper after years of renovations. These roofs, when involved in fire, will give off tremendous amounts of thick, choking, black smoke. The thicker the roof is, the harder it will be to vent. The saw blade gets bound up in the sticky mess of heated tar. In addition, a well-involved roof fire in these old structures will also readily jump over a firewall and ignite adjacent roofs. It may be necessary to keep these adjacent roofs wet with exterior streams and let the roof of the original fire building burn through.

The size of the roofs makes them less than ideal for the trench cut. It is better cutting large holes in the area of the fire. If the trench is the only alternative, it will likely be necessary to give up at least one more roof and possibly two to get the cut accomplished before the fire gets to it. Attempting to cut a trench too close to the main body of fire will usually be unsuccessful, as the required amount of cutting will take more time than the crews will have to get it done. If you are going to attempt a trench cut on one of these buildings, it is prudent to commit a large amount of manpower to the operation; at least four ladder companies. Committing any less than that will almost certainly ensure the job will not be successfully completed.

Manpower sponge tendency

A serious fire in a contiguous building will threaten not only the building of origin, but also the entire exposed row. Bringing this fire under control and ensuring that lateral fire spread is halted is both work- and manpower-intensive and will require a major commitment of resources. The first alarm companies will be "eaten up" very quickly by the many duties required in the fire building. Additional alarm companies must be requested to initiate and sustain operations in the attached exposures. These companies must be summoned early and put to work early.

A good rule of thumb to apply to manpower needs would be to check your tactical reserve. Any time you do not have at least two companies at the command post, especially if the fire is still escalating, you had better request an additional alarm. This rule of thumb works well for any type of fire, but is

The need for additional manpower also requires the need to reduce the span of control. Additional command officers will be required to meet these demands. *(Bob Scollan, NJMFPA)*

especially applicable to contiguous structures. Extra manpower will be required to reinforce the fire building operation, stretch lines into exposures, supplement the roof operation, search and evacuate the adjoining exposures and possibly other buildings in the row, pull ceilings on the top floors, and also provide relief to those companies already in the building and the exposures. If you have additional tasks to be completed and there is no one to assign them to, you are out of options. This may force incident command to lose control of the situation and quickly turn the incident from a potential winner to a loser.

The need for additional manpower will also demand the need for additional command officers to supervise the various areas of operation. Reduction of the span of control by decentralizing command is essential to a safe, organized operation. Ideally, each area of operation that cannot be directly seen or controlled by the incident commander should require that a chief officer be assigned to that area. These areas include, but are not limited to, the roof, the rear, and the attached exposures.

This rule of thumb should also apply to rehab requirements. Even if the fire is not escalating, you will still need to relieve the men operating inside the building. If you have no one to relieve them with, you have two choices. Either leave the fatigued men inside and risk the injuries associated with overworked firefighters or pull them out and give the fire the advantage. Incident commanders who resist additional alarm requests usually wind up with many fatigue-related injuries in addition to many parking lots.

In many fires, manpower is usually the difference between control and loss, and sometimes life and death. Throw as many men at the fire as is necessary and keep a tactical reserve to back them up or be available in case something unplanned for happens. Be proactive in your approach and you will rarely be on the short side of the "catch-up" disadvantage.

Difficult rear access

Buildings that are attached, whether new- or old-style, have never been rear-access friendly to the fire service. Just take a look at the rear of row houses. You will see individual fenced-in yards, clothes lines, and sometimes excessive debris behind each residence. If the fire is in the middle of the row, you might have to jump three or more fences to get to the rear of the fire building. In this case, it may be easier to use the adjoining building to get to the rear. You can even carry a ladder through the window in the front and out the window in the rear. This may be the quickest access if the apartment layout is conducive to this maneuver. Take a few seconds to check the layout before you either bring in a ladder that won't make it through or carry it all the way around the block and over fences when the easiest access was

A 6' wood fence surrounds the rear to these structures. The firefighter assigned to the rear must be flexible in his approach and report immediately to incident command any areas that cannot be accessed.

In areas where there are blocks of row houses, either you can access the rear of the fire building via the B or D exposure or it may be possible to go through the C exposure, and jump the fence. Firefighters assigned to the rear must not be victims of tunnel vision.

through the adjoining building. A few seconds of size-up may pay off in time saved.

Another way to get to the rear may be to access the yard of the building directly at the rear (the side-C exposure) and jump that fence. To quickly recon the rear, especially in "nothing showing" investigations, just stick your head out of the rear window and make a report on conditions. This will be quicker than waiting for the roof team to ladder the building and look over the back.

New-style contiguous structures are sometimes built in areas that are aesthetically desirable, thus the exorbitant price tags. In North Hudson, the Palisades' cliffs overlook the Hudson River and Manhattan. The trend of building townhouses and condos into the cliff side has made many a builder wealthy, but the access at the rear, which is a cliff with brush and trees, is virtually inaccessible to firefighting. In addition, they have also crammed many of these buildings of lightweight construction onto the waterfront. They are built European style, with very narrow streets and access on only one side. The other side is either on the river or faces a park-like green. Winds on the waterfront are constantly blowing. A fire that gets a good start in these buildings could easily result in a conflagration. Another townhouse complex is built on a pier that juts out into the Hudson River. The only access is from the front, and that is limited as the buildings face a long courtyard and green. Sides B, C, and D are on the river. This is a job for a fire boat as land apparatus will be

These townhouses built on the Hudson are constructed of lightweight wood truss. The close spacing, narrow streets, and the prevailing wind conditions on the waterfront make this area a conflagration threat.

There is no access on the B, C, and D sides of this complex. Including a fireboat response in the pre-incident plan for this complex may be the difference between a save and a total loss.

If it is unclear which door to attack through, order a team to locate the fire and determine the best access route into the fire area. This is better than having to back out a prematurely committed line from an area that is not involved.

ineffective in these areas. It is a good idea, in this complex, to have a fire boat dispatched as soon as a working fire is confirmed.

Operations as well as special responses to these areas must be preplanned in advance so that when fire strikes, contingencies that call for out-of-the-ordinary strategies and tactics will already be provided for.

Nearly as problematic as rear access is the often confusing access into the front of the dwellings. Many times, residences are built on several levels. There may be more than one entrance at the ground level, one on a short walk-up (half-story), and one at the side, all leading into different levels of the structure and different apartments. Without prior knowledge, which door leads to which area is anybody's guess. You may find the source of smoke showing from a window in one area may not be accessible from the nearest door. In this case, it is prudent to let the recon team find the best route for the hose stretch before stretching to the wrong area and having the fire extend while the line is repositioned. The best way to fight a fire in these types of occupancies is with your head. Thoroughly preplan these buildings and ensure all personnel are cognizant of the idiosyncrasies of each.

Unusual layouts

Old-style contiguous buildings, unless renovated, will usually be standard in both room size and layout. It is the new-style occupancies that will present a challenge to the fire forces. Due to the unorthodox design of many contemporary occupancies, search, recon, and hose stretching will all be more confusing and dangerous. There are many potential mantraps in these buildings. Searching firefighters should consider using a lifeline and a thermal imaging camera.

Duplex and triplex apartments will present both a unique search and fire spread problem which, if not known beforehand, can cause searching firefighters to get lost or trapped above a fire that they might have thought was concentrated on the floor below. Duplex and triplex apartments are multi-level, with open access stairs leading from one floor to another in the residence. The opening in the floor created by the duplex or triplex layout will allow fire to spread from floor to floor with frightening rapidity. There are few indicators of apartment layouts or the presence of a multi-level apartment. Checking adjoining occupancies is not reliable as each residence may be customized to the tastes of the owner. It might be beneficial to obtain a set of building blueprints from the building manager to best plan your strategy and attack.

Equally confusing in these occupancies is the possibility of multiple entrances at different levels. There may be an entrance on all three levels of a triplex, leading to problems of coordination when a member is attempting to relay his position to incident command from inside the building. It may be necessary to use a guide rope and thermal imaging camera when operating in these structures to keep from getting lost.

Another problem with the new-style structure is the presence of two-story atriums and lofts. A loft may jut out to cover half of the first floor and is usually occupied by a bedroom. Thus, at night, this is where victims may be found. A fire originating on the first floor of the duplex or triplex can quickly channel the products of combustion to this area. Searching firefighters can unknowingly crawl off the ledge and fall to the floor below. In addition, the only egress from the loft may be down through the apartment or via a window. If the rear is inaccessible (where the bedroom is likely to be) rescue may be difficult or next to impossible. Be prepared for these types of rescues.

The residences in these condos and townhomes are usually built on many levels. Often, the stairway is unenclosed and is located on the exterior of the structure, such as in a garden apartment. While many of the exterior stairways in garden apartments are made on non-combustible material such as steel and masonry, the exterior stairways in many townhouse and condo developments are made of wood similar to the planks used to construct picnic tables. A fire extending out of a lower apartment can ignite these stairs, cutting off occupant escape and rendering the attack route untenable. It may be necessary to extinguish a wood stair fire before an attack can be made on the fire area. Personnel working above the fire should be cognizant of conditions on the stairs. Incident commanders should ensure that two ways out of all areas are provided. The same problem may exist with wood balconies. Occupants taking refuge on a wood balcony on the floor above the fire can really get a hot foot. The small dimension of the wood used for the balcony can precipitate early collapse. Balconies are sometimes no more than an extension of the lightweight truss flooring. Fire may be attacking the hidden balcony void and may fail unexpectedly. Take the necessary steps to remove these victims without delay.

These exterior stairs are constructed of wood planking. Fire extending out of an apartment can ignite the stairs, trapping firefighters and blocking egress of occupants above. Always have an alternate escape route when operating above a fire.

One other area for concern that may cause a revision or modification of your attack plan is the sheer size of many of these complexes. Long hose stretches and accompanying friction loss may complicate the attack. Manifolds fed by large diameter hose and positioned in a courtyard may be the answer to a longer than usual stretch. Unorthodox routes of attack may also be devised to effectively address the unique problems presented by a large area structure. In one large complex I know of, the shortest hose stretch has been determined to be via a ground ladder to the fire floor. Through preplanning, we found that the line can be more easily stretched up the ladder into a safe, uninvolved apartment remote from the fire. The line can then be stretched into the hallway and down the hall toward the fire apartment. This was determined to be the most effective path of least resistance into the fire area in this particular building. Due to the poor planning in the design of the building, this unorthodox stretch has been determined to be even more efficient and less manpower-intensive than stretching from the standpipe, which is a ridiculous distance from the most remote areas of the building.

Dumbwaiter shafts, laundry chutes, and elevators

The new-style duplex and triplex condominiums and townhouses are sometimes equipped with dumbwaiter shafts, laundry chutes, and elevators. Dumbwaiter shafts will usually connect the kitchen with the upstairs hallway or bedroom. A laundry chute will often connect the laundry room in either the basement or first level of the unit with a room upstairs, sometimes the bedroom or master bathroom just adjacent to the master bedroom. These shafts are unprotected, and are usually nothing more than a sheet metal sleeve run through the floor. A fire originating in the vicinity of these shafts could channel fire and the lethal products of combustion to the sleeping areas. Identify these areas during preplanning visits or early in the operation and take steps to cut off any extending fire. In addition, check all areas bordering on the shaft for fire extension.

Elevators will be only for the use of the tenant and have no fire service controls. Therefore, they should not be used by firefighters. Firefighters should also note their presence and orientation in regard with the rest of the floor so that they do not crawl into the open shaft, mistaking the door for a closet and fall to the floor below. If the elevator is not at the floor where the search is being made, the door, which will open out toward the firefighter, will be locked. When forced, the firefighter may encounter an open, folding gate across the elevator opening just inside the door. If you open a door in smoke and feel this type of gate, beware, it may be an elevator shaft. Always probe for a floor in this instance and anytime you feel a change in the floor surface or an opening into another room.

Poor apparatus access

Old-style row houses will not usually be a problem in regard to apparatus access. However, in urban areas, double- and even triple-parked cars can impede our ability to get to the fire building. In addition, wires may hinder aerial operation. This is not too much of a problem as these buildings will rarely be more than three stories high, with the great majority being two stories. Ground ladders should be effective in reaching just about all areas of the building. In addition, ground ladders (properly placed at a windward exposure) will have no problem reaching the roof.

The new-style contiguous structure, often built on the outskirts of an area, on cliffs, or on waterfronts, may offer only a "one-way in/one-way out" scenario. Comprehensive pre-fire planning is required to properly position apparatus and take advantage of the best water supply to most effectively fight the fire in the occupancy. In these types of street configurations, you usually only get one chance to position apparatus properly. If it is not effectively done at the outset, the entire operation may be at a disadvantage, one that may not be able to be overcome. At the Roc Harbour fire (see page 287), the initial apparatus positioning was the single most critical factor in confining the fire to one relatively small area and probably the difference between a "stop" and a major loss.

Poor water supply

A problem with water supply in a contiguous structure fire will likely be the limiting factor that causes the complete destruction of the complex. This can occur for several reasons. In areas where the neighborhood has become run down, hydrants are often rendered useless by vandals who cut off the brass stem or threads for money or stuff debris inside the outlets. Plan for alternate water supplies or relays from operational hydrants whenever operating in these areas.

An urban city recently lost four large apartment buildings during the winter because the city had shut the hydrants in the area down during the summer due to open hydrant abuse. The hydrants were never turned back on. When the fire department attempted to utilize them for the fire attack, they were found to be dry.

New-style contiguous structures may be built on the very same ground that the old ones burned down on. Sometimes, the water available does not match the fire load of the new buildings, and consequently, the buildings are under-protected and may possibly burn down again.

Other problems include dead-end hydrants in new developments or a single main feeding a giant loop around the whole complex. Establishing many water supplies in this case will result in engine companies stealing water from each other. At the Roc Harbour fire, the water supply was more than adequate, being fed from two directions (See p.287). In addition, the complex lay at the bottom of the cliffs, supplying extra head pressure to boost the residual pressures. The main on River Road was brand new and 24". If need be, the Hudson River could also be used as a drafting source. Water was not a problem at that fire. Companies responding into these areas should not forget that swimming pools may be available for drafting if the water supply is not adequate, even if it is used exclusively for exposure protection.

This "hydrant" is actually a decorative standpipe connection. It operates exactly like a hydrant, but is in an area where there is absolutely no apparatus access. In addition, it can only supply handlines. Know your buildings.

Storage areas/garages

Storage areas not only contribute to the fire load in the structure, but may, especially in new-style contiguous structures, be located an areas that are difficult to access and just as difficult to control a fire in. These areas should be protected by an automatic wet-pipe sprinkler system, especially in the light-weight construction. A fire originating in this unoccupied area can go unchecked for a long period of time. The fire may also be difficult to locate as many times these storage areas will be located beneath the building, which can cause conflicting reports of smoke origins as well as ready avenues of fire spread. Invariably, they will be difficult to ventilate as well. The townhouse complex built on the pier on the Hudson River, already mentioned, has a storage area that is served by a corridor that is below the grade of the rest of the complex (the complex is raised up from the street level) and runs the entire length of the structure. It has no windows and each storage area is enclosed by a steel door leading to the narrow corridor. The area is sprinklered, which although it may control the fire, will cause an extremely thick and lazy smoke condition. Personnel operating in such an area must use a guide rope and thermal imaging camera to access the fire area and rely on unorthodox methods of ventilation. As the entire operation must be conducted in smoke, extra SCBA cylinders, manpower, and control by incident command must be exercised to keep the operation safe. In addition, if the living units are built above, as in this complex, aggressive fire extension reconnaissance along with line support is mandatory.

Old-style contiguous structures, such as row houses, usually will have any storage located in the cellar. This will also add to the fire load of the building, but unlike the newer construction, there will likely be no sprinkler protection. A serious fire breaking out in the cellar may not only spread to the upper floors, but also to the adjoining cellars as well. For this reason, the adjoining cellars must be checked at any fire involving the cellar of a row house.

Garages are another problem of the new-style structures. Often, they are either built as the ground floor of the living unit or, slightly better, the unit is raised on stilts and the vehicles are parked beneath it. Access to multiple parking space areas may be difficult, but if they are open to the outside air, the ventilation problem will take care of itself. If they are enclosed, apparatus as well as attack team access may be problematic. Any car fire in a garage, especially one that is enclosed or exposes the underside of the living areas must be treated as a structure fire. Lines must be stretched to protect the life hazard above as well as to extinguish the fire. In addition, both primary and secondary searches must be conducted. The complexity of a car fire in these situations may require an additional alarm response. Look at the big picture and summon the manpower to address the problems and potentialities of the incident.

Life Hazard Problems in Contiguous Structures

Early collapse

It has been well established that lightweight construction of any kind is not to be relied upon to maintain its integrity for an extended period of time under fire conditions. Failure can occur in as little as five minutes. Remember, that a contents fire is usually not attacking the structural members; however in lightweight construction sometimes just the heat of a fire, especially with the excessive amount of petroleum products in the home today and their higher Btu generation rate (16-24,000 as compared to 8,000 for wood), may be enough to cause the connection device of a lightweight truss (the sheet metal surface fastener) to pull away from the wood. Keep this in mind, especially if you do not know how long the fire has been burning. Once the compartment flashes over, the interior and roof operations should be discontinued and the fire area surrendered, for collapse will occur—either of the roof or of the lightweight trusses or laminated I-beams that hold up the floors. These structural components just do not hold up under the direct assault of fire.

In buildings of lightweight construction, depending on the building and the location of known firewalls, an attempt to cut off the fire may be made. This pincer action worked well at the Roc Harbour fire, but it is imperative that the stop be made from a safe location. Be sure when using this strategy, that if the floors and roof collapse the areas where the stand is being made will not collapse with them. If the integrity is unknown or suspect, it will be necessary to set up collapse zones and operate in a strictly defensive manner, using master streams to knock the fire down. It may be acceptable, in this case, to give up several apartment units and attempt to make the stop several exposures downwind in a "combustible-clearing" mode, removing as much of the fireload in the fire's path as possible. This will include pulling ceilings to expose the cockloft. It may even be possible to position a portable, unmanned deluge gun in the danger areas to attempt to cut off the fire. If this is not possible or too dangerous, then the entire row may be doomed. Remember, we don't unnecessarily jeopardize personnel to save a building. Buildings can be rebuilt.

The old-style contiguous structures were built with greater fire-resistant members. Unless the building is the site of previous fires or has some other known weakness due to lack of maintenance or renovation using lightweight materials, the structural integrity should remain intact for long periods of time. Even if the fire building is heavily involved, many fire-stopping activities can be safely carried out in exposures. Top floor ceilings must be pulled as quickly as possible–that means the entire ceiling. Just poking holes in the ceiling may not be enough if the fire sudden-ly breaks through from the adjacent occupancy. If the ceilings are not pulled down, effective stream penetration may not be possible, allowing the fire to extend right past the exposure crews, forcing them to withdraw. When this happens, the race will be on to see who will be in the next exposure first, the fire or the fire forces with charged lines and successfully pulled ceil-ings. My bet is on the fire in this case. The moral here is to get the area in the adja-cent exposure completely opened early so that if the fire breaks through, it will be stopped there.

Secondary collapses can be more deadly than primary collapses. Here, the impact load of a roof collapse caused all subsequent floors to fail in a pancake fashion.

Rescue/evacuation load

A primary search must be extended in the fire area as soon as possible, before conditions become untenable due to either the fire condition or the inher-ent weaknesses of the building construction. In a large building, such as the new-style contiguous structures, this may be a very difficult task, consuming a large amount of manpower. Duplex and triplex living units, due to unusual floor layouts, will create a safety hazard regarding firefighter disorientation. Many of these units will have contemporary, unorthodox layouts, which may make a systematic search ineffective and dangerous. It is best in a heavy smoke condition to use a rope and thermal imaging camera to search.

Buildings that are contiguous also present a much larger evacuation problem than other types of buildings. It will be necessary to evacuate the immediate exposures on both sides, especially at night. Then a decision must be made as to what additional units must be evacuated. If the fire is doubtful and escalating, it is prudent to evacuate all the downwind exposures and possibly most, if not all, of the upwind exposures. Wind conditions will play a big role in this decision. At least in this way, if the fire escalates, the building will already be empty, removing that headache from the incident commander's list of priorities.

Cockloft backdraft

A backdraft in a cockloft can be unforgiving, trapping firefighters below and spreading the fire into areas that were previously uninvolved. As mentioned earlier, the new-style contiguous structures will usually have a larger area for heat accumulation and subsequent distribution and dissipation than their old-style counterparts. Attached townhomes, condos, and garden apartments will usually all have poorly fire-stopped roof spaces (peaked to some degree). Adding oxygen at the wrong time and from the wrong place can lead to instantaneous ignition of an entire area. It can also lead to ignition of gases in a traveling flashover-type spread in remote areas. Properly placed and coordinated ventilation will be the difference between unwanted below-the-ceiling ignition and desired above-the-roof venting (and usually) ignition. Before the ceilings are opened, get the superheated gases out of the cockloft by opening the roof, if it can be done safely. This usually means from an aerial device or properly placed roof ladder. Fire venting through the roof is a good signal that the building is being cleared of the products of combustion and the fire is being localized. Always vent high and your problems will diminish.

This same tactic is also warranted at old-style contiguous structures. Fire in row houses will march right down the block unless some type of vertical venting is accomplished to localize the lateral spread. The cockloft of

The entire roof space is one open area. While there is more room for heat spread, the lightweight characteristics of the roof support members cannot be expected to hold up as long as their old-style counterparts.

these buildings will usually be less spacious, allowing the superheated, unburned products of combustion to heat up faster and reach their ignition temperature. These gases will mostly consist and behave like carbon monoxide. Carbon monoxide is the most abundant byproduct of incomplete burning. In the roof area, where the oxygen supply will be limited, this gas will be present in great quantities. The ignition temperature of carbon monoxide is somewhere

Thick smoke pushing out of the roof area is indicative of cockloft involvement. The darker and more pressurized this smoke is, the more dangerous the situation. To avoid a cockloft backdraft, the roof must be opened *before* the ceilings below are pulled. Coordination between the interior and the roof is critical in this situation. *(Bob Scollan, NJMFPA)*

between 1100°F and 1200°F. Gases in the cockloft will usually be above this temperature. If oxygen is added from below, a backdraft can occur, with the expanding gases and accompanying blasts of fire taking the path of least resistance. This would be directly at the firefighters opening the ceiling from below. The safest action is to open the roof first and let the products of combustion out, then pull the ceilings and attack the fire in the cockloft. Coordination between the roof and the interior top floor crews is critical. It is a good idea to place a chief officer in this area to ensure that proper and safe coordination of this operation is completed. Sometimes firefighting crews get caught up in the heat of the battle. Having a chief officer operating in the area will increase the safety factor through continuous hands-off evaluation of the fire conditions. This firefighting strategy and accompanying tactics have been repeated in this book for a reason. Properly coordinated fire operations, coupled with efficient communication, will give the operating forces the best chance of safety and success.

Accountability

Any time you, as an incident commander, are confronted with a large potential fire area (such as in a contiguous structure) the safest action to take is to reduce your span of control by decentralizing the fireground. This entails placing (preferably) chief officers in command of specific areas that would be difficult to control from the command post alone. Areas such as the rear, the roof, and both exposures are deserving of this delegation. This would provide for the safest and best control over the accountability on the fireground.

It has already been established that these structures are very manpower-intensive—a veritable manpower sponge. Multiple alarms will be required at any serious fire involving these structures. As companies report in to the command post, they should be logged in on either a command board, tactical worksheet, or whatever the department uses to control the fireground. Assignments should then be issued based on the needs of the incident and on requests received from the various divisions, sectors, or groups operating at the fire. Once in their assigned division, these companies operate under the chief officer or other supervisor assigned the responsibility for that area. Companies should work in this area and report to this supervisor until the incident is brought under control, they are reassigned, or they are sent to rehab for a rest. Once out of the building or area, the company, as a unit, must report back to the incident commander or the operations chief (if designated) so that they will be logged in as "rehabbing". When the rest period is complete, the company reports as a unit back to the command post to be reassigned to another area of the fireground.

Remember that a PAR (Personnel Accountability Report) every twenty minutes or at the request of the incident commander will help establish company location and assignment status at regular intervals.

The point to be made here, which can be referred to as an "Assignment Model" is that companies:

1. Stay together as a unit.
2. Report to the command post for assignment.
3. Report to assigned division, sector, or group supervisor.
4. Operate in the assigned area.
5. When relieved, report back to the command post for reassignment or rehab.
6. If reassigned, go back to step #3.
7. When rehab of the company is complete, report back to the command post as a unit for reassignment.

The key to safe operations on the fireground, as stated earlier, is disciplined officers leading equally disciplined and trained firefighters. These companies operate within the parameters of the action plan established by incident command. Following these simple and straightforward guides will decrease and hopefully eliminate freelancing on the fireground.

A great majority of the problems the incident commander will encounter will be in areas he cannot see. Assigning a chief officer to supervise operations in these areas will increase the safety and the efficiency of the operation.
(Ron Jeffers, NJMFPA)

Case Study

The Roc Harbor Townhouse Complex fire

The building and fire condition upon arrival. This fire struck at approximately 1:00 p.m. on a clear day. The wind was out of the west, toward the river, at about 10mph. The temperature was about 50°F. Roc Harbor is a wood frame townhouse complex located on the Hudson River waterfront in North Bergen, New Jersey. The gated complex is crescent-shaped, with the open end of the crescent facing the Hudson River. A series of attached three-story townhomes, some duplex, face an outer roadway. The roadway does not encircle the property, but dead-ends on the south and north side of the property. The rear of the buildings face a garden-type courtyard. The peaked roof is constructed of open, lightweight-wood truss construction. The floors are also of lightweight, parallel chord wood truss construction. There are open wooden stairs on the exterior that provide access to all townhouse units.

I was acting as deputy chief in command of the third division on the day of this fire. I received the alarm over the radio for a fire in the roof area of 8000 River Road. The response for North Hudson Regional Fire and Rescue on a reported fire is four engines, two ladder companies, a battalion chief, and the deputy. Roc Harbor, also known as 8000 River Road, is in the third battalion.

At the time of the alarm, I was on 10th St., in the first battalion, some four miles away. Battalion 3 Chief Nick Gazzillo was first on the scene, and having been from the former department of North Bergen, was extremely familiar with the complex as well as the difficulties in regard to apparatus positioning. As platoon training was being conducted at the time of the fire, some of the first due companies were off-duty and not dispatched on the first alarm.

Chief Gazzillo reported a working fire and knowing that the positioning of the apparatus was pivotal, had the responding companies stage at the head of the property, where there was a large parking area. The FAST team was dispatched with the report of a working fire by incident command. Fire and heavy smoke was showing from the third floor window of a duplex apartment on the southwest side of the complex. It had already extended to the combustible exterior walls and roofing material and was threatening to extend to the leeward units. Smoke was also pushing out of the peaked roof eaves on both the windward and the leeward side. A second alarm was requested, bringing another two engines and a ladder company. Again, the companies were staged in the parking lot and told to await orders. I finally arrived as the second alarm companies were pulling up.

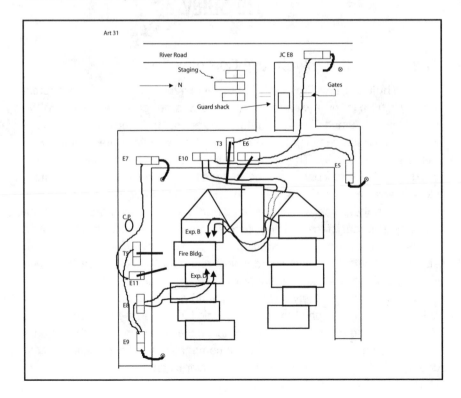

This is relatively early in the operation. Road access as well as water supply operations were difficult. Smoke is showing at the roof level at each end of the involved unit. This is indicative of the open cockloft.
(Tom Foster)

As interior positions were being established, a Telesquirt stream was placed in service to knock down the fire extending via the combustible exterior wall. In these instances, stream discipline is crucial. No exterior water should enter any interior area.
(Tom Foster)

Fire area operations. Chief Gazzillo wasted no time is setting up the fireground. A cautious offensive/defensive strategy was pursued. Lines were to be stretched to the fire area, while taking actions in both exposures to keep the fire confined to the building of origin. The biggest factors in this strategic decision was the lightweight construction of the building and the fire conditions upon arrival.

Lines were stretched up the exterior stairs, which were located and accessible from the rear, to the fire area in an attempt to knock down the fire. The first-arriving ladder company, Ladder 5, was placed in front of the building to initiate a primary search of the fire building. Ladder 4 was left in staging, and the manpower was used to assist with the search and recon operation at the rear of the building.

Next, a Telesquirt was positioned on the flank of the leeward side of the fire. The strategy was to use the Telesquirt to knock down the fire involving the combustible shingles and roofing material, which was threatening to extend via auto-exposure to the entire row of buildings. As soon as the fire was knocked down, the Telesquirt was shut down.

Lines were also stretched to the windward side of the fire. Due to the set-back of the building and access to the rear stairs, the hoseline stretches were long and arduous. A second ladder company was positioned at the windward side of the fire. Chief Gazzillo, in the first five minutes of the fire, set up three separate water supplies and positioned aerial apparatus for master stream use. In addition, the Telesquirt was being used to knock down the exterior fire. These actions set the stage for the companies to make the stop.

Exterior exposure operations. At about this time, I assumed command of the fire. I was briefed by Chief Gazzillo, who went to work coordinating operations in the Delta Exposure on the leeward side. I struck a third alarm. On the response were another two engines and a ladder truck along with an additional battalion chief. These companies were used for manpower, reinforcing the interior operations on both sides of the fire. The additional battalion chief

was placed in charge of the Delta Exposure and Chief Gazzillo, moved to the windward side, Exposure Bravo, to coordinate the operation there. By now, two lines were stretched to the fire area via the leeward exposure and two were stretched to the windward side of the fire. The fire unit was reported as untenable and fire was now showing from a small hole in the roof.

The strategy now became defensive/offensive as companies utilized a "pincer" action to attempt to pinch off the fire between the two exposures, thereby confining it to the unit of origin.

Better stream penetration could be accomplished from the rear and lines were placed there to hit the fire in the involved unit. A ladder pipe was set up at the front of the building to hit the main body of fire, but I did not as

As fire begins to break through the roof, companies were in position to apply a "pincer" action and pinch off the fire to the unit of origin. An extensive combustible-clearing operation enabled interior companies to maintain their positions and localize the fire. *(Bob Scollan, NJMFPA)*

yet want to open the exterior streams until a good part of the roof in the main fire area had burned away. The Telesquirt was also standing by.

Roof operations. Ladder companies laddered the roofs via aerials and attempted to cut holes in the leeward and windward sides to slow the spread of fire in the cockloft. At the same time, companies were working feverishly on the interior, pulling ceilings in the exposed areas. This was essentially a "combustible-clearing" operation in the attached exposures so that the fire would have nowhere to hide when it attempted to spread beyond the unit of origin. The vent teams were instructed to remain on the aerial during the venting operation and abandon the roof if conditions warranted. Both ladder companies reported heavy fire in the cockloft, but were able to provide vent holes in each roof before abandoning. It remained to be seen whether these actions would slow the fire.

Heavy fire was venting through the roof by now and I was seriously contemplating abandoning the leeward exposure. However, the report from the interior on that side of the fire indicated that there was no fire as yet in the building. The ceilings were completely down and the trussloft exposed. I decided to let them continue. Crews were experiencing some difficulty in the windward exposure, Bravo. This was due to the fact that the windward operation was actually being fought from an adjacent unit in the fire building,

not a different building. As such, conditions were difficult. Battalion Chief Gazzillo told me later that at that time, he was going to give it thirty more seconds before he pulled the plug on the operation and withdrew. But, through the efforts of the men working inside and a little help from the wind, conditions slowly began to improve.

Ladder 5 prepares the ladder pipe as heavy fire breaks through the roof of the fire building. This fire condition actually helped to localize the fire to the unit of origin. The upward draft of the venting fire pulled in fresh air from below and established a favorable venting direction. *(Bob Scollan, NJMFPA)*

The exterior streams were now put into operation, as the roof had burned away. They were positioned in such a manner that they could not only hit the main body of fire, but if the leeward side became untenable and withdrawal was required, they would be able to apply water at that point as well. The fire burning away the rest of the roof was a good sign as the fire was now released from its confines, somewhat localizing it. It was also drawing air currents into itself from the floors below, making conditions more comfortable for the crews operating in the adjacent exposures.

Sensing that the operation could still escalate on either the leeward or windward side or both, I struck a fourth alarm and requested a Telesquirt be dispatched as part of this alarm (North Hudson has three Telesquirts, one in each battalion). I also had a ladder pipe set up on the windward side in case the Bravo exposure had to be abandoned. The fire did break out of the cockloft, but there was an open stairwell that acted as a fire break, which allowed the companies on the windward side to maintain their positions. The fourth alarm brought an additional two engine companies, a ladder company, and an additional chief. I positioned the Telesquirt on the windward side next to the ladder company. They set up the Telesquirt, received a water supply from an engine that had taken a hydrant on River Road, and stood by. Now, if the windward side became untenable, we had two master streams ready. The other companies were used for manpower to reinforce the operation on both sides of the fire.

Although it looked like hell was breaking loose on the exterior, things soon began to get better on the interior. Continuous progress reports from the interior will often assist incident command in deciding whether to maintain a position or withdraw. *(Tom Foster)*

As all hands were working, I realized I had no one in reserve. I struck a fifth alarm to establish a tactical reserve and provide relief for the operating companies. Another two engine companies and a ladder company responded as well as a battalion chief from Jersey City. I utilized this chief, along with other responding North Hudson chiefs, to be my eyes and ears in and around the fireground. Remember that the most problems incident command will encounter will be from unseen areas. When companies are actively engaged in these areas, it is wisest to summon additional chief officers to keep an eye on the men, the building, and the operation. It may seem like a long period of time took place from the first to the fifth alarm, but in reality all of this happened in the first forty minutes of the fire.

Smoke color begins to change as companies begin to get a handle on the fire. The fact that the roof burned away was a significant contributor to our ability to maintain interior positions. Nevertheless, some extremely down-and-dirty, hard-nosed firefighting also made the difference. *(Bob Scollan, NJMFPA)*

Conclusion. The fire unit eventually completely collapsed, pancaking the roof and all the floors into the garage. However, the fire never spread beyond the unit of origin. Even the attached units of the fire building were saved. This was mainly due to two factors. The first was the effective apparatus positioning set up by Chief Gazzillo in the initial stages of the operation. The second was the aggressive and hard fought battle waged by the firefighting companies to keep the fire confined to the building of origin. This was, based on arrival conditions and the access and positioning difficulties, truly an incredible stop.

Basic Firefighting Procedures

Basic firefighting procedures will center around determining the extent of the fire and the most likely path of travel. Once this is determined, fire confinement and then extinguishment strategies can be applied.

CRAVE

Regarding the CRAVE acronym:

Command

- Forecast fire travel when positioning the command post.
- Develop and maintain a strong command presence.
- Decentralize the fireground by assigning chief officers to critical strategic areas.
- Demand that reports of conditions be furnished from all unseen areas as soon as possible.
- Beware of the truss!
- Be cognizant of the manpower sponge characteristics of contiguous structure incidents and summon additional resources as required.

Rescue

- Recon all sides of the building, including all shafts.
- Conduct an aggressive primary search of all tenable areas.
- Evacuate attached units on the both the leeward and windward side of the fire as conditions dictate.
- Ensure a thorough secondary search is conducted once the fire is under control.

Note the half-shaft located at the center of this building. The attached building, recently torn down, had a matching shaft, forming a narrow diamond. Firespread here will be extremely rapid.

Lines are stretched into both the leeward and windward exposures. Commit lines early in contiguous structure fires, especially for a top floor fire. Fire can spread like lightning through an old, dusty cockloft. *(Bob Scollan, NJMFPA)*

Attack

- Establish a strong primary and secondary water supply.
- Stretch lines via the safest, most effective path of least resistance to:
    a. The fire floor/area
    b. The floor above
    c. Top floor of leeward exposure
    d. Top floor of windward exposure
    e. Adjoining cellars if fire is located in the cellar of the fire building
- Coordinate attack operations with support operations.

Ventilation

- Ladder the windward exposure, when possible, to access the roof.
- Aggressive fire confinement tactics must be employed to prevent spread.
    a. Vent appropriately in relation to the location of the fire
    b. Vent at the top of the vertical artery to clear the stairwell
    c. Horizontally ventilate opposite the attack line
    d. Cut the roof at top-floor and cockloft fires
- Use an aerial device or roof ladder if cutting a roof constructed of light-weight materials.
- Coordinate with attack operations.

Extension Prevention
- Consider the paths of least resistance for fire travel:
  a. Pull ceilings completely to expose the cockloft
  b. Coordinate with the roof ventilation team
- Examine scuttle and skylight areas in row houses.
- Check all concealed spaces for extension, both vertically and horizontally.
- Have charged lines in areas of vulnerability.
- Clear combustibles in the path of the fire.

To access the roof in a relatively safe area, place ladders to the roof of the windward side exposure. If possible, place the ladder at least two roofs away from the fire building for an added margin of safety.
(Bob Scollan NJMFPA)

Conclusion

Contiguous structures offer one of the most visible opportunities for the incident commander to show how well or poorly he can manage a fire. It is hard not to notice a whole block gone. The incident commander's ability to forecast the potential fireload profile and summon the proper resources to overmatch that fireload will, to a great extent, make or break his operation. He must remember that serious fires in these structures will tax the firefighters on the scene. Proper forces must be ordered early to provide relief for those committed to the tasks of both containing and extinguishing the main body of fire in the fire building and those taking aggressive preventive steps to keep the fire from spreading to the attached areas. If this relief is not immediately available, once the first-string troops are drained and exhausted, the game is over and lost.

A roof team must be assigned to cut the roof at any top floor or cockloft fire. Roof venting will assist in localizing the fire and slowing the horizontal spread of fire to adjacent structures. Get off the roof when the task has been accomplished. *(Newark, New Jersey Fire)*

Questions for Discussion

1. From a building stability standpoint, discuss some of the reasons why old-style contiguous structures were superior to new-style contiguous structures.
2. Discuss some of the reasons that sheet metal surface fasteners are prone to failure.
3. Discuss fire control in contiguous structures in regard to the CRAVE acronym.
4. Discuss roof operations aimed at localizing a fire in a contiguous structure.
5. Discuss top-floor tactics in both the fire building and leeward exposures when fire is present or suspected to be present in the cockloft.
6. What is the Assignment Model and how does it enhance firefighter safety and accountability on the fireground?
7. Discuss some of the fire control strategies and tactics to address involvement of combustible wall surfaces at contiguous structures.
8. Describe problems pertaining to rear access in contiguous structures. Name some ways to overcome these problems.
9. Discuss why fires in contiguous structures are typically manpower-intensive and require additional alarms.

CHAPTER NINE

Taxpayers & Mixed-Use Occupancies

Mixed occupancies are structures that house both commercial and residential occupancies in a single building. This includes two-story taxpayers, although some two-story taxpayers will be occupied by commercial occupancies on both floors or by assembly occupancies such as social clubs on the second floor. These occupancies combine the heavy fire load of a commercial occupancy with the severe life hazard of a residential occupancy. In a mixed-use occupancy, the first floor is typically occupied by a commercial establishment, while the upper floor(s) house apartments. There will usually be two entrances, one for the business and one for the occupants. The entrances may be side by side or the store entrance may be on the main street while the residential entrance is on the side street, at the rear, or even in an alley between two adjacent structures.

This mixed-use occupancy houses a woodworking shop that faces the main street, while the residential dwellings occupy the two top floors. Note the residential entrance is on the side street (Side D).

This chapter will refer to mixed-use occupancies two stories in height as "old-style" taxpayers or simply "taxpayers". Those structures three stories or higher will be referred to as mixed-use occupancies. One-story taxpayers, both new-style and old-style (often referred to as strip stores or malls) will be discussed in the next chapter.

This old-style taxpayer is two stories of ordinary construction. Stores occupy the first floor, while a dance studio and a hall-for-hire (both of which are open at night) occupy the second floor after the store is closed.

Taxpayers and mixed-use occupancies will most often be of ordinary construction, although older buildings may be wood frame. They will be grouped in clusters, some as long as a block. There may be a common cockloft over the row of buildings or there may be a firewall separating the individual tenants. Without prior knowledge, this information can only be ascertained by the team on the roof, either by direct examination of the cockloft or by the presence of parapet walls located between adjacent buildings. It is wise to assume there are common cocklofts and firewall penetrations until it is proven otherwise. An inspection of the cockloft, from the vantage point of one of the exposures will show whether the cockloft is common to all buildings in the row or separated. This information must be immediately relayed to incident command. Mixed-use occupancies will require a large commitment of manpower, especially if the cockloft is found to be common to the entire row of buildings. Aggressive fire confinement and recon operations will be required.

I once operated as supervisor of interior operations at a large, two-story taxpayer in another town. There were apartments on the second floor and a business on the first. Fire was raging above the tin ceiling, which was present over both of the two large apartments on the second floor. The jurisdiction's ladder truck was out of service, and a great deal of time was spent investigating a heavy odor of smoke before a working fire was declared. In that time, the incident commander did not request a ladder company respond to assist in the investigation. I responded on the special call along with a ladder company from my old department, Weehawken. It took us approximately ten minutes to reach the scene.

By the time we arrived, there was heavy smoke puffing out of the entire roof area. It was evident that there was a heavy fire condition in the cockloft. I told the incident commander he had better call for additional alarms based on my initial evaluation from the street. My ladder company quickly raised the aerial to the roof and initiated roof-cutting operations. When I got upstairs, I saw that there were only two men operating in one of the apartments, working feverishly with a pike pole and a handline. At the time of my arrival, there were plenty of fire-fighters around the building, but for some reason only two were inside. It seemed everywhere they opened the tin ceiling, there was heavy fire. Reports from the roof confirmed this as well. Eventually, the interior operations had to be abandoned, for by the time the second and third alarm companies got to the scene, it was too late to stop the fire that had by now extended throughout the entire cockloft. As a result, the top floor, the roof, and the building were lost. The failure of the incident commander to recognize the manpower requirements for this type of structure as well as his reluctance to organize an attack or request the response of a ladder company on the initial alarm led to the loss of the building.

Let's discuss some of the difficulties of fighting fires in these structures.

Firefighting Problems in Mixed-Use & Taxpayer Occupancies

Forcible-entry difficulties

Forcible-entry problems will mirror those found in commercial occupancies. Roll-down gates, metal shutters, and access to the rear will all be problematic. Expect delays in operations whenever faced with these barriers. In addition, be cognizant of the signs of backdraft when a fire in a closed-up store is encountered.

Fires in the cellars of these occupancies will also

Roll-up gates are just some of the access difficulties encountered in ground floor store fires. Be cognizant of any indications of a potential backdraft in a heavily secured building that has been closed up for some time. *(Bob Scollan, NJMFPA)*

pose the difficulty of forcing metal sidewalk doors. Commercial occupancies such as strip malls that are built of slab construction are erected over a concrete slab foundation. There is no cellar. Mixed occupancy structures will not have this luxury.

As mentioned before, don't forget that dogs may be present in stores at night and in cellars during the day while the store is in business. Try to ascertain this information beforehand to avoid unnecessary injuries.

Light shafts

This raised scuttle covers a shaft that rises between two adjacent buildings. Windows of both buildings open on this shaft. This cover is not always present creating the danger of falling into an open shaft.

In rows of attached buildings of this type, there will usually be light shafts present between each building. These shafts, open from ground to the roof, will negate the existence of the firewall that serves as a barrier to horizontal building-to-building fire spread.

Mixed-use occupancies and taxpayers will usually be of smaller dimension and area when compared with newer commercial structures and even multiple dwelling apartment buildings. Thus, the light shafts will also be of a smaller dimension. They are often diamond-shaped, with four windows bordering on each shaft per floor. The windows facing each other in the shaft will serve apartments in two different buildings. There are usually no windows serving the shaft on the ground floor, as it takes up valuable space in the ground-floor store.

Where stores occupy the ground floor of a mixed-use occupancy, the shaft between buildings may start at the second floor, serving only the residential portion of the building. If the building is attached, the shaft will only be visible from the roof or from the interior.

Light shafts create a flue between the two buildings. The smaller shaft present in these structures will heat up faster than a larger shaft commonly found in larger multiple dwellings, causing easier ignition of combustibles bordering on the shaft. For this reason, fire venting into this shaft can rapidly spread fire into the adjacent building. Protective lines must be stretched into the exposed apartment as soon as the hazard is recognized.

Roof reconnaissance and regular progress reports are critical to fire confinement and control operations. These shafts are usually visible only from the roof. However, they may have been covered over by roofing material to keep the building more energy efficient. Recognition of these areas on the roof is critical to firefighter safety. Any area that seems to have newer roofing material is suspect for obvious reasons. One easy tip-off is that there is usually a

Beware of areas where the roofing paper does not match the rest of the roof. In this case, the patch job covers a hole created by a roof venting operation from a previous fire. These areas may not be properly supported. Note also the loose chimney bricks on the extreme right.

parapet and coping stone between the individual buildings where the firewall extends through the roof. Where the diamond-shaped shaft is present, the parapet will split off in two opposite directions to accommodate the shaft and then meet again at the other end of the shaft where the parapet wall will be continued. If the area inside the diamond is covered over, chances are it is unsupported. Don't step on it. The presence of these and other unsafe features make probing of the roof before moving into any untested area imperative.

Another area that should not be stepped on is the scuttle hatch. Covered usually only be unsupported plywood, it cannot be expected to support the weight of a firefighter. Other features that may also have been covered over are skylight openings and holes created by previous firefighting operations. Any area of a roof where the roofing material does not match that of the rest of the roof should be suspect. Caution is always the name of the game in roof operations.

Cellar concerns

Cellar fires in any building will be a major insult to the very existence of the structure and to adjacent exposures. First and foremost, the entire building is exposed to the fire, as are all the occupants. If a fire is on the top floor or in the cockloft, it only has one way to go: up and through the roof. It may take more time and effort to set up operations on the top floor, but, generally, the life hazard will be less severe. A cellar fire will require a larger commitment of manpower and apparatus as the fire forces struggle to keep the entire building from falling victim to the least resistant path of heat travel.

Another factor is the often-confusing layout of many cellars, especially cellars below commercial occupancies housed in older ordinary construction and wood-frame buildings. Cellars may be larger than anticipated as one business may use the rear half of the adjacent occupancy for storage, while the front half is used by the business above. It is not hard to get lost in a smoke-filled cellar. Other cellars may be open for the entire length of the row. For this reason, any personnel entering a cellar should not do so alone. In addition to operating with a partner, if not stretching a hoseline, a lifeline and thermal imaging camera should be used.

Speaking of life, the cellar must be thoroughly searched at any fire. It is not uncommon for store employees to sleep in the cellar under the store. A fire inspector from another town told me that they once found fifteen people living in the cellar beneath a Chinese restaurant. Their only means of egress was by way of the trapdoor leading into the store.

The contents of the cellar is also a concern. Stock may be piled high and stored in an arbitrary manner. Streams from hoselines can cause collapse of stock, trapping firefighters advancing lines in the cellar. Always maintain a clear path of retreat.

Lack of fire protection equipment

Many of these buildings are old and were built before modern fire protection codes were enacted. Thus, they do not require sprinkler protection. Modern codes may require that the normally occupied business area be protected due to the life hazard, while storage areas may not require sprinkler protection. Fires may get a good head start given the combustibility of the fireload and the building itself, especially in the cellar. While sprinkler activation may hold the fire in check, it may contribute to structural failure as contents absorb water and swell. Ensure, during inspection visits, that there is enough clearance between piled stock and enclosing walls. In addition, ensure there is adequate clearance between the sprinkler head and the height of stock. Many times, the sprinkler activates, but due to high-piled stock, the extinguishing agent does not reach the burning area. As the fire grows, it may then overtax the system, rendering it useless.

Common cocklofts

These buildings are often built in rows. There may be a cockloft common over the entire row. Fire strategy and tactics should focus on confining the fire to the building of origin. Most often, a scuttle or skylight will be present over two-story buildings of this type, while a bulkhead door will probably be present at mixed-use occupancies greater than two stories. It doesn't matter where the fire is located, whether it be the cellar, the

Once heavy fire reaches the common cockloft, the race is on to save the rest of the block. If sufficient resources are on hand and positioned early and wisely, the fire may be confined. *(Ron Jeffers, NJMFPA)*

store, the floor(s) above the store, or the cockloft. These vertical arteries must be opened as soon as possible to prevent horizontal fire spread by way of the common cockloft. The roof will generally not be cut unless the fire is located on the top floor or is in (or spreads to) the cockloft.

Once the natural openings on the fire building are opened, the natural openings on the adjoining buildings must be opened and checked for evidence of extending fire. A report of conditions must be sent to incident command from the roof position as soon as possible. Any spreading fire must be addressed immediately. It is best in these buildings to have recon teams enter the adjoining occupancies to also examine the cockloft from below. In any fire where early control is doubtful, a line should be stretched to the top floor of the leeward exposure immediately after lines are placed inside the fire building to confine and attack the parent body of fire. That way, if fire spreads into the adjoining structure, the line is already in place. The same is true if there are any windows found bordering on enclosed shafts or if there is a cellar fire and there is a possibility of lateral fire spread to the exposures via unprotected openings in the cellar. In other words, if the fire is severe, plan early for extensive operations in the leeward exposure first, but also don't forget the windward exposure. Timely reconnaissance will reveal if lines are required in this area as well.

The concept here is that if there is even a possibility that common areas for fire spread are present, it is wise to get resources in the form of manpower and equipment into these areas as soon as possible. Additional alarms may be required to cover these areas. It is the practical and effective incident commander who has lots of resources on the scene and lots of building left to show for it at the end of the operation than an incident commander who conserves manpower, but winds up with nothing but a pile of rubble to show for it.

Tin ceilings

Above many of these stores and apartments are tin ceilings. Whether these ceilings are help or hindrance will depend largely on the location of the fire in relation to the ceiling. When the fire is located beneath the ceiling, the problem may be one of high heat, but the ceiling will act as a barrier to upward fire spread. Areas around pipe chases and other utility openings such as light fixtures must be checked, as they will represent the paths of least resistance for fire travel to the upper floors.

Fires above the tin ceiling, especially in the cockloft, will present a bigger problem. The fire will be difficult to access and due to the confined and limited area and ventilation opportunities, backdraft conditions may be present. For this reason, it is imperative that the roof be opened before any openings are made from below at any serious fire in a cockloft, especially where a tin ceiling exists. Radio communication between interior teams and the roof team are critical so that roof openings are completed before openings from below are made. Once the roof is opened, the ceilings are pulled and streams are applied from below.

A fire in the cockloft will take the commitment of several ladder and engine crews, operating at least two lines on the top floor. The operation is manpower-intensive, but the ceiling must be pulled and streams applied from below. This will be the only way to save the structure. If this is not done, the fire will burn the roof off, chasing the roof team from the roof, and possibly spreading to adjacent exposures.

I have been to many fires where chief officers are reluctant to commit manpower to the top floor to operate aggressively, pulling ceilings and attacking the cockloft fire from below. They would rather allow the fire to burn through the roof, taking the top floor with it. These same chiefs then apply as many master streams as possible into the roof opening made by the fire, negating the natural venting action and pushing the fire into the lower floors. This strategy just does not make sense.

The best strategy to pursue at a top-floor fire is to

- commit sufficient manpower to the top floor with pike poles and hand-lines.
- have the roof team open the natural openings on the roof as soon as possible to alleviate any backdraft potential.
- pull the ceilings from below until there is no ceiling left to pull or the fire is exposed and extinguished, whichever is first. Aggressively pulling the ceilings will expose the cockloft and the problem.
- cut the roof as directly above the main body of fire as is safely possible, at the same time as you are pulling the ceilings.

Remember also that the presence of a drop ceiling will often mask the presence of a tin ceiling above and create an additional cockloft between the drop ceiling and the tin ceiling. This ceiling must be pulled as well and examined for fire spread.

What is being offered here is that an attempt must at least be made to stop the fire *before* the roof becomes untenable. If this strategy is unsuccessful, at least a valiant effort was

This tin ceiling is being covered over by a drop ceiling as this store is being renovated. This will create another cockloft above the drop ceiling.

made. Personnel will then be ordered from both the roof and the interior of the building. Master streams are then applied from below the roof through windows to hit the fire that is not yet venting through the roof. The fire under the roof is where the problem lies; the one that may still spread laterally. Fire burning through the roof is good as long as it is not threatening an adjacent and possibly larger structure. If that is the case, the lines should also be placed to protect exposures, the highest priority. The fire burning through the roof should be allowed to do so if at all possible.

If a major part of the roof has not yet been burned away, allow the fire to vent and apply exterior streams from below through windows or other openings. This stream angle will allow water to be directed at the still intact ceiling and roof space. A Telesquirt is an excellent tool for this task. *(Bob Scollan, NJMFPA)*

You may ask, "when is it acceptable to use master streams from above?" This would be warranted when the fire has burned a major portion of the roof away, indicating that the streams operating from below the roof were not successful in reaching the seat of the fire. In this case, the streams are redirected into the now wide-open roof area to reduce the amount of flying brands and quench the seat of the fire. A stream now operated through the window from below would shoot straight out of the building where the roof used to be, so these streams would now be ineffective at this angle. A stream operating from an elevated position in such a manner as to apply water behind the enclosing walls will have the best chance of hitting the main body of fire. It should go without saying that when the operation reaches this stage, men will already be withdrawn from the immediate area and collapse zones will be established.

Once the fire has burned away a major portion of the roof, the stream strategy should change. Elevate the master streams, keeping in mind the collapse zones, change the angle of the streams to downward, and hit fire burning behind enclosing walls. The platform is the tool of choice for this tactic. *(Bob Scollan, NJMFPA)*

Poorly fire-stopped cornices and façades

Just when you think you're done checking for firespread via the cockloft and via the light shafts, the next area of possible extension comes—the open cornice or façade. This artery may be open across the entire frontage of the row of buildings. The feature may be open by design or by the result of structural deterioration over the years. Cornices and façades may also allow unimpeded access into the cockloft. Usually found hanging cantilever-style off the front of a building, this horizontal channel may allow fire to spread past firewalls between buildings and into adjacent cocklofts. Cornices and façades must be opened from above by teams on the roof or by personnel operating from an aerial device. The device of choice would be the tower ladder, aerial platform, or Telesquirt. Not as desirable, but still effective for both opening and stream penetration into the cornice is the aerial ladder. Make sure any firefighter operating from an

Fires that originate in or extend to façades can spread fire horizontally along the exterior of the building. When it finds pokethroughs such as those created to run electrical conduits, it will spread into the interior. Ensure the area is thoroughly examined.
(FF Jeff Richards, NHRFR)

aerial ladder is safely belted into the device in case the ladder has to be moved in a hurry due to fire erupting out of the opening. Any ladder movement should either be rotation or raising, never retraction or extension as the firefighter on the aerial may get caught in the moving aerial rungs.

An unintended advantage offered by these poorly fire-stopped openings is that if the interior operation must be abandoned due to untenability, this cornice area is an ideal place to apply an aerial master stream. It still puts the

Opening up the cornice allows for stream access into the cockloft. Make sure you cut a vent hole in the roof so the stream does not push the fire into uninvolved areas. *(FF Mike Castelluccio, Newark, New Jersey Fire)*

stream below the still-intact roof, and does so at a better angle than a stream operated from ground level. Keep this in mind at serious cockloft fires in these buildings. The more tools and options in your strategic and tactical arsenal, the more chances you will have at applying the proper tactic for the situation and the more effective you will be. This tactic will also work at multiple dwellings as they too often have a cornice at the front of the building, and possibly the side if the building is located on the corner. Tactics at these fires require a little ingenuity sometimes. Victims of functional fixity do not do well in these situations.

Hazardous materials

Just as in any commercial occupancy, hazardous materials specific to the business conducted on the premises should be a tip-off as to the presence of chemicals stored, which may include those which react with both fire and water. Right-to-know information as well as Material Safety Data Sheets furnished to the department by the business will be of use here. It is unacceptable to be unaware of hazards that are supposed to be common knowledge due to disclosure laws aimed to protect emergency responders.

Some businesses that should raise a flag of caution include:

- **Dental labs / medical labs / doctor's offices** – radioactive and etiological materials, oxygen
- **Photo labs and Video stores** – Nitrocellulose, which is a component of film and video tape, burns extremely toxic and may be explosive
- **Pool supply stores** – chlorine and other water-reactive materials
- **Auto part stores** – flammable liquids and corrosives
- **Hardware stores** – a variety of chemicals including pesticides and possibly propane
- **Garden Stores** – also will include pesticides and fertilizer

- **Gun and ammunition shops** – the danger here is obvious, but these commodities may not be so obvious in a sporting goods store
- **Air conditioning and refrigeration dealers and service providers** – refrigerants
- **Dry cleaners** – solvents and corrosives
- **Drug stores** – a witch's brew of chemicals may be stored, especially in the cellar; I have seen ether stored in the cellar of drug store, which is unstable and potentially explosive
- **Exterminators** – pesticides and other poisons
- **Paint stores/home improvement stores** – flammable liquids
- **Plumbing supply companies** – flammable liquids and gases, including propane and acetylene

Remember that the primary route of exposure is via the lungs. Smoke produced by many of these burning consumer products will be extremely toxic and may not be reversible. In addition, the damage done today may not be evident for many years. This is as true standing outside the building as it is inside the building.

A chief officer, assigned to the duty of safety officer at a fast food restaurant fire, was performing reconnaissance at the rear of the building. The fire originated in the broiler, but had spread beyond the point of origin to a deep fat fryer and the HVAC system ductwork. The HVAC unit contained about twelve pounds of chlorodifluoromethane, also known as Freon-22. The heat

Many businesses use hazardous chemicals as part of their everyday operation. This fire in a vitamin manufacturer produced smoke toxic enough to warrant the response of a Haz Mat team and a decontamination operation. *(Ron Jeffers, NJMFPA)*

of the fire ruptured the sealed pressurized copper refrigerant line, exposing the Freon-22 to the open flames.

As the rear door was opened by the interior crews to help in the ventilation effort, the chief, standing approximately twelve feet from the building, was engulfed by a large volume of acrid smoke discharging from the building. The rear of the building was both opposite the attack and leeward. As the smoke cleared, it became apparent that the chief was experiencing respiratory difficulty and he immediately sought medical assistance. Two other members also experienced respiratory difficulty and were subsequently treated and released from the hospital. The chief was treated and released after two days of testing and treatment. A week later, medical units were dispatched to his home, where he was found in respiratory arrest and without a pulse. Despite the efforts of those who responded, the chief could not be revived and died an hour later. An autopsy revealed a number of complications and severe injuries to the respiratory tract, including pulmonary edema, as a result of inhaling smoke containing phosgene gas.

Freon-22 undergoes thermal decomposition when exposed to flame or hot metal and forms hydrofluoric and hydrochloric acid as well other gases including phosgene. It was believed that the acrid smoke that enveloped the chief contained the hazardous chemicals released by the combustion of the Freon-22.

Although the chief was not directly involved in firefighting efforts and was positioned in what would normally be considered a safe distance in regard to his duties, he nevertheless fell victim to the deadly products of combustion from an exterior position. This glaringly shows that the outside of the building can be just as deadly, and in this case, more so, than the interior.

The occupancies listed above represent some of the common businesses found in every jurisdiction. It is a short list, but just take a ride through your own area of responsibility with this in mind and you will see the potential for hazardous material exposure in many unlikely places. Be aware of these dangers and ensure the safety of personnel by ensuring proper protective equipment is worn, including SCBA. Allow no one in the area of the building without it, especially in buildings where the toxic byproducts of the fire may be detrimental to health.

Life Hazard Problems in Mixed-Use & Taxpayer Occupancies

Transient occupancy

Other than the staff, the occupancy load found in the commercial portion of these buildings will be relatively unfamiliar with its layout. This could spell disaster if a fire strikes. Smoke will blind and choke patrons attempting to escape. A resulting panic condition could cause needless death.

Fortunately, the area of stores located in these commercial/residential occupancies is relatively small, which hopefully will also limit the occupant load. The drawback here is that there are usually fewer exits in these stores than in larger commercial establishments. Usually, the only way out is via the front door. Other exits are usually accessed from behind a counter or by having to traverse heavily stocked and cluttered areas of the store. Most people will not seek these exits. If the fire is located between the occupants and the exits,

If fire or a human overload blocks the front door, this blocked exit may be the only other egress point.

the only thing that may stand between life and death is a rapid response and deployment of personnel to attack the fire and remove the victims.

Other problems relating to egress is that in these stores located below residences there usually seems to be a lack of the same code enforcement that is present in strictly commercial establishments. One is the placement and proper illumination of exit signs. Check out your local storefront occupancy. These are almost never present, and when they are, the lights invariably don't work. The other is a lack of functional and accessible secondary exits. You may have to negotiate a mountain of stock to get out any way other than the front door.

Another egress problem is that the exit door is not required to swing in the direction of egress, out of the structure in most of these stores. I found that

Because of occupancy load restrictions, this door is not required to swing out of the occupancy. This could cause the death of panicking patrons who pin each other, preventing the door to open.

Access to apartments above the stores in this row of attached mixed-use occupancies is located at the rear. Proper preplanning before and proper recon during the fire will discover this before it becomes a problem.

it has to do with the fact that many of these stores exit directly onto the sidewalk and an outward-swinging door would whack an unsuspecting pedestrian in the mug if it was suddenly opened. NFPA requires an outward swing in assembly occupancies and in commercial occupancies with an occupant load of fifty people or more. If you happen to be in a store with an occupancy load limit of less than fifty people, you may be out of luck if a fire starts. This really makes no sense. All it takes is for one panicking person to pin another at an inward-swinging door and there could be multiple deaths where a clear path of exit was available, but was negated because of a technicality of the code. When a fire occurs, I guess the person on the sidewalk who did not get hit with the door can watch from a point of safety as the patrons pile up behind the door and die.

Life hazard above the store

The main life hazard, especially at night, will be to the residents who live above the store. Efforts must immediately be made to access this area and conduct a primary search. Lines must be stretched to protect the search and guard against fire extension from below.

When assessing the life hazard profile in the floor or floors above the store, it will be useful to use the fire escape rule of thumb mentioned in the multiple dwelling chapter. Recall that the rule states to count the number of fire escapes on the building, and multiply by two. This will correspond with the number of apartments per floor, allowing manpower requirements to conduct search, rescue, and evacuation to then be estimated.

Also, don't forget that many people who own stores live in rooms at the rear of the store or, as mentioned earlier in this chapter, in the cellar. Primary search of all areas must be conducted so no potential victim is missed.

Another life hazard above a store may be found where these two-story tax-

Primary search must be conducted in all areas, regardless of the location of the fire. Victims located above stores are likely to necessitate rescue via use of aerial devices. *(Bob Scollan, NJMFPA)*

payers have businesses on the second floor, which typically involve groups of people unfamiliar with the exits. These include dance studios, martial arts schools, self-help groups, and social clubs. In these cases, the incident commander should treat the occupancies as if they were residential and take the same steps to safeguard life as if these areas were occupied by apartments.

Cellar access and egress

Fires in cellars of these buildings will be hot, smoky incidents. The access to these areas will also be difficult and punishing. If you're lucky, there will be a set of interior stairs at the rear of the store leading to the cellar. If there is, this is the best place from which to attack the fire as the outside entrance can be used as a vent point. However, this will usually not be the case in these old buildings. What is more likely is that there will be a trap door-like opening somewhere in the store leading to the cellar, usually via a set of flimsy wooden stairs.

This access point may be extremely difficult to get through wearing SCBA. If this is the case, the door should not be used for attack, but should be closed and protected with a line. An attack line will then have to be stretched via the exterior sidewalk entrance if one exists, or to the cellar possibly from a door in the rear. If the fire proves to be beyond the capability of attack lines to extinguish it or even reach the seat of the fire, this opening may be ideal for the use of a distributor or a cellar pipe.

At a recent fire in one of these occupancies, the initial companies could not find the cellar entrance via interior of the store. The decision was made to

This narrow trapdoor opening may be the only access into the cellar. The combustible wood steps are often weak and steep. If another entrance can be utilized, keep this door closed and guard it with a line, if conditions permit.

attempt an attack via the sidewalk entrance. When the door was forced, the stairs only led to a 5' x 5' area that was used for utility control. The access way to the rest of the cellar had been sealed off by concrete. The cellar door was found by stretching the line through an adjacent alley, turning in a 180° direction and stretching to an isolated room where there was a small stairway leading to the cellar. Be prepared for unusual situations like this in old buildings. Ladder company personnel on reconnaissance missions must keep an open mind and be ready for surprises.

The other usual access will be by way of the sidewalk doors. The problem with using the sidewalk access is that fire and accompanying products of combustion may be pushed throughout the store and possibly the building when attacked from this area. Ventilation oppor-

If the fire cannot be accessed from the interior, the only alternative will be to attack via the front entrance, usually the most ideal vent point. If the fire is attacked via this front entrance, order a crew immediately to the rear to check for rear ventilation opportunities. *(Bob Scollan, NJMFPA)*

This stairway at the rear of the store is an ideal place from which to make an attack, but its safety is compromised by the presence of these rollers.

tunities opposite this area will be limited. The best plan is to use the interior cellar entrance if it is safe enough to do so. Another problem is the presence of chutes or rollers left in place to accommodate the movement of stock into the cellar. It is one way down for the unsuspecting firefighter who steps onto one of these items. Be sure where you are stepping at all times, especially if visibility is obscured. If you cannot see the stairs due to a heavy smoke condition, make sure that a tool is used to probe the opening before heading down the stairs

Conveyors (either mechanically driven or on rollers) may also be found between floors in establishments such as dry cleaners where the finished products rotate on a track from floor to floor. These openings may be missed in heavy smoke. In addition, material hoists in the middle of the floor area may present the same problem. Remember that openings in walls usually lead to exits, while openings in the middle of the room usually lead you into trouble.

Overhead wires and utility hazards

Power entering the building may be the same as that for residential occupancies, but may be of higher voltage dependent on the nature of the business conducted at the premises. Check where the service enters the building. That wall will usually be where the electrical shutoffs and, subsequently, the largest electrocution hazard will exist. Make a note of this before entering the cellar and use extreme caution when operating in cellars.

When accessing and operating on the roof, especially in structures of two or three stories, note where utility pole-mounted transformers are located. They are sometimes located frightfully close to the roof edge. This may affect ladder placement and other operations in proximity to these features.

Always give electrical wires and attendant equipment a wide berth, at least ten feet when operating ladders. This includes both aerial and ground ladders.

Keep clear of wires when raising and lowering ground ladders. In addition, be cognizant of surrounding areas, especially when the weather is bad. One fire-fighter was killed and one nearly killed when the ladder they were using to vent windows on the upper floors of a mixed use occupancy came in contact with overhead power lines. As the ladder was being pulled from one window and maneuvered to the next, the butt slipped on ice on the sidewalk, causing the ladder's fly section to contact the wires.

Basic Firefighting Procedures

Procedures in these types of buildings should address the problems caused by the particular business occupying the structure as well as the amount of families located on the floor or floors above.

CRAVE

Addressing the CRAVE acronym, consider the following:

Command
- Be cognizant of the special hazards presented by each individual occupancy.
- Ensure that the commitment of manpower and apparatus is appropriate for the problem at hand.
- Consider a strong tactical reserve.

Taxpayer fire operations will require a strong incident command presence. Decentralizing the fireground by assigning incident command officers to supervise danger areas will contribute to a safer fireground. *(Ron Jeffers, NJMFPA)*

- Request additional chief officers to respond where operations are out of the direct control of incident command and require extra supervision.

Rescue

- Attempt to ascertain the number of apartments above the store by:
  a. Doorbells/gas meters/mail boxes
  b. Fire escape rule of thumb
- Converge on the life hazard above the fire from different approach routes:
  a. Interior
  b. Exterior – fire escapes / ground ladders
- Don't forget that the cellar may be occupied by store owners/staff.
- Search the rear of the store, as living quarters may be located there.

Attack

- Secure a strong primary and secondary water supply.
- Consider large diameter handlines in store and cellar fires.
- Ensure proper and early line coverage is provided in the residential areas above the store.

  - Make sure to back-up all lines.

Ventilation

- Use whatever means necessary to thoroughly vent cellar fires.
- Be aware of the backdraft potential when solid roll-down gates are covering the storefront windows.
- Vent storefront windows to alleviate heat condition and allow attack access.

Spreading fires often make the normal means of egress untenable to occupants. Search teams must converge on the living areas via as many paths as manpower permits.
(Ron Jeffers, NJMFPA)

Note the fire escape drop ladder at the front of the building. If fire is venting from the front show windows, it may be necessary to reroute any occupants on the fire escape to another egress point. *(Ron Jeffers NJMFPA)*

- Thoroughly vent upper floors and natural roof features when fire is in the store or cellar.
- Consider exposure problems created to adjoining buildings when venting on shafts.

Extension Prevention

- Consider extension via autoexposure from storefront windows. Consider extension into store and upper floors via an open trapdoor from cellar to the first floor.
- Check paths of least resistance to upper floors such as pipe chases through tin ceilings and light fixtures.
- Be aware of and take steps to prevent extension via shafts.
- Always check the cockloft for presence of fire.

Conclusion

The lower the fire is in the building, the greater the fire load and attendant fire control problems. Cellar and first-floor fires will involve commercial stock, creating a larger, hotter fire that will expose all areas and occupants above. The incident commander must have an acute awareness of the specific and sometimes insidious hazards associated with specific types of commercial occupancies and how they may impact any living spaces above them. A keen eye on fire conditions is needed; look out for indications that something may not be going according to plan due to the unique contents of the fire structure.

Questions for Discussion

1. Discuss the difference between a taxpayer and a mixed-use occupancy and the characteristics of each.
2. Discuss some of the problems caused by unseen shafts and how to best remedy these problems.
3. What are some of the actions that can be taken to stop the forward progress of a top-floor fire with a tin ceiling?
4. Give some examples of tenants of mixed-use occupancies and taxpayers that might cause a hazardous materials problem and what those hazards may be.
5. Discuss fire control in taxpayers and mixed-use occupancies in regard to the CRAVE acronym.
6. What are some of the ways to ascertain how many apartments may be located over a store in a mixed-use occupancy?
7. Discuss the disadvantages presented by poorly fire-stopped cornices and façades and how to best address these problems.
8. Discuss the life hazard problems found in taxpayers and mixed-use occupancies.

CHAPTER TEN

Commercial Occupancies & Strip Malls

Commercial occupancies present some of the most difficult arenas in which to wage the battle against fire. The sheer variety of building construction, layout, stock inventories, hazardous processes, and storage profiles make it difficult to establish one procedure to handle all of the occupancies of this type in a community.

This chapter will address various commercial establishments in which fires are fought. This will include business, mercantile, and industrial occupancies in addition to the wide variety of possibilities and attendant problems of fire control in each. Addressed will be occupancies that are strictly commercial;

Commercial buildings represent a major firefighting problem, especially in congested urban areas. In addition to having its own unique occupancy-specific fire problems, radiant heat, flaming brands, and collapsing walls will present hazards to the areas around the fire building.
(Newark, New Jersey Fire)

that is, the only life hazard that exists is the customer and work-staff load. Included in the chapter will be strip malls and strip stores, sometimes called "new-style" and "old-style" taxpayers as well as a section addressing United States postal property fire response.

Firefighting Problems with Commercial Occupancies & Strip Malls

Heavy fire loading

Fire load can be defined as the total amount of combustible material available to burn in a structure. This will, if applicable, include the structural makeup of the building itself in addition to the stock and other materials housed in the occupancy.

Commercial occupancies with high fire loads will burn furiously and emit large amounts of radiant heat. Note the smoking cars parked adjacent to the fire building that was occupied by a bus depot. *(Louis "Gino" Esposito)*

The contents of commercial establishments will often be highly combustible, packed closely together, and may be carelessly organized. This is especially true during holiday seasons, when stores stock up for increased demand. In addition, there may also be hazardous materials on site that will likely compound the fire problem. These features make for a fast-spreading fire, one that may be beyond the control of the fire forces upon arrival. Passive building features such as draft curtains and self-closing fire doors are intended to limit fire spread and, if operating as designed, will be an ally to the action plan.

It is best to protect these occupancies with automatic sprinkler systems, however, proper fire protection systems are not always maintained, if they're present at all. It is also not uncommon for an occupancy to change ownership, bringing in a more severe fire load, but never updating the sprinkler system. For example, an occupancy with a sprinkler system designed to protect paper or other Class A stock may now be occupied by a business that manufactures and stores plastics or flammable liquids. The old system may be quickly overwhelmed, allowing the fire to spread beyond the capability of the responding department to handle it.

The fire load in this occupancy is essentially an unassembled wood frame housing development covered by an unprotected steel truss roof. A well-designed and functioning sprinkler system might save this structure.

Some older buildings may have no protection at all. When no auxiliary protection exists, it is the responsibility of the fire department to survey the water supply profile and ensure that the hydraulic needs of the occupancy are met by the available flow in the area.

Hazardous processes/materials

The variety of products that can be found in commercial occupancies is staggering. Ranging in type from warehouses and factories, strip malls, as well as service and clerical businesses, these locations may house both hazardous processes and materials that may be deadly to the working firefighter. This danger will be compounded by the masking

Just because the building does not house any special hazards, don't overlook other dangers when on preplan visits. A serious fire at this seemingly innocent soda distributor can cause havoc when propane tanks used to power the forklifts explode.

characteristics of smoke and the ability of heat to cause some materials to change state. This change of state can make materials more deadly as fumes generated by the process of both pyrolysis and vaporization may combine in a synergistic manner to create an even more lethal combination of chemicals.

As is the case with most occupancies, the most effective method of operating safely in these buildings is to plan ahead of time. Pre-incident planning and routine visits will allow firefighters to familiarize themselves with the hazards that exist at a given property. This approach will allow a more effective solution-finding process to be designed that will allow firefighters to operate safely in an otherwise extremely hostile environment.

Forcible entry/access difficulties

A fox lock, often found in commercial occupancies, will present a difficult forcible-entry problem. Through-the-lock entry techniques are preferred. The lock cylinder will be found in the center of the door, opposite the operating mechanism shown here.

It is not unusual for the stock and inventory of a commercial occupancy to be more valuable than the building itself. For this reason, building owners and tenants substantially fortify the buildings not only at the front entrance, but even more heavily at the rear and even the roof. Sometimes, just getting through the first door will only lead you to another door and another as you move deeper into the structure. The obvious answer to this problem is to have the owner provide the keys or install a Knox Box. If building owners don't accommodate the fire department by providing these easy access features, the result is needless damage to the building. If at all possible, fire personnel should preplan the best way into a structure. Having this knowledge beforehand can save precious time and damage later.

Front. At the front, some of the barriers to access that we can expect at these occupancies are roll-down gates, both the open-type (where you can still see into the store) and the closed-type (where the storefront is completely hidden behind the gates). The gates will usually be secured with some type of padlock, and possibly many. If there is more than one gate, expect up to a dozen padlocks. The best way to open these is to use a rotary saw with a metal-cutting blade. Power saws operated at an arm's level height can be extremely hazardous, if not controlled. For this reason, it is important to control the lock with either a Halligan tool or a pair of vise grips and a chain. It will make the job of

Roll-up gates may be the closed type (left) or open type (right). The closed type of gates don't allow easy viewing or access into the fire area and may conceal the signs of a backdraft. Request reports from all sides of the building before you order this gate forced.

the firefighter operating the saw much easier and safer.

Before proceeding with the operation of cutting the gate and barging head-first into the structure, take a second to size-up the conditions that may be taking place just behind that gate. A closed-type gate may conceal, among other things, the indicators of a backdraft condition.

Backdraft situations occur as a result of stalled combustion. As a fire burns, it will consume oxygen. If that oxygen is limited, flaming combustion will cease once most of the available oxygen is consumed in the compartment. The area will be extremely hot, more than 1000°F, as the process of content decomposition due to the effect of heat continues. The fire tetrahedron is not complete. There is certainly abundant heat for combustion to occur and ample fuel present. However, oxygen is the missing element that will allow the burning process to continue. How this oxygen is introduced into the building is critical. Oxygen entering in the wrong place can have disastrous results, causing a backdraft.

Be aware that a backdraft can occur in the cockloft above a store or apartment. If superheated gases are present in the cockloft, the area must be opened from above. Opening from below before topside ventilation is complete can cause the superheated gases to expand into the newly firefighter-created oxygen-rich atmosphere. The now-ignited products of combustion will envelop the area, having fatal consequences to the firefighters below. I remember reading that a firefighter was killed in this manner. He was opening a ceiling in an apartment adjacent to the fire apartment. The fire was on the top floor. There had been a problem with the topside roof ventilation operation due to a saw malfunction. This information was never relayed to companies on the interior. When the ceiling was opened, a fire of explosive proportions filled the room, trapping the firefighter who was unable to escape the apartment due to barred windows. He subsequently died.

Communications are crucial in coordinating this topside ventilation with fire floor operations. Concealed spaces that may be prone to backdraft conditions can exist anywhere in a building, such as in a closet or even in a vehicle. The only way to properly handle this situation is to recognize the signals, properly wear personal protective clothing including SCBA, and utilize safe and effective strategy and tactics to mitigate the incident.

While a rarity, it is imperative that all firefighters recognize the indications of a potential backdraft situation. The trend toward sealing buildings up to conserve energy may make buildings more conducive to backdraft scenarios in the future. In addition to those already mentioned, other indicators include:

- Dense, dark-gray to yellow-gray smoke issuing under pressure at intervals
- Little or no visible flame
- Windows stained black
- Incipient-type fire conditions around windows due to minute air spaces allowing flaming to occur
- Buildings that appear to be sealed-up, causing confinement of heat
- Extremely hot window glass
- Rapid movement of air rushing inward when opening is made (not a good sign—don't be the one standing in front of the opening)

Taking those extra few seconds to receive reports from companies performing reconnaissance of the building at the rear, sides, and possibly the roof will round out the information visible at the front and will give the incident commander the best initial information from which to formulate an effective plan of action.

This sealed-up building is a candidate for a backdraft. The doors are sealed and what few windows there are have been bricked-up. This, incredibly, is a store where the good majority of the life load could be children.

Rear and Sides. While these areas will not be as critical from an initial standpoint, they will be important indicators of the amount of support that will be available to the attack teams inside the building. The rear of the building will often be more fortified, usually having steel doors set in metal frames, encased in some type of masonry wall. These types of doors are extremely difficult to force. Furthermore, the door will most definitely be an outward-swinging door, making such devices as the Rabbit tool ineffective. It will probably be easier to breach the adjacent concrete block or brick than to force this door.

Roll-up gates present a forcible entry challenge to firefighters. Be cognizant of backdraft indicators, especially if the business has been closed for some time. Be sure to prop up or support the remaining gate. *(Bob Scollan, NJMFPA)*

Support in the form of horizontal ventilation may be impeded by the fact that there may be few if any windows at the rear and sides. In addition, what windows are available are usually small, and may be barred or made of wired glass. This will severely hamper effective ventilation opposite the advance of the hose line. Many times the windows on older commercial structures may be bricked up, necessi-

This old renovated theatre has extremely limited ventilation opportunities. Backdraft must be a concern. Ventilation at the roof is critical to tenability of the structure.

tating a breaching operation, a time-consuming task. For these reasons, it will be imperative that swift and effective vertical ventilation along with the breaking out of the front show windows be accomplished as soon as the line is ready to commence with the attack. Remember to properly assess ventilation-unfriendly buildings for signs of backdraft before any ground level activities take place.

Roof. Another barrier may be presented by roof openings that thwart efforts to effectively vertically ventilate the structure. Skylights and scuttle opening may have been permanently closed up due to burglary problems. If the roof is constructed of lightweight steel trusses, it may be too dangerous to attempt to cut such a roof. Due to the wide spacing between the bar joists in this type of roof (up to 8 feet), a firefighter attempting vertical ventilation can either slice right through the thin top chord of the truss or make an opening in the roof that is not supported. Either condition can drop the roof team right into the fire area. For this reason, always use a roof ladder, or better yet, an aerial platform to perform such a cut on this type of roof.

Some high-value occupancies have added steel plating to the roof to discourage thieves from cutting their way in. Not only does this add to the structural load of the roof (earlier collapse), but will completely eliminate any hope of vertically ventilating the structure. The inability to vertically ventilate in conjunction with inadequate horizontal ventilation opportunities may doom the operation. Roof problems as well as obstacles to adequate support actions must be immediately communicated to incident command for evaluation and possible revision of the attack plan and strategy. Any information that can be determined beforehand by pre-fire planning and building inspections will be to our advantage. The incident commander who has prior knowledge of a problem will have a better chance of finding a solution in a timely manner.

Other barriers. There may be other obstacles to effective firefighting that will impact on our ability to gain access to and operate inside the building. The

presence of guard dogs certainly will play a very big factor in the decision to send men into the building. This information must be available prior to the fire, including not only the presence of these dogs, but also their location. A firefighter entering a building may not encounter the dogs until he accesses a certain area of the store such as the basement or where the safe is located. These danger areas should be known in advance.

There may also be barriers on the interior of the structure such as interior roll-down gates, temporary partitions, and excessive

This sign is a clear warning of an extra hazard inside the building. Often, this sign is not present and firefighters are needlessly injured.

stock impeding normal access and subsequently our ability to stretch lines to the seat of the fire. Again, prior knowledge of these conditions will save operating forces time and needless expended energy later.

This building takes up a square block, has been subdivided, and is occupied by dozens of small businesses, including some residential "artist" lofts. This is a potential deathtrap. Keep tight control over accountability and progress reports.

Another problem encountered is the amount of tenants in a large-sized commercial building. It is becoming more common to see buildings that were once occupied by one company to be subdivided, with the spaces being leased out to smaller commercial establishments. The types of occupancy will vary widely and create a maze of corridors, some of them dead-end. Not only will these buildings now become mantraps, but the forcible entry problem will increase exponentially. Whereas before the subdivision, there may have been only one or two locks to force to access the building, there may now be dozens.

I once responded with my battalion to a water flow alarm to one of these buildings. There was water streaming out from beneath a large roll-down gate, obviously a broken sprinkler head. As the weather was extremely cold that week, we had had numerous responses to this type of problem. Ice was beginning to accumulate in the street around the building, so we had to act fast. Not wanting to destroy the expensive roll-down gate, we attempted to find out who the owner was and have him respond. Neither the dispatcher or the police department could provide this information. We decided to force entry to the front door. It was simple enough, only an easily forced cylinder lock. Once inside, the problems began. There was a maze of doors. It was like being on "Let's Make a Deal". As we tried to get closer to the source of the problem, we were met with numerous locked doors, one more heavily fortified than the next. We wound up destroying several doors to find the mechanism to raise the gates and access the cellar where the sprinkler controls were. Had the owner provided us with keys or provided a Knox Box system, the damage could have been avoided.

The department should urge as many businesses as possible to provide exterior key boxes so a non-destructive means of entry can be made. There are two types of systems that work well. One is a combination key box. All the boxes in the jurisdiction should be coded with the same combination. The other is a key box which has a master key that only the fire department has. Again, all the boxes in the jurisdiction should use the same master key to avoid having to keep multiple keys. All key boxes should be mounted up high enough to prevent unauthorized tampering. They should be placed at a height that only can be accessed with a small ladder, say an attic ladder. The keys inside should be color-coded for easy use. One system that works well is to color the top of the keys. In Weehawken, we used green for entry keys, red for alarm panels, blue for basements and cellars, and orange for elevators. It was an easy and effective way of identifying which key fits which lock at a glance. If there are several keys of the same color, at least the choice is narrowed.

All commercial establishments in the area should be urged to install this system. It is quick, easy, and saves valuable time when there is an emergency. Unfortunately, not all owners comply. Most building owners get the point and install the boxes once their doors are battered down several times.

Terrazzo and concrete floors

Occupancies such as laundromats, appliance dealerships, and other businesses that must support a heavy contents floor load will often have floors constructed of terrazzo or concrete. Terrazzo, an aggregate of marble chips set in several inches of concrete, can be found in older structures, while concrete may be found in more recently constructed buildings. No matter how old the floor is, there is usually one thing in common: these masonry floors are frequently supported by wood joists. A fire in the cellar of one of these buildings may destroy the wood joists holding up the floor. The masonry, however, may stay intact, that is, until an unsuspecting firefighter steps on it. It may then collapse without warning. Twelve firefighters from New York City lost their lives this way when a cellar fire in a drug store burned the wood joists away that supported a terrazzo floor. The floor failed without warning.

Another problem may exist where masonry floors exist. Due to the desire to keep the floor level with the common grade, the wood joists may also be reduced in size. As a result, joists that are normally 2" or 3" thick and 10" to 12" inches wide may now be only 6" or 8" inches wide. This will drastically reduce their load-carrying capacity and under the assault of fire, may lead to an earlier collapse than if normal size joists were used.

Know your buildings! If a severe cellar fire is exposing an open joist cellar ceiling below a masonry floor, beware. Personnel operating both in the cellar and on the floor above the fire may be in serious danger. Recon reports from the cellar may not reveal the open-joist construction due to the smoke layer, however, a report of a masonry floor from crews operating on the floor above should raise the flag of extreme caution, even if it is not ascertained what the cellar ceiling characteristics are.

Water-absorbent stock

Buildings that house large amounts of paper, cardboard, and similar absorbent material pose a special hazard to firefighters. Water that is absorbed into these items from hose streams not only causes them to swell beyond their normal size, but also adds a considerable amount of weight to them. Rolled paper is one particular item that will cause significant problems if allowed to absorb excessive amounts of runoff water. These rolls can push out walls and collapse floors if saturated with water. Fortunately, modern building codes establish minimum clearances from walls that store these type products. These include newspaper and magazine manufacturers and storage warehouses, as well as recycling centers. Don't take for granted that this clearance is always provided.

It is critical that the incident commander be aware of the disposition of runoff water. Expanding paper will absorb water, swell, and can push walls out. There also exists the hazard of heavy machinery. A serious fire in this occupancy is a loser.

It is important to stay abreast of the housekeeping habits of such businesses.

One other area that is more insidious, however no less hazardous, is where paper and cardboard are stacked. This includes files and folders stacked haphazardly on shelves. Often, businesses keep their records and other paper-work on shelves in areas that cannot be used for any other useful purpose. This may be in a cellar, attic, or other area undesirable for office or showroom space. It is important to know where this storage is. Water from hose streams can cause this stock to collapse by either absorption or the power of the stream striking a pile of papers. This stock collapse may bury firefighters and/or pos-sibly cut off the escape route from the area. Use caution when stretching hose in these areas, especially in limited egress areas such as cellars. Poor hose management can cause a pile of stock or debris to collapse behind the hose team, cutting off their egress. It might be best, dependent of the situation and the danger present, to use the reach of the stream and keep personnel out of these precarious areas.

Pay particular attention to the final disposition of the water that is being used to fight the fire. If it is not running out of the building, it may be collecting (and absorbing) into unwanted areas. It is critical in these situations to have a chief or safety officer reconnoiter the suspected danger area to determine the extent of the fire stream runoff profile in order for incident command to make a more informed decision regarding dangerous building conditions.

Open façades

The façade, or false front, is usually found on strip malls, but may also be found on other type commercial buildings to add to the aesthetic value of the building. Usually a mansard-type roof constructed over the masonry bearing walls, the façade is constructed of wood or aluminum studs extending off the main bearing wall. More often than not, the façade will be open across the front and sides of the building. If a fire were to extend into the area, say from a vent-ing store fire, it can very easily access the entire row. The fire may also spread

This façade is open across the entire width and depth of this mall. Fire can spread throughout this space and burn into roof voids. Open and examine this space early in the operation.

into the cockloft via utility penetrations in the wall. Aggressive control tactics will be in order, with extensive pre-control overhaul a priority in heading off a spreading fire.

Roof and floor overload

Overloaded floors and roofs often result from a change in occupancy. Buildings that have changed ownership and subsequently occupancies, may have been constructed for a particular type of occupancy originally, but may now be occupied by a business that uses equipment larger and heavier than that for which the building was designed. This may lead to concentrated loads in places that were never intended to hold them. A concentrated load is defined as a load that is localized over a small area. The concentrated load is the direct opposite of the uniformly distributed load, a far more desirable condition in a building. A fire exposing the members supporting this concentrated load can lead to early collapse. During pre-fire planning visits, be sure to make note of heavy equipment located in one area. Further, during fire operations, ensure a team is sent to the floor above the fire to, among other tasks, locate any concentrated loads over the area of active fire. The presence of this load may cause an earlier withdrawal from the fire building than expected.

Concentrated loads are not only limited to floor loads, but also to roof features. This may include any roof-mounted water tanks. It is likely that the tank was designed into the original support system of the building and the roof below it will be properly reinforced. However, if the building has been allowed to deteriorate, it might be a good idea to take a good, hard look at any feature such as this that is not only a concentrated load, but, if filled with water (which weighs eight pounds a gallon), will be significantly heavier and more of a collapse hazard. Another hazard is the tank supports, which are likely to be

This roof-mounted HVAC system was added during renovations. Note the steel support intended to distribute the load of the unit in a more efficient manner. Unless proven otherwise, expect that this roof was not reinforced to support this load.

unprotected steel. Fire exposing these supports may cause early failure of the steel. Such a failure may initiate a secondary collapse of the roof and floors as the impact load of the tank crashes into the roof. Once the floors go, the walls will be next. A roof load often not originally designed into the building may be an HVAC system that was added during later renovations. Often, the roof is not sufficiently reinforced to support such a load, which can be several tons. A fire attacking the supporting members below could cause early roof failure. It is critical that roof firefighters report the presence of these and any heavy roof features to incident command immediately for evaluation.

Combustible cockloft/truss roof

This text has already gone into the dangers of the types of trusses found in these buildings. Recognition of the truss is the first and most important step to safeguarding operating personnel. Trusses can be found in almost any type of commercial building. There will be buildings that are exclusively of truss design, such as those demanding large, open areas. Bowling alleys, auto dealerships and repair centers, shopping centers, and large warehouses will fit into this category.

The best way of recognizing the truss is when it is being constructed. This will give information on the type of truss in the building, whether it is lightweight wood or steel, bowstring, or any number of other variations on the truss. If construction site viewing is not possible, then a marker on the building indicating the presence of truss construction should be mandated. A marking system used successfully in New Jersey is an orange or white triangle with a capital "R", "F", or an "RF" inside it. These stand for "Roof", "Floor", and "Roof/Floor" respectively and indicate the presence of truss construction in those areas.

The basic rule of thumb to follow regarding truss roofs in commercial structures (and any structure) is that if the fire has involved or threatened to involve the truss, interior firefighting must be discontinued and a defensive mode of operation pursued.

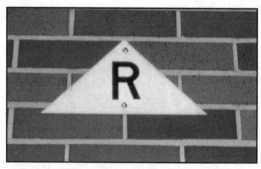

This triangular sign is used in New Jersey to warn firefighters of truss construction. In this case, the roof is of truss design. An "F" in this box signifies a truss floor is used. For buildings that use both truss floors and roof, the letters "R/F" appear in the triangle.

If the ceiling is open, as it is in many types of structures where it is not important to be aesthetically pleasing, only functional, the truss may be clearly visible. This is beneficial because operating crews inside the building can easily recognize it. It is also more dangerous as a fire at the floor level can easily spread to and ignite the trusses above. If the truss is concealed by smoke, then either pre-incident knowledge will need to be relied upon or the roof will have to be examined, preferably from the safety of an aerial device.

On the other hand, in many of these commercial structures, the presence of a suspended or directly affixed ceiling will often conceal the presence of a truss roof. If this is the case, a cockloft will be created where fire can travel and feed on the combustible wood trusses or heat the steel trusses to failure. Either way, if the fire enters the cockloft, it will again be time to pull the plug on the offensive operation, as the collapse of the building may be imminent.

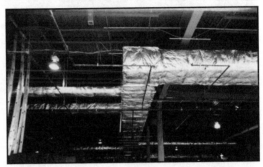

Once construction is complete, a drop ceiling will enclose these lightweight steel trusses. Fire below may still enter this space via HVAC openings and open tiles. Recognition of the truss is 99% of the game.

The cockloft in large commercial buildings also presents the danger of both backdraft and lateral firespread. This is especially true in strip malls, where the partitions between the stores may not extend to the underside of the roof, but only reach to just above the drop ceiling. As mentioned in earlier chapters, even if there are partitions between stores, the presence of utility services usually negates their compartmentalization value.

Regarding the backdraft danger, there may be situations where the gases in the cockloft are above their ignition temperature, needing only oxygen to ignite them. This is where coordination between roof crews and ground (attack and support) crews is critical. If the gases are indeed ripe for ignition, it is absolutely imperative that the roof be opened before the ceiling is pulled. Opening the roof will correctly exhaust the superheated gases up and out of the structure. Then, the ceiling can be opened and streams applied. If, however, the ceiling is opened first, due to either ignorance on the part of the operating forces or poor communication and coordination, the fire may ignite with explosive force and expand into the path of least resistance, which is likely to be the area from where the ceiling was opened. This uncoordinated action has cost

Heavy smoke at the roof level could indicate a backdraft condition is present in the cockloft. Dark smoke is more indicative of flashover, while a lighter, yellowish-gray smoke may indicate that a backdraft is possible. Proper strategy will allow for release in the least damaging manner. *(Ron Jeffers, NJMFPA)*

more than a few firefighters their lives. It is good practice, if heavy fire is expected in the cockloft, to communicate with the roof teams to both ascertain roof conditions and ensure that the roof is open and fire or gases are venting before opening the ceiling. This is especially crucial if there is a heavy smoke condition at the ceiling and roof level, but the area at the floor is relatively clear. Remember that heat rises. That's why it's hanging out in the cockloft. Don't allow it to drop back down. Take advantage of building features to eliminate it.

Presence of a rain roof

A rain roof is a new roof built over an existing roof. As the building ages, the roof may deteriorate to a point that no matter how much tar and roof paper is added, the roof still leaks. This will be especially true at the areas where the roof meets either the walls or roof openings such as skylights, mon-

The rain roof added to the top of this large bank is constructed of wood trusses. Note the small scuttle opening on the sloping portion at the right. This opening may be utilized to ascertain conditions in the trussloft. Note also the diminished height of the parapet wall.

itors, and scuttles. The older the building gets, the more of a problem this becomes. Instead of replacing the whole roof at a considerable cost, the owner has a new, cheaper roof built over the existing one. This new roof may be built with lightweight materials such as wood trusses. It may also be built using bowstring trusses. Look for the classic hump to denote the presence of the bowstring truss. Be aware, however, that if a high parapet or billboard is present, the roof crew will be the first ones to see this hump and must immediately notify incident command. Take steps to immediately withdraw companies and operate in a defensive mode outside the collapse zone if fire is found to have entered the truss area.

Rain roofs can be found on any type of building, but are included in this chapter because they are more prevalent in older, commercial buildings of ordinary or heavy timber construction with large roof areas. Rain roofs present several problems from a firefighting standpoint. First, the new roof is not as structurally sound as the original roof, which is now several feet below it. Thus, it may fail early. Second, where there was once a parapet wall surrounding the roofline, there is now none. The new roof height usually will eliminate the parapet wall, allowing firefighters to walk off the roof in heavy smoke. Third, the roof will act as a heat accumulation point, allowing fire conditions to destroy the lightweight material as well as make the space conducive to a backdraft potential. Companies making inspections of this area should discourage building owners from using the new roof space for storage. A new concave cockloft, as in the case of a bowstring truss, may be particularly inviting to storage. Ensure that all stairways that look like they lead into the roof area be examined during pre-fire planning and inspection visits.

The best way to recognize the presence of a rain roof is to observe it when it is being built. If this cannot be done, pre-fire planning visits and inspections should hopefully reveal this. If the inspection of the roof reveals an arched roof and/or parapets that have been reduced or eliminated, suspect the presence of a rain roof. Look at the roof from the interior to confirm your suspicions. If the roof joists are visible from below and are flat, while the roof exterior is arched or pitched, this should raise a flag of suspicion.

It is critical, during firefighting operations, to recognize a rain roof. It can be a lot of work to vent a fire building with a rain roof. Just cutting the rain roof will not vent the main fire area at floor level, just the new roof space. With a rain roof, once the hole is cut, when you try to push the ceiling down with a pike pole, you will be in for a surprise when you hit the roof boards of the old roof. If the old roof is several feet below the rain roof, it will be impossible to vent the roof. This condition may cause the withdrawal of the interior forces due to the lack of fire area ventilation and doom the operation. Roof crews encountering a rain roof must immediately make this fact known to the incident commander so an evaluation can be made as to its impact on the strategy.

Utility dangers

The dangers presented by utilities cannot be underestimated in these occupancies. By the very nature of the occupancy, the magnitude of gas, electric, and water may be greatly multiplied when compared with ordinary residential service. Utility control must be a high priority on the incident commander's action plan. Ignoring this hazard can lead to grave consequences later.

Gas service supplying commercial structures will be higher pressure and volume than that of residential occupancies. The shutoff to this entire multi-tenant system is below the circular regulator at bottom left.

Gas may be piped into the structure at higher pressures than expected and in greater quantities. A leak in this larger piping can set the stage for an explosion of great proportions. In addition, a fire fed by a broken pipe can expose a great deal of stock and building components. It is imperative to request the response of the utility company as soon as any fire of significance is encountered.

Water is not thought of as a major problem in firefighting except when there is not enough of it or when it can't be used for extinguishment due to the lack of compatibility with the burning product. Water from broken pipes, especially large pipes used to supply industrial occupancies, can quickly flood out an area such as a cellar or subcellar. Firefighters searching these areas should use caution when attempting to wade into a water-filled area. Remember that water will seek its level, and what appears to be a shallow, flat surface, can be disguising a service chute or steep steps. Always probe ahead of you when encountering water in low areas. In addition, the problem of electricity will compound the problem of water. Water in contact with electrical equipment can become charged and cause a firefighter to be electrocuted and then drowned, neither of which is particularly pleasant and will ruin your day.

Water flowing from broken pipes can also rob the sprinkler system of its supply. There will be times, due to access problems, that the only thing keeping the fire in check will be a properly functioning sprinkler system. Having this system knocked out of service could allow the fire to spread to major proportions before the fire forces are assembled and in place to launch an attack. Any report of flowing water in the structure must be investigated as to its source before it creates an unwanted problem.

Perhaps the largest problem is electricity. It is a good bet to surmise that more firefighters have been killed by electricity than by all other utilities combined. Large commercial occupancies, and even smaller ones, will require a substantial amount of electricity to operate. Most residences do not require more than 110 or 220 volts, however, there may be many thousand volts required to run a large business. The equipment required to support this voltage may be located in a single location, which would simplify matters. However, electrical equipment may be located in many areas of the building.

It is imperative that the fire department preplan the locations of all electrical supply areas and attendant shutoffs. In addition, the presence and location of generators that supply power to the premises when the main power is disconnected must be identified.

In a commercial occupancy, doors that open toward you often signify a utility area or a change in grade (stairs). Use caution when encountering a door that is flush with the wall, indicating an outward, "toward the firefighter" swing.

One obvious tip-off to an area of danger would be a HIGH VOLTAGE sign on a door leading to the equipment. This may not be visible in smoke. Searching firefighters must be aware of areas that contain high voltage equipment. One rule of thumb that may be followed is that doors leading to building level changes and utility closets, which not only contain electrical equipment, but other dangerous processes as well, often open outward toward the firefighter. In addition, there may be a louvered grill at the bottom

Building neglect over the years leads to many dangerous conditions. This large conduit for electrical service rotted away, exposing the wires. A feeble attempt was made to use cardboard to cover the hole. Older commercial buildings are fraught with these hazards.

of the door. Any door opening into a hallway, especially with a grill at the bottom should raise the flag of extreme caution. Even without the presence of the louvers, these doors that open into the hallway may also be doors leading to a change in grade. Rushing into these areas could get you in serious utility trouble or lead you head-first down into a lower level. Recall that gravity never takes a day off and beware the door that opens toward you in this type of occupancy.

In newer buildings, the electric service from the street may be underground and can come into the building anywhere; this must be preplanned. However, in older areas, the service will come in from a service connection located on a utility pole. It is imperative for two reasons to know the location where the service enters the building. The first is that if you are sent to disconnect or isolate power to the building, the controls will most likely be located on the wall below where the service enters the building. Second, knowing this will keep you away from this area if searching in smoke. Inadvertently following the wall during a systematic search can lead a search team to an area where high voltage equipment is located. This can lead to electrocution. Having this information beforehand will also allow the firefighter to use caution with hose streams in that area of the building.

I once responded to a fire in a block-long storage warehouse on a mutual aid assignment. We did a great deal of mutual aid from Weehawken before the consolidation to North Hudson Regional. It was a nasty and raw Sunday morning and the ladder company from the neighboring city was temporarily out-of-service. While they were switching equipment to a reserve ladder truck, the alarm came in. As a result of the mutual-aid agreement at the time, my company was dispatched on the first alarm.

We arrived on the scene to find a good deal of light gray smoke issuing from the building, which was three stories and constructed of heavy timber. After forcing entry, we attempted to locate the source of the smoke, which was evidently being generated by a fire being kept in check by operating sprinklers. The cold smoke was moderately thick on the first floor, but we could see where we were going so a search rope was not required in the huge structure. We found no operating sprinklers on the first floor and now concentrated our efforts to finding the entrance to the basement. It took quite a while to locate the entrance as it was in an area that was not only on the exterior, but appeared to be in a different building. The smoke in this area was thick as there were few ventilation opportunities other than cutting the floor or taking out the cellar windows which were boarded up with plywood and tin sheeting. This task was being accomplished by the second-due ladder company, which was now on the scene. However, it took a great deal of time to make the cut as the heavy timber flooring was very thick, being constructed of a subflooring made of 1" x 4" planks laid on edge covered by a finished floor laid flat in a tongue-and-groove fashion. To complicate matters, the area directly over the fire was covered by storage, so the hole could not be made in the most advantageous area.

The storage in the cellar, which was as large as the building, was piled up to the ceiling in an arbitrary manner. There were dead-end corridors and maze-like conditions created by the storage, creating a disorientation potential. With this in mind, we utilized a rope to attempt to find the fire. As the area was difficult to negotiate at best, I told the hose team to wait at the entrance until we could find the fire. It was difficult enough having to redirect the rope when we met a dead end. Having to redirect a hose line would have been a nightmare. As I had seen where the holes were being cut in the first floor before entering the cellar, I tried to orient myself in their direction. I also listened for the sound of either the saw cutting the floor or the sound of water flowing from a sprinkler. Normally, in heavy smoke, the fire can usually be found by moving in the direction of increasing heat. This was not the case here, as the sprinkler had apparently removed the heat and accompanying steam from the fire area before we located and entered the cellar. All that was present was the nuisance of cold smoke being created by the discharging sprinkler system.

Finally, we found the head discharging water. It had succeeded in controlling the fire in the stock below, but it could not reach the area above it, where the ceiling was ignited and still burning. As it was a very tight space, I called for the line to be passed forward to my crew and we finished the job of extinguishing the fire. It was really a "nothing" fire, but once the smoke cleared, I saw that our position was about five feet from a whole wall of buess bars, apparently live. In the smoke, we could easily have grabbed these bars and been killed.

Once outside, I saw that the wall where the buess bars were located was directly below the service connection where the electric supply entered the building. The lesson to be learned here is that if you are going to be operating in the cellar or anywhere else for that matter, it is a good idea to check where the electric service enters the building, especially if you've never been in the building before.

This brings up the controversy surrounding electrical shut-off. Should the entire building be shut down for a small fire? The answer is no, unless the area of operation is located near the electrical equipment and an unnecessary risk is presented to operating personnel. It is always safer to leave building services intact if at all possible. However, once the utilities present a hazard to the fire forces, either shut them down completely or isolate them to exclude the

The electrical shutoffs in many older commercial occupancies are located on the wall directly beneath the point where the service enters the building.

area of operation. It goes without saying that in a heavy fire situation, it is best to have the utility company cut the power to the building from the street.

Another utility service that must be controlled is the HVAC (heating, ventilation, and air-conditioning) system. This system can spread smoke and fire to uninvolved portions of the building. They must be shut down immediately.

There will be times, however, when HVAC systems not only spread smoke throughout a structure, but are the cause of it as well. I responded as a captain to a variety-type store on a smoke condition. When we arrived, there was a haze evident at the ceiling level. At first, it was thought to be a ballast problem, but there were no indications that any ballast was malfunctioning. Next, we used a ladder to check above the drop ceiling. There was nothing showing there; as a matter of fact it was completely clear. I was perplexed. The chief arrived, took one sniff, and ordered the HVAC system shut down. He told me to take my crew to the roof and open the motor cover on the HVAC system. He said we'd find a burnt belt on the motor housing. We followed his orders, opened the motor housing cover, and there was the burnt belt. The products of combustion being generated by the burning belt were permeating the entire store by the operating HVAC system. Experience, especially someone else's, is a great teacher. I never forgot that smell or the lesson. It is like food burning on the stove. Once you smell it, you never mistake it again.

Large, open areas

Buildings whose dimensions can best be described as "big by bigger" will demand a strategy that includes strict command and control. Ventilation opportunities may be limited, hose stretches long, and sprinkler-induced cold smoke may mask the seat of the fire, making even routine operations more complex and dangerous.

Large, open areas should automatically raise the flag of caution to the officer performing either a preplan or size-up. These widespread areas are often indicative of the presence of trusses. Newer buildings will often be of non-combustible construction with an open web of lightweight-steel parallel bar chord trusses. Lightweight wood trusses are usually limited to residential occupancies such as townhomes and condos, but this is not an absolute. Either way, the collapse threat may be realized in as little as five minutes. The main difference between the two types of lightweight trusses is that the steel trusses will sag prior to failure while the lightweight wood trusses will not. They will usually fail without warning. Older buildings employing truss construction with an open-floor design will likely be a heavy timber truss such as the bowstring. These are not guaranteed to provide any more fire-resistance than their lightweight counterparts, as many of these so-called heavy timber trusses are nothing more than a bunch of 2" x 6" joists connected together by through bolts or other means. The concept here is that no truss should be trusted to maintain its integrity once exposed and/or involved in fire. While the open floor design is not always going to be covered by a roof that employs a truss construction system, it is a good enough rule of thumb to warrant a closer look when sizing up the building. As indicated earlier, if the fire has involved the trusses, it is time to operate in a defensive manner, establish collapse zones, and protect exposures.

The open-floor design offers no barrier to fire extension and is usually an environment subject to unrestricted air movement. Therefore, rapid and extensive fire spread can be expected. Open floor areas will include large showrooms, display

The connection point of this "heavy timber" truss is actually five small dimension boards connected by an unprotected steel through bolt. Note also the spaces between the top and bottom chord timbers of this truss assembly. Fire can attack all sides of the wood at one time.

floors, and even cubicle areas. Cubicle areas will often be partitioned and quite congested at the floor level, but will be wide open to the ceiling above the height of the individual cubicles.

Wet-pipe automatic sprinkler protection is the most effective method to prevent fire spread before the arrival of the fire department. In the absence of

such a sprinkler system, remember that open floor spaces usually mean open ceiling spaces. As products of combustion rise, the upper regions will be subject to the most heat build-up. It is imperative to ventilate vertically as directly over the fire as is safely possible. This will help localize the fire and prevent spread along the lateral run of the ceiling. When heavy fire and high heat conditions are encountered, it will be extremely difficult to advance to the seat of the fire. For this reason, structures of larger than normal area should prompt the officer to stretch and operate with larger diameter handlines such as a 2^1/$_2$" line with an 1^1/$_4$" or 1^1/$_8$" solid bore tip. This will provide both reach and penetration far superior to the 1^3/$_4$" line, even with a solid bore tip. If this larger handline doesn't provide the extinguishing power to control the fire, then you must accept the fact that it is not your day and the fire is going to get its way for the time being. Prepare defensive positions and monitor and evaluate the reports from the interior for it may come time to withdraw and you must be ready to continue the firefight with the least amount of reflex time in the offensive-to-defensive transition.

This bowstring truss roof has skylights on each side. These should be vented from an aerial device and the truss area examined for fire. Strategy will then be modified as conditions (involvement) in the truss dictate.

Drop ceilings

Drop ceilings, also called "hanging" ceilings, present many problems to the firefighting force. Problems created above the drop ceiling include the ability for a fire to start in the drop ceiling space due to the numerous ignition sources present (such as light fixtures and wiring).

The area above the ceiling is also called the "plenum" because it is sometimes used as a return for the HVAC system. This area may not only be several feet high, but may also be one of several hanging ceiling spaces in the same area. As buildings are renovated, ceilings are added to enhance the aesthetic appearance of the room, as well as provide for the conservation of energy due to less space below the ceiling

Drop ceiling tiles, if properly in place, will offer a barrier to fire spread into the ceiling. Above them may be maze of wiring and building support systems, all of which may be a source of ignition. Once the trusses are exposed, expect failure. *(Capt. Alan Ballester, NHRFR)*

to heat. A fire can originate, spread into, and hide in any one of these ceilings. All must be opened to examine for extension. When you can see or feel the structural supports of the roof or floor above, whether they are wood joists or lightweight-steel trusses, there is no need to probe further upward. Remember that a fire exposing the lightweight steel trusses can cause failure in as little as five minutes. Ensure proper reconnaissance reports are issued by operating personnel as soon as possible. If there is heavy fire above the drop ceiling and steel trusses are present, expect early collapse.

Most of the building's nerve center (such as the HVAC systems, electrical wiring, and alarm system components) will be located in the drop ceiling. Thus, the ceiling may be open over a large area. Where partition walls are present between adjacent occupancies, poke-throughs may have been made to accommodate this wiring. It may or may not be properly protected against lateral fire spread. Assume it is not and take steps to head it off.

There is also a danger of backdraft conditions existing in these spaces above the ceiling, especially if the roof has not been properly vented and there is a great deal of extreme heat present. It is imperative that topside ventilation be completed before the ceilings are opened from below. Introduction of oxygen in the wrong place will cause the gases to expand in that direction. If the direction is downward toward an unsuspecting firefighter instead of upward into the atmosphere, the results could be tragic. It is crucial that interior and roof operations be coordinated whenever a fire in the drop ceiling space is present or suspected.

The other danger is the collapse of drop ceilings, which can trap firefighters inside the store. Many times, these ceilings are supported by unprotected steel rods or even wiring that can easily fail in the heat of a fire. Dependent on the age and construction of the ceiling, the firefighter may or may not be able to extricate himself from the ceiling. Some older buildings may have plaster and lathe or a wood framework as the drop ceiling. In this case, the ceiling will be extremely heavy and even more difficult for the firefighter to break though on his own. Newer drop ceilings are made of a lightweight steel grid with removable lightweight panels. Firefighters should have no problem breaking through this ceiling.

Using the reach of the stream to punch through the ceiling from a protected area such as a doorway or from the exterior will prevent ceiling collapse entanglement. A Telesquirt is an excellent tool for this strategy. *(Ron Jeffers , NJMFPA)*

The best way to protect firefighters against these ceilings is to recognize their existence and take steps not to be trapped under them. This entails using the reach of a large caliber stream from either a handline or master stream device to blow the ceiling apart and extinguish fire ahead of the stream. Another is to ensure that the crew on the roof pushes the ceiling down when ventilating the area from above. Firefighters engaged in pulling ceilings on the interior must ensure they keep their back to their exit point, never stand directly below the pull, and pull all material down and away from them. It is the responsibility of the company officer to supervise their assigned personnel in recognizing and avoiding these dangers.

Mantraps

This is one of the major reasons to preplan what I like to call "weird structures". A weird building can be defined as "a structure that, whether by its occupancy, geographical location, layout, or other unusual feature or condition, will create an incident that will challenge all area personnel and resources". Some may call them "target hazards", however, many buildings that would not normally fall under the category of target hazard will pose the biggest life safety problems for fire personnel due to the inherent hazards in their contents and layouts.

Some of the danger areas to use extreme caution in are cellars and sub-cellars. Most modern commercial occupancies are constructed without a cellar, making access and ventilation efforts more easily completed. However, many old factories, warehouses and similar occupancies were built with below-grade areas that afford very little in the way of access and ventilation. It is very easy to get lost in these areas. For this reason, a lifeline and thermal imaging camera or hose line is urged when entering these areas. In addition, firefighters should never, under any circumstances, enter these areas alone. Sometimes, the only access point is also the only ventilation point, especially in subcellars, which are completely below grade, sometimes as much as two or three floors below grade. These areas will have no ventilation opportunities. A fire extending beyond the capability of the attack team can channel superheated gases that have the potential to ignite when they reach open air. This open air may also be the only egress point. Firefighters operating below grade may have to fight their way out of such areas. For this reason, it is critical that at least an additional handline be positioned at the entrance to these below-grade areas.

There are also numerous areas that are not below grade that expose firefighters to disorientation hazards. These include cubicles. Some businesses are a maze of these cubicles, which make for a "rat in a maze" condition at floor level while exposing the whole ceiling area to the products of combustion. Even businesses that are primarily process-oriented will often have an area where the administrative segment of the business is attended to. It is important to know where these areas are, for the life hazard here is more prone to entrapment than in the wide open floor areas. In addition, the potential for firefighter disorientation is

Search bags must be available and should be utilized whenever encountering an area that presents the hazard of disorientation. Rope-guided search operations must be practiced if they are to be effective on the fireground.

greatest in these areas. It is best to search and operate in these areas using the rope-guided search method. All members need not stay on the rope, but must stay in close proximity to it. Tethers can be used to extend the main search line into adjacent areas. Usually, cubicle areas will have a center aisle off of which the cubicles are accessible. Extending the search rope down the main aisle and then extending the cubicle search from this main line is far less confusing than attempting to snake the rope in and out of each cubicle. For one thing, the entanglement problem will be difficult at best. Another even more critical hazard is that if a victim is found or a firefighter is running out of air or has an SCBA malfunction, the rope has to be followed back through the maze it negotiated on the way in. Stretching the rope down a main corridor will allow the most direct route to the egress point.

Another hazard most often associated with industrial and manufacturing occupancies is open floor spaces with arbitrarily located level changes. These open spaces are most often dotted with heavy machinery and sometimes floor

level dip tanks, acid baths, freight elevator shafts, and other unprotected openings in the floor with or without guard rails. Even when guard rails are present, they are sometimes pipe-type with the lowest bar being about a foot and a half above the floor. In smoke, a firefighter can easily crawl beneath this bar and into an unprotected opening in the floor. For these reasons, the rope-guided search and a thermal imaging camera is an absolute

An opening in the middle of the floor is almost never is a good sign. Although there is a guard along the floor, a firefighter crawling in dense smoke can still fall into this sewerage plant grit removal pump.

must when searching these large areas. It is best to have some knowledge of the area beforehand. Pre-fire information with floor plans should be available to the incident commander. It should go without saying that a thermal imaging camera is essential when operating in these areas.

In situations where employing a rope-guided search is necessary, the rope must be played out in as exact the same manner as the paths covered as possible. Slack in the rope should be assured as it is laid out in the direction of travel. The rope must not be pulled taut during the operation. Pulling the rope taut will move it into the path of least resistance, which may be in the direct path of some of the above mentioned hazards and mantraps, which were not traversed by the original search path. Firefighters attempting to follow the rope out may follow this new path into trouble. One method that works well is to tie off the rope with a knot such as a figure eight on a bight at each change of direction. This certainly keeps the rope from moving into undesirable areas, but if a non-verbal communication system between the search team and the exterior is used that involves tugs, the tug will be lost at the knot no matter which direction it originated from. A better system is to clip carabiners onto the rope and a stationary object at each change of direction. This will not negate the tug system when it is needed most. Each member searching can readily clip a half-dozen carabiners to his turnouts to be used during the search.

As always, knowing as much as possible about the building, the processes, and the layout will afford the incident commander the best opportunity to make informed, intelligent decisions on behalf of the fire suppression forces.

Deep-seated fires

In many commercial type occupancies, especially large area buildings such as warehouses and factories, pallets of stock may be piled high and tight. Stock piled several pallets high cause several difficulties in fire extinguishment. When hit by hosestreams, the stock may collapse causing casualties and trapping firefighters. When sprinklers are operating, waterlogged stock, if unable to collapse because of adjacent stock, may list and lean on other pallet loads. During overhaul, when pallets are being moved, a progressive collapse may occur. In addition, a fire in the middle of a tightly packed pallet area may be difficult to access, causing operational delay and subsequent fire extension. Another problem with stacked palletized stock, is that even if it is tightly packed, the open-joist construction of the pallets will allow air to circulate in buried sections of

High-piled, tightly packed stock may lean over and collapse when struck by hose streams or overhauled. Circulating air caused by open-joist pallets may also spread fire that will be impervious to stream penetration and sprinkler coverage. *(Hartz Mountain Industries)*

the stock, inviting fire extension. This fire extension may be influenced by wind and by mechanical ventilation.

When fires are deep-seated, final extinguishment may be difficult due to the depth of the load. Use of mechanical aids to move the stock may be necessary. Ensure the main body of fire is knocked down before attempting to move any stock. Moving stock by hand is both manpower demanding and dangerous. It may be best to use forklift machinery to more safely and efficiently move stock so final extinguishment can take place. Manually operated forklifts are preferable to propane powered ones.

Overhaul will be extremely manpower intensive. Where stock is tightly packed and it can be safely accomplished, remove it to a safer area and break it open to finish extinguishment. A charged line must be ready to operate in this area.
(Hartz Mountain Industries)

Assign an officer, preferably a chief officer, to the interior to supervise and coordinate the forklift-assisted overhaul operation. Visibility may be limited, so operations must be strictly controlled to ensure the safety of operating personnel. The officer supervising the overhaul operation should have a thermal imaging camera handy to direct the pallet-moving process. If the smoke is thick, the camera will be mandatory.

I operated recently at a fire in a sprinklered warehouse at about 2:30a.m. The sprinklers were successful in controlling the fire, but the stock was packed so tightly, it was difficult to see what was still burning. Luckily, the stock was palletized, about twenty feet high. I had a thermal imaging camera and could see where the fire was still burning, but could not tell how far back the pallet was located. The smoke was so thick I could not see where one pallet ended and the next pallet began. A firefighter with forklift experience was removing the pallet stacks to get to the fire, which was still being held in check by the sprinkler system. The only way I could tell that the ignited pallet was being moved was that the fire I could see in the camera screen started to move. Without the camera, I could see very little. It would have been extremely difficult to conduct safe operations.

Keep the initial attack line at the fire area ready to knock down any flare-ups when stock is moved. Also, have a charged line ready in the area where the stock is being moved to so the pile can be opened and overhauled as required. The exterior of the building is the best place for this.

Rack storage

With the advent of the super warehouse (occupying many hundreds, thousands, and even millions of square feet of storage space) has come not only the problem of large, open spaces, but also the presence of rack storage. Rack storage can present a tremendous fire load. Racks can hold everything from paper (business records) to furniture to boats. Ceiling height sprinkler systems are often ineffective against fire traveling inside the rack spaces, thus fire can spread throughout the rack system relatively unchecked.

The major problem regarding rack storage is its construction, usually of light-weight unprotected steel. Recall how fire and heat affect unprotected steel. Then add the weight of the stock, which may also be water absorbent and the collapse potential is evident. When operating in proximity to rack storage, collapse zones must be established and maintained. Just like an exterior collapse zone, the interior rack storage col-

This heavy stock is supported by unprotected steel. Note also how the toilets sit on a downward angle. Fire weakening the supports or the power of a hose stream could cause these items to collapse.

lapse zone should be at least the height of the rack and the entire horizontal distance of the rack system. If this cannot be maintained, flanking will be necessary. If flanking is not possible due to the orientation of the area, withdrawal may be the only alternative.

The only defense against such a huge fire load resting on such an inherently weak support system is a properly placed and supplied automatic wet sprinkler system. One of the first actions to take regarding engine company operations is to ensure that the fire department connection is supplied. If the sprinklers are not effective due to limiting factors such as improper design or obstructions, the fire will probably become substantial and the building will be destroyed. The use of large caliber streams from a safe distance may not be effective against an advanced fire in rack storage.

As is always the case in a large area structure, a strong incident command presence is required. Interior supervision is critical. A large area warehouse can be likened to a high-rise laying down. An expanded incident command organization will make the operation easier to manage and safer. An operations officer should be assigned as well as an interior division commander. If possible, especially if the building area is expansive, a resource post should be established in a safe area. Possible sites for this post should be preplanned in advance. Resources and personnel should be far enough from the main area of operations (the Hot Zone if you will), but close enough to be effective and quickly deployed. A possible site for this post is on the safe side of a firewall. Due to the size of the structure and the complexity of the operation, it is wise to request more than one FAST team at these structures. Consult with management personnel or building fire safety directors as required.

The life safety problem for firefighters is not so much the huge fire, but the fire that is being controlled by sprinklers, hose streams, or both. The smoke condition may be severe, causing disorientation and difficulty in finding the seat of the fire. Thermal imaging cameras will be useful here. Carbon monoxide levels will be high. In addition, all firefighters in the operational area should either be on a hoseline or a lifeline. A lighted path of egress should be established.

Firefighters operating hoselines should also be aware of the potential for stock collapse caused by the power of a high-caliber stream. Fire damaged stock can also collapse off the racks, injuring or trapping firefighters. Hoselines being dragged into a building can also cause piles of stock to collapse. Hoseline management is critical wherever the potential for stock collapse is present.

Depending on the severity of the fire, the search effort may be severely handicapped by the large area and the smoke condition. The civilian life load will depend to a large degree on the type of storage occupancy. Some rack storage facilities are completely automated, with machines handling the stock. Ensure the automated system is shut down to prevent injuries to firefighters operating in the area of these machines. Few employees will be found in these areas. Unfortunately, this can lead to a "needle-in-a-haystack" problem. Thermal imaging cameras coupled with common sense search procedures will afford the search team the best chance of success in these types of search operations.

Other rack storage areas will be occupied by employees only. Propane-fueled forklifts will be prevalent. The problem here is of the potential for more victims and the possibility of a BLEVE threat from the propane cylinders. Inquire as to where the spare propane cylinders are kept and try to protect that area if necessary. Utilize the warehouse manager and/or the timecards to determine if anyone is missing and where they may have been working.

Still another type of rack storage area is where access to the public is permitted. Many home improvement and furniture outlet stores have large rack storage areas open to the public. This will compound the search problem. Accountability of civilians in the affected area will be difficult at best. Fire department/management cooperation coupled with a frequent and effective inspection program will aid in minimizing the risks to firefighters, the workforce, and the public.

Sprinkler-induced cold smoke conditions

Many commercial occupancies, as well as many mixed-use occupancies and taxpayers, have sprinkler systems. Sprinkler systems save life and property, there is no doubt about that. Activated sprinkler systems, however, cause a multitude of firefighting problems, causing incident command to operate in a somewhat different manner than in an unsprinklered structure.

Activated sprinkler systems not only quench fire, but also cool the temperature of the smoke and fire gases. Water applied to the fire also creates more smoke due to the inefficiency of burning. This also creates more carbon monoxide. Downward action of the sprinklers also aid in pushing the products of combustion toward the floor. The cool gases also sink toward the floor. SCBA is an absolute must.

Cold smoke may create a severe visibility problem in the building. Firefighters can get lost in the smoke even in a small building. The name of the game must be control of personnel. No one should enter unless on a lifeline or carrying a hoseline. A thermal imaging camera will also be extremely valuable in orientation, but the camera may malfunction, so a lifeline should still be followed.

In sprinklered buildings, an expanded command organization will be necessary. Rotation of personnel into and out of the fire area must be effectively coordinated. A separate officer may have to be assigned to monitor firefighter entry and exit times. It is best to feed companies into the fire area and relieve them from the area as task forces. This will make rotation and time keeping easier. An interior commander must be assigned (preferably a chief officer). The safety officer must also be no stranger to the operations. In addition, it may be wise to request an additional FAST team to the scene. Accountability of personnel through pre-established systems must be adhered to. PARs must be requested by incident command at regular intervals.

After the fire condition is definitely under control, the sprinklers may be shut down. The firefighter at the controls must stay there in case the system has to be re-charged. The HVAC system, if present, may then be used to clear cold smoke. It is best to consult with building personnel before attempting this. Fire department fans may also be used, but both these operations must be strictly monitored and constantly evaluated. The slightest sign of a problem must cause the stoppage of the air-moving equipment.

Strip Mall Characteristics

Modern strip malls, sometimes called "new-style" taxpayers, will exhibit some of the same characteristics that all commercial occupancies will display. This includes large, open areas for fire to spread, difficult access at the rear, heavy fire loads, and roofs supported by lightweight steel parallel bar chord trusses. They are usually one story in height, although some may have an additional half or full story above the first floor that may used for storage. There will be no cellar in these structures, so the threat of cellar fires does not exist. However, the threat of fires that originate and/or spread to the cockloft has caused the destruction of many of these types of structures. If superheated gases accu-

The great majority of strip malls will be non-combustible. The main fire load will be in the contents of each occupancy. Occupancies like Mr. Pool (chlorine-based products) and Benjamin Moore (paints and other flammables) are warnings to potential fire control problems.

mulate in the cockloft, the potential for a backdraft will exist. Ventilation at the highest point prior to the pulling of the ceilings from below will effectively vent the cockloft and assist in localizing the fire, reducing the chance of lateral fire spread and backdraft. To aid in this ventilation, there will usually be skylights, scuttle hatches, or automatic ventilation hatches available. This will allow roof teams to vent the building without resorting to cutting the metal roof, a very dangerous task.

As always, these buildings must be preplanned so firefighters may become familiar with the associated fire load of each occupancy as well as the location of fire department connections, available hydrants, and utility shutoffs.

In a fire in these strip malls, there is a very real possibility of losing the whole row of stores. It is not uncommon for the cockloft to be open over the entire row of stores. For this reason, it is critical that the strategy pursued be one of offensive-defensive, that is, aggressively attacking the main body of fire while taking steps to cut off the fire in adjacent exposed stores. Lines must be stretched into exposures and ceiling spaces examined. This will require a considerable amount of manpower to accomplish. There will often be no time to play catch-up at these buildings. Hit the fire hard and aggressively cut it off. However, it is also imperative that the incident commander be aware of the limitations of the construction type, in most cases, non-combustible. A heavy body of fire exposing the steel roof members must force the incident commander to consider the feasibility of an interior attack. Be aware of what the fire is doing to the building and react accordingly.

"Old-style" taxpayers also fit into this commercial category. Old-style taxpayers were the predecessor of the modern non-combustible strip mall. The owner of a parcel of land found that it was cheaper to construct a single building that housed a row of stores, all who would pay rent and subsequently, the building's taxes, to the owner. The owner would make a substantial profit from

Most new-style taxpayers will have a space above the drop ceiling. Consider this area to be open over the entire complex. Note the heavy, black smoke issuing from the stores on the right. This area is about to flashover. (Bill Tompkins)

using the land in this manner. Thus, the concept was well accepted and used all over the country.

Commonly one story in height, these buildings are usually built of ordinary

Old-style taxpayers will have a combustible interior and roof as well as a cellar. The structure will add to the fire load. Note the cellar windows at bottom right. This opening will be inadequate in venting the cellar. Like the common roof, the cellar should be considered to be common until proven otherwise.

construction with individual commercial establishments occupying one area of a large building, just as in the modern strip mall. However, the major difference in the new-style and the old-style is that the construction of the old-style will add to the fire load. In addition, most old-style taxpayers of ordinary construction will have a cellar where storage is kept. The cellar problem here will mirror that of the mixed-use occupancy. In fact, many jurisdictions refer to two-story buildings with a commercial occupancy on the first floor and a residential occupancy on the second floor as taxpayers. Fires often originate in the cellar and will pose a major fire control problem due to difficulties encountered in access and ventilation. In addition, the roof will be constructed of combustible wood joists and wood planks covered with tar paper and, of course, layers of tar applied to its surface over the years. The problems present in this roof are basically the same as in a multiple dwelling with a flat roof. The common cockloft problem as well as the cockloft backdraft potential is present.

Old-style taxpayers will fall into two categories, the one-story discussed above, and the two-story taxpayer. The two-story taxpayer will usually have an area above that is rentable just like the first floor, but may be used as an assembly occupancy, such as a meeting hall. Other two-story taxpayers will house residences, usually one or two apartments above the business occupancy. This is technically a mixed occupancy and was discussed in the last chapter.

Taxpayer Differences

New-Style
- Non-combustible construction
- No cellar present
- Steel truss roof
- Drop ceilings

Old-Style
- Ordinary construction
- Cellar present
- Wood joist roof
- Tin ceilings (& drop)

Life Hazard Problems in Commercial Occupancies

Transient occupancy

Most of the occupant load in many of these structures will be clientele. As such, they will have very little, if any, knowledge of the layout of the building. The staff cannot be relied upon to lead shoppers unfamiliar with the building to the closest exits. Especially in larger stores such as department stores or large public warehouses, there will be (by law) many exits. However, most people will usually seek to use the door they came in as an exit, bypassing clearly visible, more accessible, exits. This is classic tunnel vision, and is one of the reasons for the potential for mass flight and panic of a crowd.

Accountability

Because businesses open to the public are primarily occupied by a transient population, it will be extremely difficult to ascertain whether or not the building has been fully evacuated. For this reason, a primary search must be conducted as rapidly and thoroughly as conditions allow. Even when the primary search is complete, it is not a guarantee that all patrons

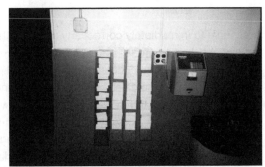

Occupancies such as warehouses and factories are usually only occupied by the workforce. Timecards are a way of accounting for workers, a sort of crude tag system for employees. This information may be helpful in ascertaining who may be missing.

have been evacuated. This can only be confirmed by a thorough secondary search. It s a good idea to question staff employees to ascertain if any employees are missing, but as far as customers are concerned in a heavy, rapidly spreading fire situation such as an explosion or arson, there will be no way of knowing who was in the store at the time of the fire and who was not. It may take quite a long time to determine if any patrons are missing.

Blocked exits

The accessibility of exits is one of the most important life safety issues in regard to commercial occupancies. This is especially critical during holiday seasons when stock is excessive to meet the demands of the public. In fact, not only is the stock load increased, but so is the human load as more customers visit the store on a daily basis, store hours are extended, and extra sales staff are hired to meet these seasonal demands. "Temporarily-stored" stock may interfere with exit facilities. It is imperative that the fire prevention bureau stay on top of this problem, but it is also the responsibility of the line fire

Somewhere behind this debris is an exit door. A fire started in this outside rubbish could easily spread to the interior. Note the heavy fire load such as the mattresses and the stock piled floor to ceiling in the window on the second floor.

personnel to immediately correct any situation that is unsafe, such as blocked access and egress areas.

I remember going to the same supermarket on more than a few occasions to find the roll-down gates closed over some of the exits because the store was not that crowded and the manager was afraid it was an easy way for shoplifters to just walk out of the store. The store had open-type roll-down gates and was part of a mall. The gates did not lead to the exterior, but to the mall atrium. The same store also made it a habit of stacking soda and cookie displays against the wall just inside the store entrance. This, in effect, cut off half of the available area for egress. After several warnings and a subsequent hefty fine, they saw the light. Store managers and owners must be made to understand that the fire department means business when these infractions regarding exits are found.

I remember my department fining a popular restaurant on the waterfront five thousand dollars on several occasions for blocking exits with tables and storage. Unfortunately, many times, by the time that exits are discovered blocked or inoperable, it is often too late. Other times, inventory is ordered removed, only to be put back in the unsafe area after the fire personnel have left.

Panic

Panic is the result of rational thinking gone astray due to extraordinary circumstances caused by life-threatening and potentially life-threatening situations. All of the situations regarding life safety mentioned above can easily result in a panic situation. Panic can also be contagious, causing usually rational people to become irrational as the "mob" mentality takes over. This irrational thinking is the reason why people jump from high-rises to their deaths during a fire. The overwhelming fear of burning to death causes them to lose their ability to reason. This same lack of reasoning allows them to override the certainty of death due to the leap from the upper floor. There is really not much the fire department can do in the face of true panic by patrons except either provide escape or put the fire out.

Basic Firefighting Procedures

The actions taken must be concomitant with both the potential area of involvement and its impact on firefighter safety. Consideration must be given to the hazards associated with the inherent fire load associated with the specific use of the occupancy. Time of day will play a major role in the life safety profile of the incident. During business hours, the major focus will be on both civilian and firefighter safety. When closed, the focus may shift primarily to fire-

Fires that occur at night may get a good head start. Additional alarms will be required to surround the building and to address exposure problems. Additional chief officers will be required to establish a proper span of control and decentralize the fireground.
(Ron Jeffers, NJMFPA)

fighter safety and may be a deciding factor on the offensive/defensive strategy decision.

CRAVE

Addressing the CRAVE acronym, operations should be based on the following:

Command

- Be cognizant of the special hazards presented by each individual occupancy.
- Ensure the manpower commitment is sufficient with the potential involvement; ensure a strong tactical reserve of manpower and apparatus.
- Request extra chief officers in large and/or particularly dangerous operations to allow for effective decentralization of incident command and proper span of control.
- Take aggressive fire confinement measures when attached exposures are involved.
- Provide for firefighter safety:
  a. Consider the presence of guard dogs
  b. Consider requesting the response of two FAST teams, especially in large area structures
  c. Consider the use of 1-hour air cylinders for FAST team personnel

Rescue

- Be prepared for panic.
- Consider the use of lifelines and thermal imaging cameras when searching.
- Don't forget the possibility of security staff on the premises during non-operating hours.

Attack

- Establish a reliable primary water supply.
- Ensure a strong hydraulic reserve. Supply auxiliary appliances early in the operation.
- Consider large diameter lines when faced with heavy fire or large building area.
- Coordinate attack operations with rescue and support operations.

Ventilation

- Consider building construction and inherent weaknesses and limitations when venting.
- Anticipate the fire's direction of spread and take steps to confine it.

- Ventilate in coordination with attack and rescue operations.
- Use existing building openings whenever possible.
- Be aware of the indications of backdraft conditions.

Extension Prevention
- Open up in the paths of least resistance to expose hidden fire.
- Check for multiple drop ceilings when conducting both pre- and post-control overhaul.
- Coordinate operations with engine company support.

United States Postal Service Property Fires

While not technically a commercial establishment, the federal government owns postal buildings, vehicles, and property. Offered here are some guidelines to use in postal property operations.

Not every jurisdiction in the nation is home to a dedicated postal property such as a post office or mail distribution center. It is, however, almost certain that every jurisdiction is traversed by postal vehicles and an even more certain that there is at least one postal mail box in every jurisdiction in the United States.

Postal property is not classified as a commercial occupancy, falling more under the governmental occupancy. However, the hazards associated with commercial properties are generally the same as hazards associated with postal properties. There are however, some glaring differences regarding postal property response when compared with other commercial property. First and foremost, the regulations and guidelines regarding security when operating at postal properties are extremely strict. Second, and this applies to not only postal properties, but also to postal vehicles and mailboxes, the postal service is the largest carrier and stockpiler on a routine basis of hazardous materials

Every jurisdiction in the United States has or is passed through by post office property. From mailboxes to bulk storage facilities such as the one above, departments should adopt an SOP regarding response guidelines at all postal properties.

in the country. Each day, an unknown quantity of dangerous materials are mailed or shipped by, and stored, at the nation's postal property. For this reason, fire department personnel responding on these types of incidents should use extreme caution when attempting to mitigate an emergency situation. Personnel should be prepared for water-reactive material, explosives, toxic chemicals, contraband, munitions, and a witches' brew of other commodities that may present a hazard to responders.

To the best of their ability, both logistically and legally, fire departments should preplan and become familiar with these and other governmental agencies in their jurisdictions. Make note of restricted areas and other areas where hazardous materials are usually stored. To this end, the following are a few guidelines to use when handling incidents involving United States postal property.

General guidelines

Standard operating procedure should prompt dispatch to immediately notify the United States Postal police at all incidents involving postal property. This notification must be made whether the property is on the street, mobile, or at a fixed facility. The incident commander should take all appropriate action to ensure and maintain scene security.

Damage done to United States Postal property such as forcible entry, ventilation, etc. should be in direct proportion to the requirements of the emergency at hand. Every effort should be made to protect U.S. Mail and postal property.

Regarding entry into areas designated "Restricted Areas":

a. No less than two department members should make entry in any such areas.

b. As conditions allow, a police officer or authorized U. S. Postal employee should accompany fire department personnel

A thorough cause and origin report should be made at all fires involving United States Postal Service property.

Operational guidelines for incident stabilization – postal mail boxes

Response considerations. A single engine company usually handles this incident, however, the security concerns may warrant the response of a chief officer. The incident commander must ensure dispatch has notified the postal police of the incident and that they are responding to the scene.

Extinguishment considerations. Unless conditions demand, use of water is generally discouraged, although water may be used to cool the metal postal box. It is better to use a clean agent such as CO_2. Dry Chemical can also be used, but CO_2 is preferred as dry chemical extinguishers make a mess.

Salvage considerations. Every effort should be made to protect and preserve the contents of the mailbox. Unless conditions demand, forcible entry is discouraged. It is better to discharge the above-mentioned extinguisher into the box and stand by. If the fire does not go out, repeated applications of the extinguisher may be necessary. The box must be opened to ensure extinguishment is complete. Unless circumstances dictate otherwise, and they shouldn't, the incident commander must wait for an authorized U. S. Postal representative to arrive on scene before the box can be opened. It is very unprofessional and legally unethical to leave the scene without properly ensuring extinguishment, which can only be attained through inspection of the contents. Only an authorized postal representative can provide access to the contents of the box.

Operational guidelines for incident stabilization — postal vehicles

Response considerations. The dispatched fire department response should be appropriate for a reported fire for the vehicle type involved. Postal vehicles run the gamut from cars, the familiar postal truck or jeep, and tractor-trailers. It is the responsibility of the first-arriving officer to enhance the response as required to meet the needs of the incident. In addition, any additional companies required to mitigate the incident should be requested as soon as possible.

Extinguishment considerations. As with all vehicle fires, check for and perform those functions that preserve life first. This includes the life of fire personnel. Ensure that no power lines are involved; check for spilled fuels, especially in the case of a motor vehicle accident with fire. Also, before any action takes place, it is imperative that the wheels of the vehicle be chocked to prevent movement.

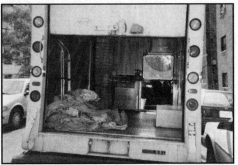

The white mail "box" trucks carry paper mail and some small boxes. This was taken at the end of the day. All of the mail has been delivered and the fire load is now just the truck.

The dark brown UPS trucks will carry packages of all sizes. This shot was taken at the beginning of the day. There is a substantial content fire load. All vehicles may contain dangerous cargo. The size of the package has no bearing on the potential hazard.

When fighting a fire in a postal building, all of the critical fire-fighting objectives must still be met; however, the third-property conservation takes on an added significance. *(Louis "Gino" Esposito)*

Discretion should be used when choosing an extinguishing agent. All appropriate actions should be taken to minimize water damage to mail contents. It may be best to move some of the contents to another area of the vehicle while extinguishing what is burning. If at all possible, try not to move contents from the vehicle without supervision and authorization by appropriate U. S. Postal personnel. It may also be possible to use combination attacks utilizing water and CO_2 or dry chemical in appropriate areas. At all times, if given the choice of an extinguisher, it is best to choose a clean agent such as CO_2 or even Halon if available. However, use of SCBA is imperative, as extinguishing agents such as these tend to remove oxygen from the area, creating an asphyxiation hazard. In addition, the byproducts of the Halon extinguisher are toxic and should be avoided.

Prior to and during fire attack, look for signs of hazardous material presence such as an unexpected reaction to the application of water, unusually colored smoke, or anything else that would raise a flag of caution.

Salvage considerations. As with all U.S. Postal property, appropriate actions should be taken to preserve and protect the mail and its contents. Generally, excessive use of water for extensive overhauling is discouraged unless conditions demand such action be taken. If possible, overhaul of a vehicle might be delayed until authorized personnel remove the contents of the vehicle, both salvageable and burned. Vehicles should not be removed from the scene without consent from authorized U. S. Postal representatives.

Operational guidelines for incident stabilization — postal buildings

Response considerations. A response appropriate for a standard structural fire should be dispatched. Any additional information available should be relayed to companies while en route to the scene. The best hydrants and apparatus positions should be established beforehand by pre-fire planning visits.

Extinguishment considerations. For a working structural fire, fire attack and attendant support operations should be consistent with department standard operating procedures. If possible, keys should be made available once on-scene to limit and reduce forcible entry damage. In addition, if some mail, including trucks waiting at loading docks can be safely moved, it would reduce the fire load and potential damage. This should be done under the direction of fire department personnel working in conjunction with postal representatives.

As with all postal property, all appropriate actions should be taken to minimize water and other secondary damage.

Salvage considerations. Salvage operations should be initiated as soon as possible. Every effort should be made to protect the building and its contents from secondary damage. To better control this operation, the incident commander should assign a salvage group/sector as early as possible during the operation. The salvage group/sector officer should liaison with U. S. Postal representatives as to the best way to protect mail and other related contents and parcels. This can include establishing a dedicated area where postal contents can be moved and secured. This security detail is best performed by an authorized postal representative or a police officer. Fire personnel should be cognizant of the fact that absolutely no contents should be removed from the building or from any other U.S. Postal property without consent from and under the supervision of an authorized United States Postal representative.

Terminating a United States postal property incident

Due to security protocols, incidents at United States Postal property should not be considered terminated until:
1. The situation is stabilized by fire department personnel.
2. The scene is turned over to either:
   a. An authorized U.S. Postal representative
   b. The United States Postal Police
   c. The police department of the jurisdiction
3. Proper notification is made to dispatch that the incident scene has been transferred to an appropriate outside agency.

Conclusion

Fighting fires in commercial occupancies present some of the most dangerous situations a firefighter will face. Be prepared for manpower-intensive operations due to a heavy fire load, large, open areas, and long hose stretches. It is the prudent incident commander who recognizes the inherent problems with each type of occupancy in his jurisdiction and is prepared for the emergency before it occurs.

Questions for Discussion

1. Discuss fire control in commercial occupancies in regard to the CRAVE acronym.
2. What are some of the indications that a backdraft condition exists?
3. Discuss proper strategic coordination when backdraft conditions are encountered.
4. What are some of the forcible entry challenges encountered in commercial occupancies?
5. Name some of the indicators that the building is absorbing water rather than allowing it to run-off.
6. Discuss some of the mantraps that may be encountered at fire incidents in commercial occupancies.
7. Discuss some of the problems caused by the utility services that are found in commercial buildings.
8. What are the advantages to using task forces at the fire scene?
9. Discuss the role of the incident safety officer at the fire scene.
10. Discuss the offensive/defensive strategy utilized in strip mall fires.
11. Discuss some of the operational considerations that must be taken when United States Postal property is involved in a fire response.

CHAPTER ELEVEN
Hazardous Materials Incidents

The First Responder's Role

You are dispatched on a single company response to a report of a strange odor in a fire-resistive high-rise office building. The building is occupied by an international telecommunications company. As you arrive, you are met by a nervous security guard that tells you that the odor is emanating from the ground floor in the vicinity of a division of the business that supplies and processes paper for the company. You notice that some renovations are being conducted in this area. There are about thirty employees milling around the entrance, several of which are lying on the ground and in respiratory distress. Many do not speak English and are trying to tell you something, but you do not understand what they are saying. Suddenly, two police officers who entered the building prior to your arrival stagger out. They are gasping for air and vomiting. The entire building is occupied with about three thousand employees. There are reports of breathing problems and headaches. EMS is not yet on the scene.

The representatives of the company are adamant about not evacuating the building, as a shutdown will have a far-reaching impact on the region and argues that the problem is only occurring in an isolated area of the building. One of your men who had entered the building now becomes violently ill.

Hazardous materials are found everywhere. This trailer, placarded with a "poison" label was located in a parking lot in a rural community. No jurisdiction is immune from the threat of an incident.

You also notice that the silver metal framing around the entrance door appears to be turning a weird yellow color. People are beginning to stream out of the building now in a panic. They are yelling at you to do something. What can you possibly do to control this situation?

As a firefighter, I am sure you have responded to your share of extraordinary calls. People tend to call the fire department when they are unsure of whom else to call. This includes hazardous materials incidents. At most hazardous materials incidents, the fire department is usually the first one requested and, with the exception of the police department, first to arrive at the incident. Keeping in mind that almost two-thirds of all victims are would-be rescuers, fire personnel should choose cautious assessment over aggressive action as the first step in incident stabilization.

Hazardous Material Response is from usually out of the jurisdiction, causing some reflex time. Fire Department personnel must exercise operational discipline and not get involved in an emergency they are not trained or equipped to handle. *(Pete Guinchini)*

Most fire departments do not employ the service of a full-time Hazardous Material Response Team (HMRT). Even those departments that do, usually respond at least an engine company with the Haz Mat Unit. It is critical that members arriving before a dedicated HMRT recognize their limitations and work within those boundaries.

Virtually all fire department members are trained to recognize the dangers of the hazardous materials release, yet time and time again we go too far in regard to our own involvement. When faced with a release, without the proper equipment and training, we are no better off than civilians.

The same goes for terrorist incidents, trench rescues, high-angle and technical rope rescues, confined space, and other exotic incidents. Know the limitations of yourself and that of your equipment.

Most firefighters are required to be trained to the First Responder Awareness Level and First Responder Operations Level of competency. Some firefighters are trained as Haz Mat Technicians, while relatively speaking, a rare few are trained as Haz Mat Specialists.

Personally, I am trained, like most firefighters, to the Operations level. I am by no means an expert at handling hazardous materials incidents, having been involved in only about a dozen or so in my career. Another name for this chapter, taking into account the expertise and level of involvement mandated for most firefighters (myself included) may be "Haz Mat for Dummies".

Let's take a look at the various levels and the limitations of involvement in a hazardous material incident.

First responder at the Awareness level

√   Trained to initiate a emergency response notification process
√   Secure the incident site
√   Recognize and attempt to identify the materials involved
√   Notify the appropriate agency

Awareness level responders are limited to a non-intervention mode of operation.

First responder at the Operations level

√   Trained to protect nearby persons, property, and/or the environment from the effects of the release
√   Defensive operations may include, but are not limited to:
  •   Ignition source control
  •   Vapor cloud suppression and/or dissipation
  •   Exposure protection
  •   Container cooling operations
  •   Confinement operations including:
    1.   Diversion: Controlling the movement of a substance to an area where it does no harm.
    2.   Diking: Establishing a temporary barrier preventing passage of the material; intended to buy time.

Operations level responders are limited to those operations that are either non-intervention or defensive in nature. They should make no direct contact with the offending material.

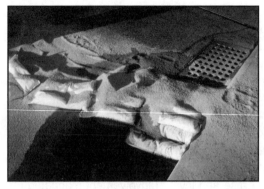

Without getting in too deep, there are many support activities that can be performed by fire department personnel. Diking ahead of the spill is a defensive operation and not intended to expose response personnel to direct contact with the product. *(Pete Guinchini)*

Personnel trained to the Technician Level of hazardous materials response are qualified to operate in an offensive manner. Direct contact with product may be necessary. Proper protective equipment is critical to safeguard personnel. *(Bob Scollan, NJMFPA)*

Hazardous materials technician

√ More directly involved in stopping the release
√ Can take offensive operations to mitigate the incident

Hazardous materials specialist

√ Provide support to hazardous materials technicians by acting as site liaison
√ Expertise may be product-specific

It is important to remember that firefighters and officers, including chief officers, are usually only trained to the Awareness and Operations levels and are limited in engagement to those activities that fall into the defensive and non-intervention modes. Attempting to operate in any manner other than that for which they were trained may result in exposure, injury, and death. This is mostly because these responders are both untrained and ill-equipped to operate in any fashion other than the defensive or non-intervention modes.

Keeping this in mind, it is absolutely critical that the incident commander realizes he is extremely limited in his ability to handle the incident without some type of expert advice and/or response.

Most fire departments do not have the luxury of the response of a hazardous materials technician or specialist in the initial stages of the incident. Thus, we will concentrate our efforts on those measures that the first

Hazardous materials incidents often occur in unexpected places. The oxygen cylinder(s) in this vehicle create a BLEVE threat as well as a heavy fire potential. Hopefully, a fire will not burn this sign up before you've had a chance to see it.

responding companies can take to attempt to stabilize the incident, reasonably provide for life hazard threats, protect exposures prior to the arrival of the "experts", and survive the experience.

Incidents that occur in the middle of nowhere, for example, on an isolated stretch of highway may warrant a total non-intervention mode of operation. In this case, depending on the nature and the disposition of the product (contained, spilled, leaking, ignited, etc.), companies may stand by in a safe area while the incident runs its course. This operational mode may be acceptable at a fire in a transportation vehicle in an isolated area where the burning of the material (thermal elimination) is the best incident mitigation action. Departments whose response districts include these desolate areas should have their degree of intervention planned in advance.

Unfortunately, in built-up urban areas, total non-intervention may not be an option the incident commander has available to him due to the enormous amounts of people and property exposed. Steps, therefore, must be taken to reduce this exposure hazard.

The scenarios in the workbook related to this chapter will focus on those incidents where the first responder can, while using appropriate caution and proper respect for the product involved, make a positive impact on the outcome of the incident. This will usually be when the product can be positively identified. We will also focus on supplemental actions that can be taken when there has been a release of "methyl-ethyl-bad-stuff," especially that which the first responder cannot readily identify. Finally, in all cases, potential fire department intervention on behalf of victims and potential victims must be weighed against the risks to the fire personnel.

Indications of the Presence of Hazardous Materials

"Copological" indicators

Approach should be on the "up and up" (upwind and uphill). Apparatus should stage a safe distance from the incident until ordered into action by incident command. This attitude of non-aggression is counter to the traditional fire department personality. Uninformed tactics may lead to unnecessary casualties. *(Ron Jeffers, NJMFPA)*

A "copological" indicator is one of the first signs of definite trouble and a rescue problem. Police personnel are inquisitive by nature. It is a large part of their job. Many times, they get in over their heads. This "copological" indicator occurs when an improperly dressed and trained member of the police department decides to investigate the product without first taking notice of the existing condition and hazards, and is subsequently overcome. A cop lying in or close to the offending material is a good bet that something nasty is afoot. Fire department personnel should take every precaution to avoid falling victim to this unacceptable action and condition.

Look for people running away from the incident

This is another dead giveaway to a problem. A driver or passenger running away from a vehicle or building without stopping to address an approaching emergency vehicle is not normal. At this point, the apparatus should stop where they are, get out of the way if the officer thinks they are too close, and attempt to get some information from the driver if he can be caught. Don't forget that someone running from a vehicle that does not flag down the apparatus as it approaches, especially a rented vehicle, could mean the potential for a terrorist act. Stay out of harm's way and prepare for something bad to happen.

Look for odd-colored liquids or smoke

This seems to be a no-brainer, but too often, this sign of abnormal conditions goes unheeded until it is too late. Take all precautions to protect your company and isolate the areas until definite product identification can be made.

I remember seeing a video very early in my career of a tank car leaking product and a chief officer of a big city department actually reaching out, taking a sample of the leaking product on his hand, and either smelling it or tasting it. How many of us would do that today? Frightfully, despite what you may think, the answer is too many. This was, as the video stated, "an act beyond comprehension".

At another incident, a tractor trailer was leaking a green-colored slime one hot afternoon. The fire department was dispatched and a Hazardous Materials Response Team was summoned. Control zones were set up and personnel were put in fully encapsulated entry suits. They approached the back of the trailer and opened the doors. What was found was rotting watermelons. However, the response to the incident was correct. The incident commander must ensure that all necessary precautions are taken to protect personnel, the public, and the environment.

CRAVE

We will address the CRAVE acronym in regard to the first responder's role in a hazardous materials incident.

Command.
- Find out as much as possible about the release prior to arrival.
- Request wind direction and speed from dispatch.
- Direct companies on safest approach.
- Locate command post in a safe area (upwind and uphill).
- Establish command post with anticipated incident escalation in mind.
- Attempt to identify product (or at least product classification; *i.e.*, poison, oxidizer, etc.).
- Set up preliminary control zones based on the most pessimistic information available.

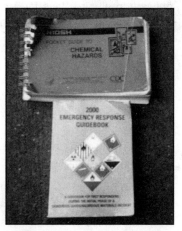

Incident command should make use of reference material to assist in the hazard/risk analysis process. This information will be instrumental in the formulation of a strategy and action plan.

Control zones

There are four control zones. They should be clearly identified. All responders should be made aware of the zone boundaries.

The Hot Zone. This is the area of the actual incident. The Hot Zone covers the area that is immediately dangerous to life and health. The size of the Hot Zone is contingent on such factors as the material involved, the topography, and the weather. The Hot Zone is generally located from the point of the incident to a point deemed safe for personnel to operate without product-specific protective clothing. The boundary line between the Hot and Warm zone is designated the Hot Line. Only personnel who are directly involved in incident stabilization should operate in the Hot Zone. Hot Zone entry requires product specific personal protective clothing and equipment. The only entrants should be those properly trained and equipped personnel assigned to a specific task. It is best to set up a decontamination station before entry into the Hot Zone is made. Fire department personnel trained to the Haz Mat Operations level should not enter the Hot Zone for any reason.

The Warm Zone. The Warm Zone is the area of forward control of operations outside of the Hot Zone, and is used to provide support directly to those operating in the Hot Zone. The operations chief as well as the Haz Mat chief are located in this area. The Warm Zone begins at the Hot Line and extends outward in all directions for a distance as required by the incident. Personal protective equipment required includes structural firefighting clothing as well as hazard-specific personal protective equipment as dictated by conditions. The number of personnel operating in this area should be restricted to the absolute minimum required to support the Hot Zone operation. The decontamination operation as well as essential equipment is staged in this area.

The Cold Zone. This area is reserved for those responders and agencies who have some function germane to the incident, but whose operation does not require a level of protection required in either the Hot or Warm Zones. The Cold Zone is an area deemed safe to operate with minimum protective equipment. The command post as well as the primary equipment and apparatus staging areas are located in the Cold Zone. This is also where EMS and police operational areas may be located in addition to any agency representatives whose participation is critical to the outcome of the incident.

The Public Zone. This area is reserved for the press, civilian onlookers, public officials, and other persons who have no function in regard to incident stabilization. A secondary apparatus staging area can be established in the Public Zone.

Ensure sufficient resources are ordered early and in quantity tantamount to the expected need. Ensure decontamination equipment is available prior to any direct intervention in the Hot Zone and that documentation protocols are followed. Make sure that everyone wears full PPE and SCBA. Firefighters who are improperly dressed are nothing more than highly informed bystanders; in other words, they are next to worthless. They are almost guaranteed to become part of the problem.

Rescue.
- Consider evacuation versus protection-in-place.
- Weigh victim rescue against fire-fighter risk (hazard-risk assessment).

Attack.
- Ensure a continuous primary and secondary water supply is established
- Consider extinguishing agent vs. product compatibility.
- Use special extinguishing agents as required.
- Preplan special agent sources and keep contact phone numbers updated.

Ventilation.
- Consider exposures when venting.
- Be aware of ventilation equipment as ignition sources.

Extension prevention.
- Movement or protection-in-place of exposures.
- Migration of product.
- Disposition of runoff extinguishing agent and/or product.

This fire in a vitamin store and manufacturing facility was reported to be issuing extremely toxic smoke from the burning contents. A Hazardous Materials Unit was requested and a Decon setup. *(Ron Jeffers, NJMFPA)*

Suggested Hazardous Materials Response Strategy for First Responders

1. Establish Command in a safe area
2. Position upwind and uphill
3. Isolate the area
4. Deny entry to the area
5. Attempt to identify the products if it can be safely done from a distance
6. Set up preliminary control zones
7. Let the experts handle the situation

Conclusion

I do not profess to be an expert in hazardous materials incidents. Most firefighters aren't. If you take a few seconds to properly size-up the situation and then make a rational judgment based on that assessment and then a common sense approach in regard to risk analysis, you will not place your personnel in situations that they are not trained for or equipped to handle. Following this simple guide will give your personnel the best chance of returning to the firehouse unscathed.

Questions for Discussion

1. Discuss the scope of most fire departments in the handling of hazardous materials incidents.
2. Discuss the levels of hazardous materials training and the scope of intervention of each.
3. Discuss the CRAVE acronym in regard to first responder actions at hazardous materials incidents.
4. Discuss some of the cues that are indicative of a hazardous materials response.
5. What are the seven steps suggested as a strategy for first responders at hazardous materials incidents?
6. Discuss how the three fireground priorities impact a hazardous materials incident.
7. What are the four control zones that need to be established at a hazardous materials incident and what are the characteristics of each?

CHAPTER TWELVE
Operational Safety

This chapter will address actions that can be taken to make the fireground a safer place. We will attempt to cover the fireground experience in a chronological fashion, from response preparedness to incident termination and beyond. In addition, several methods of alerting the entire department to unsafe situations (which includes prior to, at the fireground, and after the incident) will be discussed. Finally, fireground critiquing and how to make it more effective will be addressed.

Planning for Unusual Responses

Most departments use pre-fire plans, fire inspections, and building surveys to best ensure operating personnel are aware of hazards that may be encountered on the fireground. These preplans and surveys are usually limited to target hazards such as schools, hospitals, industrial complexes, large shopping malls, and other occupancies that will create a major problem, both in response and operations.

Sometimes, companies will come across a building or condition that warrants the attention of the rest of the department. These conditions may include such conditions as dangerous fire escapes, the presence of unfriendly dogs

Unusual responses will include occupancies such as the Lincoln Tunnel that connects Northern New Jersey with midtown Manhattan. Comprehensive preplanning and multi-agency cooperation is mandatory.

in a cellar, the presence of a rain roof, or hidden building conditions such as an attached cellar. Whether discovered during inspections, routine responses, or civilian complaints, it is imperative that all department members be made aware of these conditions. In a career department, there may be as many as four shifts. Due to unforeseen circumstances, information transfer may not be as efficient as it should be and some of the information is neglected to be transferred on a daily basis. Therefore, crucial information gets lost. In a volunteer department, members not on the alarm when the condition is discovered may never be made aware of the condition. The best way to relay this information is by making it part of the department's database. If the department uses a Computer-Aided Dispatch System (CADS), the information can be automatically teletyped into each fire station as the alarm is received. If a CADS sys-

This sagging floor condition will not be visible in heavy smoke. Personnel must be aware of this condition before the incident. CADS reminders and department-issued memos are a good way of keeping members informed about hazardous conditions.

tem is not available, the hazard or condition can be broadcast over the air to all companies as they are on the response. The problem that arises is that during the response, the firefighters may not hear or absorb this information due to the noise and excitement of the response. It is best that they be made aware beforehand and then reminded en route.

One way to accomplish this information without the use of a computer is by the adoption of an "Exceptional Response Form". This form was used in my former department and was very effective in passing on vital information regarding dangerous conditions and other building and area-specific information. Affectionately known as "The Weird Building Report", submissions were forwarded through the chain of command to the office, where the condition would be investigated. If the condition warranted, the information would be sent to the dispatcher to enter into the computer database and a copy placed

WEEHAWKEN FIRE DEPARTMENT
EXCEPTIONAL RESPONSE FORM

ADDRESS: 122 & 124 Dodd st.

CONTACT PERSON? _____

☒ NO ☐ YES - NAME & TEL NO. _____

DATE 6/30/98 **NAME OF OFFICER** captain Lemonie
GENERATING REPORT

THIS INFORMATION WILL EXPIRE ☒ NO ☐ YES - EXPIRATION DATE _____

**IF NEEDED, USE A CONTINUATION PAGE
FOR ADDITIONAL INFORMATION OR MAP.**

DESCRIPTION OF EXCEPTION:

In the rear of 122 Dodd St. & 124 Dodd St. there are single family homes (122R and 124R).

EFFECT ON OPERATIONS:

These homes are only accessible through the alleys of 122 Dodd St. & 124 Dodd st. from the Dodd St. side only. This will require a longer than normal attack line hose stretch.

SOLUTION TO PROBLEM:

Bring the water thief into the alley of 122 Dodd St. or 124 Dodd St.

PAGE 1 **OF** 1

on a bulletin board in all stations. An entry was also made in the company journal regarding the report. Companies coming on duty would read the company journal and be directed to the "Weird Building Bulletin board". They were urged to familiarize themselves with the building or condition themselves during their tour of duty.

The Exceptional Response Report can be used for both permanent and temporary (transient) conditions. It can be used to alert personnel to temporary conditions such as street detours, construction projects that affect response, or auxiliary systems that are temporarily out of service. The report can also be used for permanent conditions previously unknown to most responders. These include homes "hidden" behind homes, unusual response routes and conditions, access problems, and other critical/unexpected conditions.

The report has a section on a description of the condition, whether it is permanent or temporary, what effect the condition has on operations, and most importantly, a proposed solution to the problem created by said condition. It is important that fire personnel be both problem-finders and problem-solvers, however it is best to be more of the latter than the former. Personnel who consistently find problems without solving them are operating contrary to the mission of the department.

The Exceptional Response Report is a simple method of passing hazard information in an efficient manner to the entire department. Knowledge gained before an incident, no matter what it is, can only benefit the response.

Apparatus Positioning

Apparatus positioning is probably the most important initial action on the fireground. It is akin to a football team using the right formation to run a play. If the team lines up in the wrong formation, the play is likely to go nowhere. It is extremely difficult for the incident commander to put strategy into motion if the apparatus is not positioned to match the strategy.

The principal objective of apparatus positioning is to position each piece of apparatus to take advantage of its capabilities while allowing for tactical flexibility of uncommitted apparatus.

Apparatus function is the determining factor in apparatus positioning. With very few exceptions (which should be known to all responding companies beforehand), the ladder company should position at the front of the building. Attack engines should generally be positioned just past the building so the stretch is not too long, but also to avoid blocking ladder company access.

On narrow streets where wires are present in front of the building, the aerial may have to position on an adjacent street, out of the way of wires. On wide streets where overhead power lines are located in front of the building, it may be necessary for the ladder company to alter its position. The apparatus can either straddle the center of the street or median, or position in the lane opposite the building so the aerial can reach over the lines to the roof.

When aerial use is not possible at the front of the building due to obstructions such as overhead power lines, creative positioning may be necessary. Apparatus operators must be aware of the capabilities as well as the limitations of their equipment. *(Bob Scollan, NJMFPA)*

In this instance, engine companies may have to back into position. Proper and timely communication along with prior knowledge and training are a necessity here.

Engines assigned the water supply responsibility should look to position at the most strategic hydrants, while allowing access to the scene by later arriving apparatus.

It is extremely important that initial arriving companies announce their arrival along with their function. That way, later arriving companies will be aware that these responsibilities are already taken. For example, the first arriving engine should announce that they are "attack" or "have the building," while the water supply engine should state that they are "water supply" or "have the water". Specific department SOPs will dictate variations to this positioning, but the communication is still critical no matter how your apparatus are positioned.

The first-arriving ladder company should also announce that they "have the building" so additional aerial apparatus are aware that a ladder truck is already committed to the building and they should stage uncommitted or take a different position as per SOP. This is also important because companies that are out of their response area may not be first-due at an alarm for which they would normally be the initial arriving engine or ladder truck. Remember that first due does not necessarily mean first arriving. Once a company announces scene arrival and function, the other companies should position accordingly regardless of who those companies are and what their "normal" due status is. Proper communication will avoid a logjam of apparatus at the front of the building.

Companies that are not assigned one of the initial operating positions such as the front of the building, attack, or water supply should stage uncommitted, preferably at a cross street. This will allow the incident commander the opportunity to position apparatus in the most advantageous and flexible manner possible given the conditions. For ladder companies, the cross street will allow for positioning at the rear of, in a flanking position, or in any number of a variety of strategic positions as dictated by conditions or the orders of the incident commander.

Once a large piece of apparatus like a ladder company is in position, it is very difficult to move it. It is often easier to request another ladder truck than it is to move a poorly positioned, already committed ladder company. For engine companies, other than the initial-arrivers, it is best to back into position. That way, if repositioning is necessary, it is easier to drive forward to the new position than it is to back out of a block. It should also go without saying that once hose is dropped, the rig is pretty much "dead in the water" and repositioning is out of the question.

The best way to address apparatus positioning is by establishing a Standard Operating Procedure regarding initial scene assignment. The SOP should address the routine response. Special responses can be covered by their own SOP. For instance, high-rise response may require totally different apparatus positioning than a multiple dwelling. Most high-rises, with the exception of the company responsible for standpipe supply, will require companies to stage and report to the command post on the interior. Other occupancies such as large industrial complexes or shopping malls may require a specific response, standpipe supply, and positioning SOP. This response should be based on potential problems that may be encountered in the fire building, its exposures, and the surrounding area. This knowledge and subsequent effective positioning can only be gained through proper pre-fire planning.

Some actions to avoid in regard to apparatus positioning include the following. (Thanks to Captain Steve Winters for his work on this subject.)

Avoid travel against the flow of traffic

Unless an SOP specifically addresses this, traveling against traffic almost always creates poor apparatus positioning and is an example of apparatus freelancing. If there are cars in the street, they will be stuck between the apparatus and may block access to the building. If there is no other choice but to respond into a block against traffic, communications to the other responding companies is essential so the proper adjustments can be made.

Avoid "bumping" an engine already committed

"Bumping" is a term used when supply and attack engines switch responsibilities and positions. Bumping occurs most often on a narrow street when
the first two engines arrive in quick sequence before the ladder company. If the
first engine is uncommitted, that is, has not stretched any hose yet, the second engine can "bump" them. This means that the first engine will assume water
supply duties while the second engine has the attack. This method of fire attack
is extremely effective on narrow streets.

Attempting to "bump" an engine that has already dropped lines and
begun an attack will cause the front of the building to be blocked by the now
out-of-position engine. This will most likely prevent the ladder truck from
properly positioning in front of the building. It is best for the second-arriving
engine to listen to the radio for the first-arriving engine's Preliminary Size-Up
Report and initial actions. It is also a good idea to size up the block before committing to a position that may hamper the entire operation. Again, proper and
timely communications is the best way to avoid this problem.

Avoid blocking an intersection or two-lane road

This may be hard to avoid if overhead power lines dictate the position of
the ladder company, but every effort should be made to create a free lane of
traffic. It makes everyone's job easier and minimizes apparatus congestion.
This is also applicable to engine companies connected to hydrants, which are
often on corners. Engine chauffeurs should make every effort possible to
supply water to the fire scene while allowing other apparatus to pass.

Avoid positioning the engine directly behind
the ladder company

In many departments, the third engine backs down to the ladder company. This positioning gives the incident commander some tactical flexibility. The
engine can be used to feed an aerial master stream or can be used as a second attack engine to stretch lines into the fire building or an exposure. In the
latter case, if a hydrant is not nearby, it will be necessary for another engine to
provide a water supply. The big problem occurs when the engine is positioned
too close to the back of the ladder truck. This positioning eliminates access to
the ground ladders at the back of the ladder truck. Always leave at least 30'
between the engine and the back of the ladder truck. This is because the longest
ground ladder is usually the 20' roof ladder or the two-section 35' ladder, which

is 20' long, bedded. Taking ladders from another ladder truck because ladder access is blocked at the front of the building will cause a delay in the required tactical action. The consequences created by this condition may be severe.

Avoid positioning the ladder truck where the aerial device will be ineffective

When wires are present at the front, the side or the rear of the building may be used for aerial access. If two ladder companies are on the initial response, one can position at the front of the building to provide tools, lighting, etc. The other can position at the side and access the roof. *(Mike Borelli, FDJC)*

Overhead wires often crisscross streets and inter-sections. The aerial must be positioned in such a manner that allows the turntable and the bed ladder a clear shot at the building. In addition, especially on mid-mounts and tillers, the position must also allow the stick to be raised out of the bed and rotated into position in line with the turntable. If only the turntable clears the wires, but the end of the aerial device does not, the aerial may be unable to be raised. The same is true on narrow streets where utility poles may prevent aerial rotation. It may be best for the officer or the jump man to get out of the cab and direct the chauffeur into the best position.

Ladder personnel should always be aware of the best places to ladder a building. Sometimes, a building will be a mess of wires at the front, but clear at the sides. This will usually apply to corner buildings. It may be better to ladder the building from the side where the wires are not present. If two ladder companies are on the response, let the first ladder company take the front of the building while the second ladder company hits the roof from the side street.

Avoid committing additional alarm apparatus without orders

This type of commitment leads to loss of control on the fireground. Companies not assigned on the initial alarm, unless specifically ordered by incident command, should stage uncommitted in a position that affords for maximum tactical flexibility of the apparatus. Once improperly committed, it may be very difficult to reposition. This may lead to delays in the accomplishment of some critical task.

Avoid radio silence

Radio silence upon arrival is just as bad as radio blabber. It is imperative that the incident commander knows where companies are positioning at the outset of the incident. This will allow incident command to take a positioning inventory to determine which positions are yet to be taken, which may require reinforcement, and most important, which positions are not available due to obstructions such as wires and setbacks, barriers to response such as traffic, double-parked cars, blocked hydrants, and other conditions that will negatively affect proper and effective apparatus positioning. This early intelligence must be reinforced as required with information regarding conditions, progress made, or support needed. There is no way that the incident commander can evaluate the effectiveness of an operation without timely and concise reports from the working companies.

Dead-End Street Response

If not planned for in advance, there are few responses that may congest the scene and frustrate operations more than a dead-end street. Not only may dead-end mains affect the water supply, but the ability to quickly secure an alternative, reliable water supply may also require some unorthodox tactics. Dead-end street positioning and tactics require quick thinking and even quicker communications on the part of the first-arriving companies to avoid congestion and confusion. A standard operating procedure should be in place for dead-end street response. Dead-end streets generally fall into three categories.

Dead-end streets often cause apparatus access problems. Timely communication and structured coordination on the part of the first arriving companies is critical. A good SOP on dead-end street response will help reduce confusion.

1. Long dead-end street with no hydrant in block.
2. Short dead-end street with no hydrant on the block.
3. Dead-end street (long or short) with hydrant(s) on the block.

If at all possible, the ladder company should position at the front of the building. Factors such as width of the street, height of the building, and proximity to the intersection will influence the decision to position the ladder truck on the cross street or direct it to a priority position in the fire block. *(Pete Guinchini)*

In almost all cases, the ladder company must get the front of the building. Dead-end streets may require the engine companies wait at a cross street until the ladder company arrives. Once the ladder truck is in the block, the attack can be organized. An exception to this may be where the ladder truck cannot make the turn into a narrow dead-end. If this is the case, the ladder truck may have to position on the cross street and equipment carried into the block. This will allow the engine to enter the block first. If a Telesquirt is available, it may be best to position it in front of the building due to the aerial capabilities. If the building is three stories in height or less, then only ground ladders may be necessary to access the upper floors and the roof. They can be taken off the engine or carried into the block from the ladder company. Preplanning and response area familiarization will best provide this information.

For dead-ends with no hydrants, the attack operation will most likely depend on the length of the street and the location of the fire building relative to the intersection.

For long dead-ends with the fire building some distance from the intersection, the attack engine can either wrap a hydrant on the way in or simply proceed to the fire building and begin the attack. Remember to position in such a manner that allows ground ladders to be removed from the rear of the ladder company. The second-arriving engine will then have to provide the water supply by connecting the wrapped supply line to their discharge or to the hydrant, depending on department procedures and/or the length of the stretch. The other alternative is to either back down the block, drop a supply line and proceed to the hydrant or hand-stretch a supply line down to the attack pumper from the intersection. This decision will also depend on the proximity of the hydrant to the intersection.

For both short dead-ends or long dead-ends where the fire building is close to the intersection, once the ladder truck is in the block, the first-arriving engine can either develop their own water supply by hand-stretching to the hydrant or take a feed from the second-arriving engine. Communications will play a key role here. If the first-arriving engine can supply themselves, they must announce over the radio that they "are the attack engine and have their own water". The second-arriving engine can then assist with the water supply operation and the attack line or the stretching of a back-up line.

If the department utilizes a reverse, fire-to-hydrant lay, once the ladder truck is in the block, simply back down to the building, leaving room for the ladder company to get the ground ladders off, drop attack and supply lines, and proceed to the water source. It is best for other initial alarm apparatus to stage on the cross street and walk to the command post or fire building.

For either long or short dead-ends with hydrants, it will be critical that the responding companies know the hydrant grid system. A good hydrant inspection program should provide this information before the incident. Dead-end hydrants should be avoided if at all possible. If a hydrant on a dead-end main must be used, pump operators must continuously monitor the water available for the attack. This will begin with the hydrant or static pressure. Once water is flowing, the pump operator must take notice of the percentage of pressure drop in the compound gauge on the engine pump panel. Called residual pressure, this pressure can be used as a guide in determining attack line supply capability. If, after the first handline is supplied, the residual pressure drops by 5% or less, then three more lines of equal diameter can be supplied from that hydrant. If the pressure drop is 10%, then only two lines of the same diameter may be supplied. If the pressure drop is 20%, only one more line can be supplied. Any drop greater than 20% should indicate that no more lines can be supplied by this hydrant. This rule of thumb should be used at all fire situations, but it is extremely important when supplying water from a dead-end main

To give an example, suppose a hydrant has a static pressure of 100psi. After the first 1¾" handline is supplied, the residual pressure drops to 90psi. That is a 10% drop. This hydrant can supply only two more lines of the same diameter, in this case, 1¾".

There are times, however, a hydrant on a dead-end is not the end of the grid. The grid may continue to adjacent blocks. It will be perfectly acceptable to use the hydrant in this instance. If this is the case, the engine should enter the block first and develop their own water supply at the nearest hydrant. This may require assistance from the second engine company. It might be best, if the hydrant is some distance from attack position, for the second engine to enter the block before the ladder company and "bump" the first engine to water supply duty.

These are operational suggestions to coordinate a situation where the potential for problems is great. The point to be made here is to plan in advance for each type of dead-end condition and make sure that responding companies are aware of their responsibilities. Remember that even the best plan is likely to fail if communication between the companies is lacking.

Functional Fixity

The above situations and many others like it on the fireground will test both the flexibility and ingenuity of the officer and his or her assigned firefighters. While there are many ways to accomplish an objective, there will come

This is the stuff that legends are made of. Firefighters must be able to adapt to the situation. The firefighter unable to improvise will often be inefficient at the most critical time. *(Ron Jeffers, NJMFPA)*

a time in every firefighter's career where the situation will require an ability to improvise on the scene. Salvage and forcible entry operations are prime areas where this can best be applied. The ability (or lack thereof) of a firefighter or officer to adapt to the situation is called functional fixity. A firefighter is a victim of functional fixity if this ability to adapt is inadequate or non-existent. For instance, a firefighter has to remove a screw as part of an operation, but does not have a screwdriver. He does however, have access to a butter knife or perhaps a dime in his pocket. If he cannot reason that these items can be used to at least attempt the task, he is a victim of functional fixity.

One of the best improvisational tools there is in regard to forcible entry in a non-emergency situation is the butter knife. It can be used to jimmy a window or door. Likewise, one of the best improvisational aids in salvage work is plastic sheeting, plastic tie wraps, and duct tape. Many water-related problems can be solved temporarily by manipulating these three items to resolve the problem. Water leaking through a ceiling due to a leak in a roof can be temporarily routed to a sink by using a makeshift chute quickly fashioned from plastic sheeting duct taped to the ceiling around the leak. The sheeting is then molded to route the water to the sink. The plastic tie wraps can be used to aid in the molding of the plastic at the narrow end of the chute. This technique eliminates the need to leave expensive fire department equipment such as salvage covers at a scene.

Proper training will guide the firefighter in the use of the "right tool for the right job". Only with this prior knowledge can firefighters improvise in a safe and effective manner. Without this proper training, no frame of reference can be established, and, as a result, safety may be jeopardized. Through experience, an arsenal or skill portfolio of these improvised techniques can help the firefighter be more versatile on the fireground.

Recently, a boat fire in one of the marinas on the Hudson tested the responding companies' ability to improvise. The boat, a pleasure craft, which was nearly fully involved upon arrival and threatening to extend to boats on either side, was about 100 yards out on a pier stringer. The marina was equipped with fire-hose boxes, which contained several hundred feet of 2½" single-jacket hose. This required a significant amount of manpower to stretch the hose to the fire area. A second alarm was struck. Ladder companies made their way out to the pier and involved boat to provide recon ahead of the attack operation. When the hose was finally stretched and a water supply established via a standpipe-type supply operation, many of the lengths burst, as they had not been used in years.

To keep the fire from spreading, the ladder company on the pier used chains from the extrication equipment and pike poles as makeshift mooring lines to keep the burning craft from going adrift, having burned through its mooring ropes. In addition, more rope was used to pull and guide the exposed boats away from the danger area. As the lines were bursting, it was clear that the need to get some water on the burning vessel was crucial, as the danger of a gas tank explosion was a real possibility. Again, thinking quickly, dock lines, no bigger than garden hoses were used to wet down the exposures and several other adjacent lines were connected together to reach the fire area and apply water to the now sinking vessel and cool the gas tank area. This strategy proved successful, as the fire was soon brought under control and the damage was limited to the craft of origin.

DON'T BE A VICTIM!!

Incident Scene Management

The "what-ifs" of decision-making

There is an infinite amount of "what-ifs" in the fire service. As an instructor, you can get "what-ifs" all day long. "What-if" dangers and accompanying decisions should be based on judgment rooted in experience and knowledge. In addition, a healthy dose of common sense and gut feeling also plays a large part in the decision.

Priorities must be assigned to the "what-ifs". The likelihood of the "what-ifs" becoming reality should be based on a hazard analysis which is designed to allow you to prepare, either through pre-fire planning or rapid mobilization of forces, to address the worst case scenario as conditions dictate. The priorities of the "what-ifs" fall into the following categories:

- **Definitely won't happen**
- **Probably won't happen**
- **Probably will happen**
- **Definitely will happen**
- **Is happening** – you don't ever want to be here if you had time to prepare, but didn't.

There is a fine line between adjacent "what-ifs". Things happen on a sliding scale. Rarely does something go from one end of the spectrum to the other instantaneously, except for those who failed to recognize the signals being given by a particular situation. Remember that the definites at the top and bottom are extremes and are more rare than the probabilities. The trick is to take steps to prevent "definitely" from become "probably". If you do this, the "is happening" may not happen to you, and if it does, you will be prepared for it.

Proper and continuous size-up is the key to controlling the "what-ifs". However, spending too much time on the "what-ifs" is a mark of decision-making impotence. Second-guessing is a disadvantage on the fireground. If you worry all day about "what-ifs," you never get any-

This mangled ground ladder is a possible result of a mis-read on the "what-ifs" of decision-making. A sloping roof can be seen to the left, possibly indicating a bowstring truss. Recognition of the collapse potential should influence incident command to slide closer to the "definitely will happen" side of the scale. *(Ron Jeffers, NJMFPA)*

thing accomplished, which will cause the scale to slide into the dreaded "is happening and I'm not prepared" dilemma.

Operating with tactical worksheets

Tactical worksheets help the incident commander keep his head in the game and make sense out of a dynamic and often frenzied situation. The most important factor regarding the tactical worksheet is that it must be kept simple. There is enough going on at the fire scene, especially in the early stages, without having to fill out a bulky, complicated, and confusing document that does not give you back what you put into it. Equally as critical to simplicity is the ease of transferability. The document must allow incident command to be transferred with the least amount of confusion.

The form should act as a "tickler" file as well as a crude accountability form. Two forms that work well are the "fill-in-the-blank" and "checklist" types of forms. A combination of the two may also be advantageous. The form should be laminated so that it can be used over and over. A grease pencil or dry erase marker work best. This minimizes scribble and allows companies to be tracked without cross-outs.

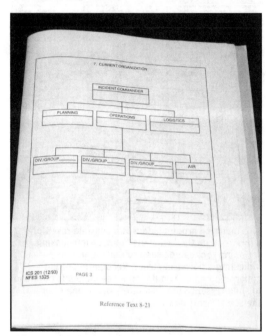

Reference Text 8-21

ICS form 201 is an excellent way to organize the operation and ease the transfer of incident command. The four-page form is simple and versatile, and helps reduce the CHAOS (Chief Has Arrived On Scene) factor.

ICS Form 201 used in the Incident Management System is one form that is generic, simple to use, and easy to transfer from one officer to another. It is a four-page document, which allows the current incident commander to include all of the germane information required to provide a comprehensive briefing should command be transferred. Page one is a place where a sketch of the incident can be drawn, which should include a crude building drawing and orientation, apparatus positioning, water supplies established, and any other information required for "at-a-glance" referencing. Page two is a place to write a brief narrative, summarizing current actions. Page three is a generic incident command chart with assignment boxes already drawn in. Page four is a resources summary sheet that allows the incident commander to keep track of resources ordered, time of arrival, on the scene status, as well as location and assignment. The form is also serrated so that the pages can be separated as the incident is decentralized and additional personnel are assigned to command and support positions. Many departments use a hybrid of this form to suit their needs.

Worksheet #1 is a checklist. It is simple and to the point. It actually operates as a "tickler" form to prompt the incident commander to provide for essential operations, request reports, and contact appropriate outside

---

NORTH HUDSON REGIONAL FIRE & RESCUE
FIREGROUND TACTICAL CHECKLIST

FIRST ALARM
- ☐ Size up -- given by first-arriving Officer
- ☐ Situation report -- given by first Chief Officer
- ☐ Status reports -- from interior companies
  - ☐ Attack team
  - ☐ Interior team
  - ☐ Roof
  - ☐ OVM
- ☐ Apparatus positioning
- ☐ Mode of operation   OFFENSIVE   DEFENSIVE
- ☐ Water supply secured
- ☐ Primary search extended
  - ☐ Complete
- ☐ Placement of hose lines
- ☐ Ventilation accomplished
- ☐ Attack coordinated
- ☐ Building laddered
- ☐ Outside agencies contacted
  - ☐ PSE&G
  - ☐ EMS
  - ☐ MSU
- ☐ FAST Team in place

SECOND ALARM
- ☐ Reinforce attack
- ☐ Secure secondary water supply
- ☐ Protect exposures
- ☐ Control utilities
- ☐ M.A.C. Cards requested
- ☐ Command Board set up
- ☐ Operations Officer assigned
- ☐ Additional agencies requested (where applicable)
  - ☐ Fire Investigator
  - ☐ Gong Club
  - ☐ Red Cross
  - ☐ Construction Official / Building Dept.
  - ☐ Medical Examiner
  - ☐ Health Dept.
  - ☐ DPW

PROGRESS REPORTS (every 15 mins)
- ☐ #1
- ☐ #2
- ☐ #3
- ☐ #4
- ☐ #5

nhrfr/sops/fg-chk.doc

Worksheet #1

---

agencies. This form can be easily modified to accommodate a high-rise operation. As a matter of fact, this is a great form for the operations officer to use at a high-rise operations post, while the incident commander uses the more comprehensive Worksheet #2.

Worksheet #2 and #2A are opposite sides of the same laminated form and is the actual tactical worksheet used by the North Hudson Regional Fire & Rescue. It is more comprehensive, but it is still relatively simple. The numbers above the building diagram at the center of the page are the company designations of the apparatus. Crossing out the number when the company arrives on the scene helps the incident commander keep track

Worksheet #2A

Tactical worksheet V12

FIRE BLDG:	APT:	EXPOSURES:
OCCUPANT:	APT:	PHONE:
SUPER:	APT:	PHONE:
OWNER:		PHONE:

OWNER S ADDRESS:

FIRE FLOOR: SMOKE ALARMS: yes no   ACTIVATED: yes no

AREA OF ORIGIN:

CAUSE:

SAFETY

ABANDON _____ TIME _____   ROOF COND. _____

ROLL CALL _____ TIME _____	BUILDING INTEGRITY	
MAC CARDS	LADDERS RAISED	
REHAB	LIGHTS	
E.M.S.	FANS	
UTILITIES	SALT	

NOTIFIED	ARRIVED	NOTES
GAS		SEARCH/EVACUATION
WATER		
ELECTRIC		
POLICE		**FIRE FLOOR** Attack stair Evac stair
E.M.S.		OPERATIONS
INVEST.		RESOURCES
GONG CLUB		REHAB
RED CROSS		
D.E.P.		
D.P.W.		
EMER MANG.		
CONST. OFF.		COMMAND POST
HEALTH OFF.		

Worksheet #2

STORY _____ LINES STRETCHED _____ OPERATING _____ FL.

FR. ORD. N.COMB. H.TIMB. F.RES.   TRUCKS: OPENING UP   FORCING ENTRY   SEARCHING

RESID. COMM. RETAIL MIXED   PRIMARY IN/PROG. _____ FL.   SEC. IN/PROG. _____ FL.

LT. MED. HVY. LOCA: _____   C _____ D _____ B

FIRE DOUBTFUL. P.W.H. U.C.   EXPOSURES: A

ALL HANDS ___ E ___ T WORKING   CONDITIONS: IMPROVING   UNCHANGED

1ST ALARM ASSIGNMENT	2ND ALARM ASSIGNMENT	3RD ALARM ASSIGNMENT	4TH	5TH
E	E	E		
E	E	E		
E	T	T		
E				
T				
T				

FAST CO: _____   COMMAND CO: _____   SAFETY OFF: _____

BENCHMARKS	E1	E2	E3	E4	E5	E6	E7	E8	E9	E10	E11	E12	E13	REPORTS
PRIMARY					T1	T2	T3	T4	T5					ATTACK
2ND ARY														SEARCH
VENT														ROOF
EXPOS														O V M
SPRKLR														EXPO
WATER														DIV
M/STR														
C/LOFT														

ROLL CALL

```
      C      D
   B     A
```

XXXXXXXX	INTERIOR	ROOF	DIV. A	DIV. B	DIV. C	DIV. D
OFFICER						
TRUCKS						
ENGINES						

A=ATTACK   B=BACKUP   E=EXPOSURE   S=SEARCH   R=ROOF   V=VENT   W=WATER of which companies are at the scene and which are not. From this information, staging area resources can be ascertained as well as provisions for task force planning and further resource requirements.

Remember that no tactical worksheet can be used successfully if chief officers do not use it and continuously update it. It is like a computer. If garbage is input, garbage will be the result. This will likely be evident, as it will translate to the fire scene, which is likely to be out of control.

A tactical worksheet must be used at the routine incident if it is to be effective at the major emergency. *(Ron Jeffers, NJMFPA)*

Span of control

The incident commander must establish an incident command organization that is directly proportional to the emergency at hand. Simply stated: big incidents require big incident command organizations. The incident commander who refuses to delegate will find himself quickly overwhelmed. This not only creates a coordination problem, but also a safety hazard. The best way to provide for the safety of operating personnel at the scene of a large emergency is to decentralize the fireground. This entails the designation of an operations officer and individual sectors (also called divisions) to address the various areas of operation. The question must arise as to when this decentralization and delegation should take place. The answer is as soon as the span of control of the incident commander is exceeded or is predicted to be exceeded in the very near future.

The ideal span of control, most sources will tell you, is five subordinates to one supervisor, with this span ranging anywhere from three to seven subordinates per supervisor. When additional alarms are struck and companies start reporting to the command post for assignment, it is time to decentralize the fireground. Designating task forces, which are a combination of resources assem-

When companies are operating in a safe area such as out of the collapse zone during a defensive operation, span of control may be increased. Here, one officer is used to supervise six firefighters on two exterior lines. *(Bob Scollan, NJMFPA)*

bled together for a particular tactical need, is an excellent way of making the fireground more manageable. An example of a task force would be two engines and a ladder company sent to operate in an exposure under the supervision of a single officer. The incident commander or his designated operations officer only speaks to the either the task force leader or the chief officer (division supervisor) assigned to that particular area. This reduces radio traffic, redundant orders, and assigns accountability for a particular area to one person.

Span of control parameters should be dependent on the nature of the incident. The more danger in the particular operation, the narrower the span of control should be. For an extremely hazardous operation (such as a confined space rescue), the span of control may be limited to a one-to-one supervisor-to-subordinate span of control. This is due to the limited area of operation as well as the inherent danger of confined space operations. On the other hand, a single officer supervising three three-man teams operating exposure lines out of the collapse zone at the rear of a fire building may be acceptable. The span of control here is nine-to-one, however, the strict control over the position of the lines allows wider span of control. When in doubt, decentralize and delegate. Being overwhelmed on the fireground is no fun.

Battalion chief position and duties at fires

Departments that have the luxury of two chief officers on the response of a reported fire or a second chief officer showing up at the scene in the initial stages of the operation should utilize this upper level of fireground management to provide early decentralization of the fireground. It is important that this "second-in-command" member be utilized properly.

At smaller or escalating incidents, incident command may transfer several times as companies arrive on the scene. The initial-arriving officer should surrender command operations to the battalion chief, who should, in a still-escalating incident, subsequently hand it over to the deputy chief. Once command operations have been properly transferred, the battalion chief can be of great value to the incident commander. Unless the block is burning to the ground, the battalion chief is much better utilized doing something productive. I have seen some battalion chiefs stand with the incident commander at the command post. This is a useless place to be standing. He is no more than a highly paid aide. The battalion chief can be of great assistance to the operation by acting as the eyes of the incident commander in areas other than the command post.

The battalion chief should be assigned what I like to call "Roving Recon". At exterior, defensive incidents, this should entail going to problem areas such as the rear, the roof of exposures, or into the exposures themselves. In this way, the burden of reporting conditions on the personnel operating in those areas of reporting conditions is somewhat lessened.

On interior, offensive incidents, the battalion chief should don protective equipment and head into the building. The battalion chief can be of great assistance to the incident commander by operating as the interior division

The "Roving Recon" concept provides decentralization and allows company officers to better supervise their subordinates and focus on their assigned tasks, increasing the safety of all personnel.
(Ron Jeffers, NJMFPA)

For offensive structural fires, battalion chiefs, when relieved of incident command operations by a higher-ranking officer, are best utilized in the control of interior operations. A first-hand account of conditions from a "hands-off" member can be issued to incident command. Face-to-face communication is best. *(Ron Jeffers, NJMFPA)*

supervisor, providing early information on the location and extent of the fire, continuously evaluating the attack operation, and relaying information back to the incident commander. In multiple dwellings members tend to bunch up on the stairway in an attempt to get into the fire area. The battalion chief can help coordinate this area and keep the stairway free of too many personnel. The battalion can be most effective by standing back, seeing the "big picture", and, based on his observations and on reports issued from other areas of the building, get the best "feel" for the fire. This interior position close to the action, but not in the way, can also allow the battalion chief to act as an inside safety officer as well as the interior division supervisor

At very large incidents, an effective assignment for the battalion chief may be that of the operations officer. This will take the burden of directing tactical actions off the incident commander so he can focus on more critical strategic decisions. The battalion chief assigned to operations officer duties runs the incident while the incident commander builds and manages the organization designed to mitigate the incident and support the strategy. Here, due to necessity based on the need for proximity, he will operate close to the command post.

The safety officer

The safety officer monitors and assesses safety hazards and unsafe situations and develops measures for ensuring personnel safety. The safety officer has the emergency authority to alter, suspend, or terminate any operation he deems hazardous to the operating personnel. His authority is on par with the incident commander. Because the incident commander has the responsibility for everyone and everything that happens on the fireground, it is critical that the department or the incident commander assigns someone who is reliable, safety-conscious, and fireground experienced. He can delegate the authority for the safety officer to operate, but he must retain the responsibility for any actions of the safety officer.

The department should adopt standard operating procedures that address the duties, responsibilities, and the authority of the safety officer so that no ambiguity exists as to the scope of his power. The scope of these roles are addressed in NFPA 1521 Standard for Fire Department Safety Officer, developed by the National Fire Academy.

All personnel must be made to understand that the role of the safety officer is not that of a "safety cop". He should not be made to micromanage a fire scene. This means that obvious safety actions such as the proper wearing of personal protective equipment and operating within the parameters set forth by department SOPs should not have to be addressed on the fireground. Disciplined officers supervising disciplined firefighters will make the safety officer's job more legitimate, which in turn will allow him to focus on greater safety issues, making the fireground an overall safer place. How can the safety officer, who has to spend most of his time telling the men to get dressed properly, watch for signs of building failure or other dangerous conditions? It would be a crime if the safety officer were telling a company to put their masks on while the building is collapsing on personnel in another area of the fireground.

One tool available to the safety officer to assist him in creating a safer fire scene is the Fireground Safety Officer Checklist (see exhibit 3, p. 406). This form is similar to a tactical worksheet, but is modified to fit the concerns of the safety officer. On the sheet are some of the factors that the safety officer should be addressing when assessing such conditions as the operational effect on the fire building, both on the interior and the exterior, as well as overall fire operations. As with the Fireground Tactical Checklist, it can be used as a "tickler" file to prompt the safety officer to check certain conditions and allow him to do a more comprehensive job. This form and notes written on it can also assist in the safety officer portion of the post-incident critique.

FIREGROUND SAFETY OFFICER CHECKLIST

BUILDING:
- [ ] ROOF CONSTRUCTION
- [ ] FACADE
- [ ] MACHINERY, HEAVY EQUIPMENT, AIR CONDITIONERS
- [ ] BALCONY, OVERHANG, ETC.
- [ ] OVERALL BUILDING CONSTRUCTION
- [ ] CRACKS IN WALLS
- [ ] LEANING WALLS
- [ ] FOUNDATION
- [ ] FIRE ESCAPE CONDITION
- [ ] WINDOW LOCATION AND CONDITION
- [ ] FRONT, REAR, SIDES OF BUILDING
- [ ] STEPS
- [ ] INTERIOR STAIRS

FIRE OPERATIONS:
- [ ] FAST TEAM IN PLACE AND READY
- [ ] ACCOUNTABILITY AND MAC CARDS
- [ ] ACCESS TO AND FROM ROOF
- [ ] ACCESS TO AND FROM BUILDING
- [ ] WATER RUNOFF
- [ ] VENTILATION
- [ ] COLOR OF SMOKE
- [ ] BACKDRAFT POTENTIAL
- [ ] BREATHING AIR SUPPLY
- [ ] LIGHTING
- [ ] REHAB
- [ ] BUILDING LADDERED
- [ ] ROLL CALL

EXTERIOR:
- [ ] FIRE LINES
- [ ] POLICE AND TRAFFIC CONTROL
- [ ] CROWD CONTROL
- [ ] STREET CONDITIONS—ICE, SNOW,
- [ ] CLIMATE PROBLEMS—WIND, COLD, RAIN

Exhibit 3

The FAST team

In incidents where there is an unusually large commitment of manpower on the scene, consider requesting two or more FAST teams. Consider also having the FAST team don one-hour SCBA cylinders in a large area building. The rationale here is that if the firefighters requiring rescue are wearing 30 minute cylinders and they can't get out, how will the second team, with the same cylinders get them out in time without falling victim to the same predicament?

The FAST team, in addition to standing by at the ready, can perform specific tasks aimed at increasing firefighter safety. Such actions include, but are not limited to:

- Sizing up entry and exit points from the building
- Staging building construction-specific tools for firefighter rescue applications
- Ensuring the area is well lit
- Raising ground ladders at the egress points on upper floors where searching members may need to make a quick exit
- Assisting in the set-up and monitoring of the command board
- Forcing doors not opened by the initial entry team
- Making available a dedicated FAST team hoseline and staging same in an easily accessible area

The FAST Team must have access to equipment. Dedicated compartments are one option for quick access to FAST Team equipment.

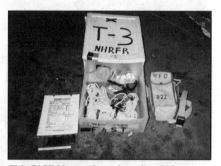

This FAST kit consists of armbands for identifying FAST Team members as well as rope-guided search bags and tethers. In addition, there is a Hazardous Area Access Form and clock mounted to a clipboard to provide for safe and organized control of a rope-guided search operation (see exibit F, p. 409).

FAST Team equipment should be gathered and staged in one place on the fireground. A dedicated FAST tarp is one way of organizing this equipment. Mandatory equipment can be listed on the tarp in permanent marker.

While these tasks are essential to ensuring the quickest ways of entry and egress are assured, the FAST team must not get sidetracked and involved in firefighting. Once these safety-oriented tasks are complete, the team must go back to the command post and stand by. It is also imperative that the FAST team maintains an ongoing interest in the command board operation so that the most up-to-date company location information is known.

The FAST team should liaison with both the safety officer and the incident commander to best have their fingers on the pulsebeat of the operation. In addition, monitoring of the portable radio, thecommand board, and any tactical worksheets used is mandatory for FAST teams. *(Bob Scollan, NJMFPA)*

Perimeter control – use of barrier tape

Much like control zones in a hazardous materials incident, operational work areas must be established to control access into the emergency environment. A system must be established to signify what dangers exist on the fireground. It is important for the department to support and enforce this and all safety procedures and that all firefighters are made aware of these procedures. It will likely be a duty of the safety officer to set the parameters for these zones based on current and forecasted conditions. If a safety officer is not designated, it will be the duty of the incident commander or his assigned designee.

Barrier tape has long been used to keep people out of the way of fire department operations as well as isolate dangerous conditions that the public should be aware of and steer clear of. This same barrier tape can be utilized to identify areas that firefighters should not enter. The manner in which the barrier tape is strung should also have special meaning to operating personnel. A suggested system follows.

HAZARDOUS AREA ACCESS FORM
DEPARTMENT: _____

DATE: _____

LOCATION OF INCIDENT: _____

CONTROL MAN: _____

ENTRY LIMITATION POINTS: 30 MIN. CYL. = 10 MIN.
                                        60 MIN. CYL. = 20 MIN.

MAIN SEARCH TEAM: (DEPT./COMP.) _____

	SCBA	SIZE (CIRCLE)
LEADER _____	30	60
MEMBER _____	30	60
MEMBER _____	30	60
MEMBER _____	30	60

ENTRY TIME: _____

ENTRY LIMIT POINT TIME: _____

    NOTIFICATION MADE: _____ REC'D _____

                    CHECK OFF   CHECK OFF

EXIT TIME: _____

BACK-UP SEARCH TEAM: (DEPT./COMP.) _____

	SCBA	SIZE (CIRCLE)
LEADER _____	30	60
MEMBER _____	30	60
MEMBER _____	30	60
MEMBER _____	30	60

ENTRY TIME: _____

ENTRY LIMIT POINT TIME: _____

    NOTIFICATION MADE: _____ REC'D _____

                    CHECK OFF   CHECK OFF

EXIT TIME: _____

SEARCH RESULTS: _____

_____

_____

_____

ROPE-GUIDED SEARCH
SOPRGSCH.DOC

Exhibit F

Fireground Strategies

There are more civilians than emergency personnel in front of this fire building. Incident command must ensure a perimeter is established to keep unnecessary persons out of the area of operation. The authority to designate this perimeter should be delegated. *(Bob Scollan, NJMFPA)*

Unrestricted firefighter access. A single length of tape at chest level (approximately 4' or 5') allows firefighters to enter the area, but prohibits civilian entry. This is the most common type of perimeter control used on the fireground. This is akin to the boundary designating the Public Area from the Cold Zone at a Haz Mat incident.

FIRE LINE DO NOT CROSS FIRE LINE DO NOT CROSS FIRE LINE

Limited firefighter access. Parallel dual lengths of tape indicates a limited access or entry point. It may indicate an area of limited collapse potential or other limited personnel condition. This is used when the hazard necessitates that only personnel essential to the operation enter the area. All other personnel must remain out of the area. A controlled access/entry point, similar to the "gatekeeper" position at the Hot Line at a hazardous materials incident, must be established as the only access route into the area inside the barrier tape. The safety officer or his appointed designee must be stationed continuously at this access point. Some type of accountability system should be in place to keep track of who is in the Hot Zone and how long they have been operating there. Operations may still continue outside the barrier tape, but personnel must recognize that a safety zone has been established for a specific reason.

FIRE LINE DO NOT CROSS FIRE LINE DO NOT CROSS FIRE LINE

FIRE LINE DO NOT CROSS FIRE LINE DO NOT CROSS FIRE LINE

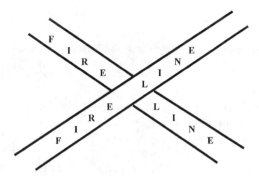

Prohibited firefighter access. Double lengths of barrier tape run in a crisscrossed fashion indicates an area where no personnel are permitted under any circumstance. This condition will most likely be utilized in a collapse zone where building collapse is possible, likely, or has already happened. All operations must be conducted outside the established boundary line. The establishment of this zone should be announced over the air to all companies as an emergency transmission.

Building search considerations

Bread and butter operations are the backbone of the fire service. Primary search, ground ladder raises, hose stretching, and other routine Firefighter I tactics are the everyday tasks by which the job gets done. It is, therefore, not surprising that they are also related to the most fireground casualties. Outside of cardiac-related deaths, more firefighters are probably killed and injured performing these standard operations than all other casualty-causing factors combined

I was fortunate enough to be involved with some very good instructors in teaching the Ladder Company Operations Primary Search station at the Hands-On Training (HOT) sessions at FDIC in 1999 and 2000 in Indianapolis, and at FDIC West in Sacramento in April, 2000. Briefly, the training was set up as follows. Headed by Lead Instructor Mike Nasta, a Captain for the Newark (NJ) Fire Department and an instructor at the Bergen County (NJ) Fire Academy, we were given dedicated buildings to modify at our will. Using saws, sledgehammers, nails, and lots of diabolical creativity, we assembled a mask confidence course that was a challenge to even the most experienced professional.

Our dedicated buildings were abandoned, so we set out to recreate a vacant building condition. At the mask confidence station, we created numerous firefighter traps inherent to this type of building. Included were loose and hanging electrical wire and mini-blind obstructions, holes in the floors and walls, missing stairs, collapsing ceilings, and an attitude-adjustment device for those firefighters who were less than receptive to what we were trying to accomplish. We had a great time both creating and delivering the course.

Students wore full protective gear, including SCBA. The firefighters were staged outside the building and given minimal instruction other than caution about abandoned building dangers. Members negotiated the course alone, not in pairs.

The attitude adjustment device used during the mask confidence course was unleashed upon those who were less than receptive to the objectives we were trying to accomplish. It's never a good idea to mess with the instructor, especially if there is a hood over your head.

The objective, they were told, was to get out, not to find a victim. This was a "rat in the maze" scenario and the only way out was forward. To complicate matters more, we pulled their protective hood over their face piece, completely obscuring vision. Students were not required to go on-air, just to wear the SCBA. This was because the length of the course almost guaranteed that no one would finish with any air left. After the evolution was complete, the students were guided through the course by an instructor, pointing out and explaining specific key points we wanted to reinforce. It was also an opportunity to answer any questions the students had.

After delivering this course to a couple of hundred firefighters, we noticed that the same mistakes happened over and over no matter what part of the country members were from. The objective of this section is to point those out and explain some key points about each that may help in overcoming these problems in the future. Some are "obvious" fireground procedures, as we also thought. However, it is apparent, that once you pull a hood over someone's' face and put him or her into an unfamiliar environment, sometimes common sense safety practices are forgotten and unsound practices are the result.

Firefighters must not only know safe search procedures, but they must be able to overcome difficult situations with confidence and skill. The following are remedies and common sense procedures to the common mistakes observed. I have included them in this book because they are critical to fireground safety.

Wear your protective equipment properly. This is something that should go without saying. In the course, anyone who was wearing gear improperly suffered. One of my pet peeves is wearing the chinstrap around the back of the helmet. I see this all the time and can never understand it. I have observed that most firefighters who wear their helmet this way have to constantly readjust or pick it up when it falls off. When crawling, either it falls off when you look down due to gravity or it falls off when you look up because the SCBA hits it. These guys spend more time fixing their equipment than getting any actual tasks done. Anyone wearing their helmet in this manner had it continuously poked off his or her head by a Halligan hook. Eventually, if the lesson was not learned and the chinstrap not properly adjusted after this harassment, the helmet was knocked into a hole in the floor where it wound up on the floor below. Anytime you lose your helmet, you are both a liability to yourself and to your fellow firefighters.

Another condition we observed was the omittance of fastening the SCBA waist strap. We had a makeshift lasso in one course that caught the strap. The harder the firefighter pulled, the tighter it got. They "died" right there. We also grabbed it and had the student drag us around until he realized that he was caught. As soon as he stopped, we dropped the strap. If they were smart and fastened the strap, they learned the lesson. If they were not and left it dangling, we continued the pestering. Minimize your problems by wearing your protective equipment properly.

Whether a firefighter is operating inside a structure or outside makes no difference. Proper protective gear must be worn at all times. Here, both firefighters are without SCBA and helmets. Incredibly, the firefighter operating the saw has no gloves on either. This is asking for trouble. *(Bob Scollan, NJMFPA)*

If you can't see – crawl. This one sounds like a no-brainer, especially if there is a hood over your mask. In Indianapolis, the course was laid out on the second floor of a row house. To access the second floor, the student started at the stairs. I cannot count how many firefighters I had to tell to get down on their knees before attempting to climb the stairs. There were a few I could not tell fast enough. They wound up in a hole created by two missing steps in the first flight. One firefighter actually had to be extricated from under the stairs; only his feet were sticking out. I think they're still looking for his tool.

Probe ahead of you. That is what the tool is primarily used for once you are inside. I saw some students just dragging the tool around. If you are not going to use it to your advantage, leave it outside. Then, stay outside with it because a ladder man without a tool is useless. I saw other students swinging it around like a baseball bat. Don't turn a rescue into a recovery by burying a tool in someone's body. They generally don't take too kindly to this treatment. For axes, hold the head and probe with the handle. With Halligan tools, probe with the fork end; hold the adz and spike in your hand. Use it to both extend your reach and to check stability.

Immediately upon passing the obstacle and redonning the SCBA, the firefighter must not move forward until he has regained possession of the tool and probed ahead for hazards. A premature move here can cause a firefighter to fall through this floor.

Sound all areas for stability. All firefighters who are not carrying hose must carry a tool. A firefighter who enters a building without a tool is nothing more than a highly informed bystander. Not only should you carry a tool, but you should use it as an extension of your reach. Whenever you enter a new area, which may be recognized by feeling a doorway or a change in floor covering or elevation, the area must be sounded for stability. The same goes for every time you enter a window. Part of the course included a mock window leading to a simulated peaked roof such as a porch roof. Many firefighters climbed onto this incline without first sounding for stability. We explained that it could be their last act as a firefighter

Don't lose contact with your tool. Try breaching a wall with your hands or feet. It can be done, but it is always easier and safer with a tool. This is especially true once you pass an obstacle such as one requiring a reduced profile maneuver. Keep the tool in close proximity to yourself at all times. The tool can be the difference between getting out and staying trapped. In the rooms where the floor was removed, any firefighter who did not show respect for tool possession lost it in the hole. Gravity never takes a day off. If the tool is the first thing to go, guess what's next?

Maintain contact with the wall and use a consistent search pattern. Once you leave the wall, you will rapidly lose your bearings. Even if conditions are decent when you enter an area, it is no guarantee that they will stay that way. Staying on the wall will allow you to keep your bearings. Openings in a wall are usually a way out of the area (doors, windows, balconies, etc.). Any openings in a room that are not on the wall usually spell trouble (holes in the floor, dip tanks, unprotected shafts, etc.). Once you have learned to stay on the wall, you must use a consistent search pattern. If you start out on the right wall, stay right. The same procedure should be followed with a left wall start. Changing your search pattern in the middle of the search can confuse you and cause you to not only move into already searched areas, wasting precious time, but cause you to become disoriented and lost.

Count walls as you move. This is a way of orienting yourself with your surroundings. Most rooms that I know of have four walls. If you are looking for a way out and you wind up on the fifth wall of what seems like a square room, chances are you missed an opening. This is also a good rule of thumb to keep in mind when you find a victim. If the victim is found, knowing which wall you are on will influence your decision on whether to keep going forward or to retrace your steps back the way you came. Finding a victim on the first or beginning of the second wall may make it quicker to retrace your way back to the door. However, if the victim is found further into the room, say on the third or fourth wall, it may be more efficient to keep going until you get back to the opening you came in from. In this case, going back the way you came in may take too much time, wearing you out and minimizing victim survival time.

Know the difference between a wall and a door. The obvious difference is the presence of hinges and a doorknob, along with some sort of molding or door stop. Another difference is the amount of resistance you receive when coming into contact with the item. Generally, a wall is solid and has no give whatsoever. A door, on the other hand, even the most solidly reinforced doors, will have a little "give" when pushed. If you feel this give, feel further for the hinges and doorknob. This guideline will work even in the thickest smoke.

Make use of cues to orient yourself to the room. There are many cues you can use to orient yourself. If disoriented in a room that has a wood floor, feel for the seams in the floor. Following the seams will lead you to a wall, most likely a non-bearing wall since the flooring is laid perpendicular to the floor joists. Many non-bearing walls are where window and doors are located. In multiple dwellings, the bearing walls are those opposite walls that are closest to each other and usually have the longest length. The shorter walls, at the front and the rear are non-bearing and will usually have most of the doors and windows in them. In the suburbs, where homes are larger, the opposite may be true. The bearing walls are usually the walls that face the street, so you may have to travel parallel or across the seams to reach a doorway. In either case, following the seams will lead you to a wall, which will help you orient yourself.

Other cues to take notice of are molding on walls and floors. These usually indicate a room change. A short protrusion found on a wall could mean that you are in an archway and about to go into another room. If you are searching with a partner who is on the wall on the other side of the room, this is where you can lose him. Reach out as far as you can, using the reach of your tool, to see if you feel the other side of the arch. Another very reliable sign of a room change is when you go from rug to wood, tile, or other floor covering or vice-versa. Be aware of what you are crawling on.

Doors that open toward you often lead either into closets or to an elevation change, such as this door leading to a basement. Often these areas will also have a change in flooring, even before the stairs drop off. Always probe ahead of you.

As far as doors go, doors that open toward you when you are inside a residential structure usually mean that you are either about to enter a closet or are about to change level. Cellar and basement doors, as well as doors leading up to attics will all swing toward the firefighter. Most doors that open away from you lead into other rooms, but always stay alert for surprises.

Make sure that the wall you choose is breachable. If you are in a jam and it is necessary to breach a wall, make sure that you don't do a lot of work for nothing. Take your tool and blast it all the way through the wall right up to the handle. Do this four times to mark the outline of your hole. If you can do this and do not hit anything, go ahead and breach the wall. If you hit something, move over a little. There may be a dresser, cabinets, or something large and heavy on the other side that will prevent your escape. Take a couple of seconds to check this out. You will be glad you did.

Know how to manipulate your SCBA to pass obstacles. Firefighters must know how to get past obstacles and free themselves when snagged on something. Knowing the reduced profile maneuver for limited clearances and the quick release for SCBA entanglements can be a lifesaver. Two things are critical here. The first is that the firefighter should never lose orientation with the SCBA no matter to what degree he removes it. In certain instances, it may be necessary to take the entire unit off and slide it ahead of you through a reduced clearance area. If shoulder strap orientation is lost, the unit may be re-donned upside down, backwards, or even worse, may pull the face piece off. The specific strap (right or left) which you must maintain contact with will vary with the manufacturer. Generally, the strap on which the low-pressure hose

SCBA manipulation procedures, such as the reduced profile maneuver, must be practiced so that they will become second nature when needed, reducing the chance of a costly mistake.

is attached to the pressure reducer is the strap you must maintain contact with. Second, take only as much of the unit off as is necessary to pass the obstacle or clear the snag. The more you remove, the more likely you are to encounter a problem re-donning. Keep it as simple as possible given the situation and conditions. These maneuvers must be practiced.

Not all exits are at floor level. It is important to recognize this fact when searching, especially if conditions deteriorate. Not only can windows be used for escape, but also for orientation. Even in relatively heavy smoke, once you get near it, you can sometimes differentiate a window from a wall. Sunlight during the day and apparatus lights at night may show the presence of a window. In addition, knowing where a window is in relation to a door can help you get back to that door. However, if you are crawling and feeling the wall only to head level, the presence of a window will not matter because you will miss it every time. In one room in the course, the only way out was through a window mock-up. Some students kept going around in circles in this room until they were told to feel up toward the ceiling instead of back and forth across the wall at the two-foot level.

Use a "swimming motion" to clear hanging obstacles. We observed that when a firefighter got caught on the numerous wires strung around the course, they would usually back-up to clear the obstacle and then continue through only to be re-snagged either on their helmet or their SCBA cylinder. The key here is to reduce your profile by turning your cylinder down toward the floor. Back up until you no longer feel the snag. Drop low, sliding your

hand along the floor while turning your body sideways in a swimming motion. This moves your cylinder and harness to the side, toward the floor. Then reach up as you would when following through on a swimming stroke, and push the obstacles behind you while "swimming" forward. Be sure to sweep wires away with the back of the hand. This will prevent the "grab" reflex from grasping live wires.

When using the "swim" motion to pass an obstacle, note the relationship of the cylinder with the floor. There is much less chance of snagging the SCBA if the body is turned sideways. The upper hand should not be grasping the wires. The wires should be swept out of the way with the back of the hand.

Don't go anywhere headfirst if you don't have to. The head is the "brain bucket". If it is injured in any way, the chances for survival are severely minimized. Passing obstacles, entering and exiting windows, and going down stairs are all best negotiated feet first. This direction of movement allows the area to be tested with the foot first before continuing forward. Remember to keep most of the weight on the rear or non-leading foot while checking for stability. In addition, test with your tool first, and then test again with your foot as the area is entered. We saw many firefighters passing between wall studs headfirst and losing their helmets or nearly strangling themselves with their SCBA harness or low pressure hose. Both ways (head and feet first) should be practiced, but always opt for the safest method as conditions warrant.

Carry and use a "magic rope" when required. The "magic rope" was a term we used when I was a firefighter working on the ladder company. It was simply a 25' length of small diameter utility rope, stuffed in the turnout coat pocket. Later, as we got more "sophisticated," snap hooks were placed on the end of one side of the rope to make the operation more flexible. The "magic rope" was not used as a "bailout rope," but utilized to assist in the many functions required of ladder company personnel. The use of the rope was only limited to the user's imagination. One use, applicable here, was when searching alone as the OV (outside vent) position. It was very useful to tie it off to a fire escape and use it as a lifeline into a smoke-filled apartment. The rope can be used as an umbilical to the exterior. Other applications were to tie tools from the roof to vent top floor windows, holding doors open when required, and tying off ground ladders at an upper floor window. There is no room for "functional fixity" if you want to be an effective ladder man.

If you have to bail out onto a peaked roof, use your tool as a lifeline. This is another reason for holding onto your tool. Suppose conditions were to deteriorate and you found you had to leave via a window. Many upper floor windows will open onto a peaked roof such as a hip, porch or shed roof. If you properly sized-up the structure before you entered it, you would know this. If luck is on your side, there will be a ladder there. However, the ladder may be at the end of the peaked roof. These roofs, if they are slate or tile, can become extremely slippery, especially if wet. In addition, all roofs can be icy. I remember one time at a winter rooming house fire almost sliding right off a roof from the third floor. As I exited the roof of the fire building onto an adjacent roof, I began sliding on the "black" ice down toward the roof edge. Luckily, I was able to grab the top of the fire escape ladder just before I was about to go over the side.

When coming out a window, the tool can be hooked over the sill and used to control the descent of the firefighter as he slides down the roof to the ladder. Even if there is no ladder, a firefighter can hold onto the tool and remain on the roof, lower than any venting flames for a relatively extended period of time, at least until a ladder is raised to get him or her down. Having the tool rather than a hand hooked over the sill will allow the firefighter to wait for rescue at a safe distance below heat and danger.

Maintain your composure. Several firefighters panicked and ripped their masks off when they became entangled. This was a controlled environment. I doubt very much that the fireground will be sympathetic to a firefighter who loses his cool like that. Relax and think. Control your breathing.

The handle of a tool can be used as an improvised lifeline when forced to bailout a window onto a low roof. The working end is hooked over the sill, allowing the firefighter to wait in an area of relative safety until a ground ladder can be brought to the area. This is a last resort operation and does not replace the requirement for laddering all sides of the building.

Once your composure is lost, the next thing may be your life.

Get in shape. There were many firefighters who were completely out of breath less than a third of the way through the course. Others had to stop and take a breather or take off their mask before finishing the course. How can a poorly conditioned firefighter expect to save a victim when it is likely he will become one himself? Firefighting is a job for those in excellent cardiovascular condition. If you are not, you are not only putting a burden on yourself, but also on your fellow firefighters. Heart attacks are the #1 killer of firefighters. Don't let yourself become a statistic.

Thermal imaging camera considerations

Thermal imaging technology is the latest craze in the fire service. It has many beneficial attributes aimed at increasing firefighter safety. This technology has been around for years in the military and the law enforcement agency. It can be used for many purposes. Other than locating downed firefighters, its most valuable function, the camera can also be used to locate fire victims, detect hidden fire during both recon and overhaul stages, search light fixtures for a defective ballast, and a myriad of other functions. It can also be used in a missing person incident. An emergency responder with a thermal imaging camera can scan a large wooded area much faster than a team of men canvassing this same area. If this can be done from a surveillance aircraft, the search can be narrowed tremendously in both time and manpower.

Personnel using these devices must understand that they do not replace the requirement to practice established and safe operating procedures. Many firefighters who are uninformed think that this camera frees them to wander around the building without limitations. This is a potentially deadly routine to fall into. These cameras, like any other piece of equipment, are subject to failure. Imagine using the camera in the middle of a large, open, smoke-filled area. Visibility of the area with the device is excellent. Imagine the instrument malfunctions. The firefighter will now be disoriented and out of touch with a wall, his best path to safety and egress from the area. It is imperative, even with the camera operating, to maintain common sense search procedures. This entails staying on the wall, setting up lights at the entry point, or using a guide rope.

Firefighters must avoid a false sense of security just because they can "see" in a visibility-obscured, IDLH atmosphere. The camera reads temperature differentials, causing warmer objects to appear bright. The hotter the object, the brighter it appears on the screen. The

These missing steps may not be recognized through a thermal-imaging camera. Although you can "see" through the device, common sense safety precautions in areas of poor visibility must be observed.
(Lt. Joe Berchtold, Teaneck, New Jersey Fire)

question must be asked, "What about holes in the floor and missing stairs?" Does the camera pick these danger spots up as well? If they are not emitting heat, they may not appear on the camera and a firefighter who is not observing safe search procedures may fall through the floor or stairs. Unless the camera has the ability to allow the firefighter to fly, he may be trapped. Don't let your guard down just because you have the latest technology. The Titanic used the latest technology too and still managed to hit an iceberg.

Progress in the fire service in the field of technology is invaluable, however, it must be tempered with caution and investigation. Whenever a new device or method is introduced to the fire department, a comprehensive fact-finding mission as well as a review of current procedures should be implemented to ascertain how it will fit into the department's operating scheme. A trial period should be identified and training should be scheduled for all personnel. Input should be invited on how to best utilize the equipment and update standard operating procedures.

Master stream considerations

So, your initial strategy did not go quite as planned. Don't take it personally. You didn't start the fire. The time has come to switch your strategy to defensive and utilize master streams. The incident commander must realize that just because personnel are operating in defensive manner outside the building, they are not totally safe from the hazards of the incident. Master stream operations have killed and injured firefighters and caused millions in damage to firefighting equipment. Firefighters must be aware of the dangers associated with operating master streams. Some common dangers of defensive operations are as follows:

Announce strategy changes to all personnel. This may be part of an emergency transmission (see next section) or a progress report. In either case, the mode of operation should be made known to all the participants. It is dangerous and an indication of a loss of control to have men operating streams in the same area on both the interior and the exterior of the fire building.

Ensure control zones are established. Like the strategy change, ensure all personnel are aware of where they can operate and where they cannot. Make use of barrier tape as discussed previously. Reinforcing safety zones with radio announcements will best ensure that personnel are aware of safety boundaries at the incident. This may be the duty of the safety officer or other designee of the incident commander.

When shifting from an offensive to a defensive strategy, discipline and coordination is critical. All personnel must be removed from inside and around the building, accounted for, and reassigned.
(Bob Scollan, NJMFPA)

Establish and maintain collapse zones. This control and safety measure must go hand-in-hand with the establishment of control zones. In fact, the control zones must be established with collapse zones in mind. It has been well established that a collapse zone is considered to be at least the full height of the dangerous wall. It can be argued that because debris can bounce when it hits the ground, a collapse zone of one and one-half times the height of the wall is safer. Either way, the full height of the wall is the absolute minimum and should not be compromised at any time, for any reason by the operating forces. In addition, the horizontal run of the wall must also be considered as some walls are reinforced laterally. This is especially true of parapet walls mounted on top of steel I-beams at strip malls.

Many times, the apparatus is positioned in a safe area, but the aerial device is operating inside the collapse zone. When operating aerial devices, not only must the apparatus be positioned out of the collapse zones, but the aerial device itself (bucket, ladder tip, etc.) must also be clear of the potential collapse area. The device must be positioned away from the wall for a distance at least equal to the height of the wall located above the platform or aerial tip. For example, if a tower ladder bucket is operating in an area where the facing wall is 20' above the platform, then the device must be kept at least 20' away

This aerial device is properly positioned in regard to the wall collapse zone. If the device cannot be raised above the wall, it must be kept away for distance at least equal to the height of the wall above the device.
(Ron Jeffers, NJMFPA)

from the wall. That way, if the wall begins to fall outward, it will not hit the device. In addition, if narrow streets prevent apparatus from positioning out of the collapse zones (i.e., a 50' wall threatening a 30' wide street), apparatus must be positioned to flank the area. Never let your guard down.

Beware of secondary collapse. Be cognizant of what could be compromised when a portion of building fails. Many times, a wall, roof, or other area can fail and strike another structural element, causing a secondary collapse due to the impact of the falling materials or by ripping a structural feature off the building. Beware of signs, power lines and utility poles, exposure roofs lower than the walls of the fire building, and trees. A firefighter was killed when a collapsing wall hit a tree that fell and crushed him.

Remember that collapsing roofs and floors will likely push out bearing walls or make them so unstable that their total collapse must be anticipated. Incident commanders must be prepared for this. If there is no room to establish a proper and safe collapse zone, streams should be operated from a flanking position or be unmanned.

Thin, unprotected steel rods support these signs mounted atop this row of stores. Fire issuing from the storefront or burning through the roof may cause the rods to lose integrity, causing a collapse. *(Bill Tompkins)*

The fact that the floors are collapsing in this heavy timber commercial structure should be a warning that the walls will go next. No personnel should be operating anywhere near this building. With this fire condition, collapse zones should have been established long ago. *(Ron Jeffers, NJMFPA)*

Don't start the master streams until everyone is confirmed out of the area of application. When switching from an offensive to defensive mode, it is imperative that the incident commander confirms all members are out of the area prior to starting the stream. This may require a roll call at the command post or a PAR request from dispatch. Make sure all personnel are well clear of the building and outside established control zones before the stream is applied. Many times, the initial operation of the stream knocks building parts loose as the stream operator adjusts the aim of the nozzle. If members are just coming out of the doorway or still inside the immediate perimeter of the building, they may be struck by falling debris.

Keep firefighters still on the interior out of the operating area. There are times when it is acceptable to allow firefighters to remain in the building when a master stream is operating. This may be the case if a quick knockdown from the exterior is desirable due to factors such as heavy fire, inaccessibility, or a fire accelerated by a strong wind blowing into an apartment. In this last case, the advance into the face of this intense heat may not be possible. Using a master steam from the exterior may be the solution to diminishing some of that heat. In this defensive-offensive strategy, companies must be withdrawn to a safe area before the stream is started. Some places of refuge may be the stairwell below the fire floor or on the floor below the fire. Another area may be behind a fire door on the fire floor. It is critical that the incident commander carefully coordinates defensive-offensive operations. Proper communication between either the incident commander or operations officer on the exterior and the sector or division commander on the interior will avoid master stream related injuries.

Consider where the water is going (or not going). Always take into consideration where runoff water is going and/or accumulating. Master streams dump as much as 1000gpm on the building. That's four tons every minute. Many fires require the use of three or four master streams. That would be sixteen tons every minute. Multiply that by sixty minutes or more and that is a lot of weight for a building that is being ravaged by fire to support. As the incident commander, you should be very interested in the final disposition of this water.

At the height of this fire, approximately 11,000 gallons of water per minute were being discharged into this structure. That's 44 tons per minute! Fortunately, a good bulk of it was draining into the Hudson River. How many buildings could withstand this barrage of water if the drainage profile was not efficient? *(Bob Scollan, NJMFPA)*

Water will behave like fire, heat, and smoke, except it will travel in the opposite direction and take the path of least resistance downward instead of upward. Older buildings, especially heavy timber commercial buildings, have very well built floors. Water accumulation here could be significant. Don't wait for it to be flowing over the windowsills to establish a proper collapse zone. Parapet walls could create a swimming pool condition on the roof, especially if drains and scuppers are clogged. This can cause roof collapse, pancaking floors in a secondary collapse, pushing out walls. Also, be aware of water accumulation inside canopies and marquees. These building features are often hollow. They can be quickly filled with runoff water, precipitating a collapse.

Consider the effect the runoff water may be having on the foundation of the building, especially if the soil in the area is loose. Buildings mired in deep mud and water can list like a ship when water undermines the building. Another problem with mud and runoff are the stability of retaining walls located on sloped land. The runoff water will flow in the path of least resistance, downhill. At one fire, this flow collapsed a block retaining wall several properties downhill of the fire building. The wall fell outward into the street at a 90° angle. Luckily, no one was standing in proximity to it when it came down.

Speaking of lowest level, don't send anyone into the cellar of a building that has been battered with master streams. Different levels of the cellar may not be apparent as water will always look level. One wrong step can lead to disaster. Wait until the water recedes.

It is also important to be aware of the occupancy. Many occupancies are conducive to water absorption due to the nature of the stock. Paper goods, rolled or stacked, textiles, recycling centers, and other such occupancies can absorb water to the point of floor and wall failure.

Another major consideration is the formation of ice. Consider where is it accumulating and what effect it can have on the structural integrity of the building. Ice forming on fire escapes, and cantilevered signs create an increased eccentric load on the wall, which may pull it down. In addition, ice from master streams that coats power lines may cause them to fall. When faced with a large fire on a freezing day, make it a point very early to have the power shut down to the block.

Keep the aerial device away from the fire. This is obvious when operating an aerial platform, but not always observed when directing unmanned elevated master streams such as ladder pipes and Telesquirts. Aerial ladders are usually constructed of all steel, which is very heavy, or a steel and aluminum alloy. The aluminum gives the ladder a lighter weight while the steel gives it strength. All the ladders are treated with a heat-retarding coating for protection. Remember, however, this is still virtually unprotected steel and will react similar to a steel I-beam when directly exposed to flame. Flame will not only destroy the temper of the steel, causing it to weaken, but also may cause it to expand and warp. Warped and expanded ladder sections may not retract or extend. Aerial devices that have been impinged on by fire may malfunction. Fire and heat may negatively affect hydraulic systems, preventing crucial ladder movements. Be aware of where fire is likely to vent and take steps to keep the aerial device well away from it.

The same is true when placing a ladder to the roof for access. If possible, place the ladder to the roof away from windows that may vent while companies are operating on the roof, potentially blocking their escape route. It is for these reasons that two means of egress from the roof are critical.

Use halyards when possible on ladder pipes. There is no reason for a firefighter to be perched on top of an aerial ladder directing a master stream. For one thing, the range of motion of the ladder pipe is limited. It is easier to rotate the turntable than it is to wrestle with the ladder pipe control arm at the tip of the aerial. Second, unlike an aerial platform, there is no continuous air supply on the tip of the aerial. Once your SCBA is empty, you have two choices. Either breathe the smoke, which is unacceptable, or climb back down the aerial to get another cylinder. It is safer and easier to operate the aerial from the turntable. Halyards will allow the nozzle to be operated in the same manner as the firefighter on the tip of the aerial. The difference is that the operator is now on the ground. I have heard firefighters say that you cannot see where the stream is going unless you are on the aerial. This is nonsense. With all the smoke, you can't see anything anyway. You are better off on the ground. In addition, the ladder pipe is used once most of the roof is burned away and the attack is directed from above. It is pretty hard to miss the fire blowing out of a giant fire-

What is more dangerous than one firefighter on a ladder? Two firefighters on the same ladder. This is almost criminal neglect of safety. One firefighter is literally hanging off the side of the ladder.
(Bob Scollan, NJMFPA)

It is much safer to utilize halyards than it is to place a firefighter on the tip of the ladder. While most halyards are operated from the street level, these were operated by personnel on the roof of an exposure building who had a better view of where the stream should be directed. *(Ron Jeffers NJMFPA)*

created roof hole. Other than for a quick recon of the fire area, keep the firefighters on the ground.

Use the proper nozzle and supply it at the proper pressure. Unless you are involved in vapor dispersion or a pressurized tank cooling operation, a master stream from an aerial device should always be a solid bore tip. Ladder pipes and deck guns are usually designed to supply 600gpm, so the tip should be 1½" in diameter. A platform delivering 1000gpm should have a 2" tip. The solid tip is more desirable not only for the advantage it offers in the way of reach and penetration, but also because it requires only 80psi nozzle pressure as compared to 100psi for a fog nozzle. In addition, to reduce friction loss, it is best to keep the supply lines short. A supply line of one hundred feet is best. This is not quite as drastic with large diameter hose, but it is still a good practice to keep the supply lines as short as possible due to excessive friction loss in long hose stretches. I have been told that 5" LDH can flow 1000gpm through 1000' of hose without any friction loss. Using this diameter and larger supply hose is like bringing the water main up on the street. Remember when calculating pump pressures, do not forget to account for friction loss due to elevation, which is the same as for a building, 5psi per 10'. In addition, if there is an appliance involved such as a ladder pipe siamese, add 15psi or whatever the manufacturer suggests to the final number.

There is a ladder pipe rule of thumb that I learned when I first got on the job. It is called the "75-80-85" rule of thumb. The numbers stand for the following: 75° angle for the ladder, 80psi nozzle pressure (solid bore tip), and 85' of elevation. These numbers are the most efficient way of operating a ladder pipe as they put the least amount of stress on the aerial device.

Let's take a look at an example. A ladder pipe with a solid bore nozzle is 85' in the air. The ladder pipe hose is 3". What should the pump pressure be for an engine one hundred feet away supplying the ladder pipe with 5" LDH?

There is a gate valve at the bottom of the aerial that allows the ladder company to control the water and drain the pipe when complete.

We will start off with 80psi for the nozzle. We then add 5psi per every 10' for elevation. That comes to 42.5psi. We will round it to 40psi. Add to that the friction loss in the 3" ladder pipe, which is roughly 35psi. Finally, add 10psi for the gate valve. The friction loss in the 5" LDH flowing 600gpm is negligible. We will therefore ignore it. Adding these figures results in a pump pressure at the Engine of 165psi to deliver 600 gallons per minute to the ladder pipe at the proper pressure.

A final consideration in master stream delivery is matching the tip size to the water available. If you have a 2" tip and only 800 gallons per minute available, the stream will not be as effective if you changed the 2" tip to an 1¾" tip. This smaller tip will match the gallons per minute and produce a more effective stream. An improperly sized tip will provide an ineffective stream.

Anchor down ground-mounted master streams. This is a safety consideration. I have seen firefighters take a ride on an insufficiently secured deluge gun. There are usually some types of securing devices for the appliance, either a skid plate that the appliance is screwed into or metal spikes that are hammered into the ground. There also may be other ways of securing the device. Make sure you are familiar with the set-up and operation and that all safety precautions are taken before charging the appliance.

Some deluge guns are designed to be supplied opposite the discharge direction. Others are designed from the same direction. Unless the appliance is specifically designed to take its supply from the same direction as the stream discharge, it is best to connect your supply lines into the device from a direction opposite the nozzle so the forces of the supply pressure and the nozzle pressure equal out. There are several styles of hose layouts that are utilized to minimize these forces as well. Ensure that assigned firefighters are familiar with the operation of each type of deluge gun your department carries. Do not take the operation for granted, as that is how injuries occur. Take every precaution to ensure the appliance is properly set up before discharging any water.

Hit the fire only. Just like on the interior of the building, master streams should be aimed at fire, not smoke. Once the fire is knocked down in one area, either reposition the stream or shut it down. If you are not hitting fire, you are just adding weight to the structure.

Unless being used for exposure protection, master streams (and any exterior stream for that matter) should be directed inside the building. Avoid splattering the streams off walls, roofs, and other building elements. If visibility is poor, the sound of the stream can be used as an indicator as to whether or not the stream is penetrating the building. The sounds the stream makes are very distinct in regard to how far away actual stream impact is. If the impact sounds close, you are probably not hitting anything but the wall. This will cause a "splattering" or "hammering" sound. Your suspicion may be further confirmed by water running off the side of the building or splashing back toward the apparatus. A master steam penetrating the building's exterior will produce a more distant sound as the stream strikes interior building features. When the stream is moved from window to window, you will hear the different sounds as the stream strikes the wall and then penetrates the next window space. While maintaining control over the aerial device, move the stream as quickly as possible between the openings to avoid unnecessary wall damage. Once streams versus building sounds are recognized, the stream operator will be more efficient in placing the stream, even in limited visibility.

Be careful around parapets, chimneys, and slate/tile roofs. Master streams are high volume, high impact weapons against fire. Not only is the point of impact concentrated, especially at close range, but the mass of water being delivered has significant weight, compounded by forward velocity. Remember that force is equal to mass multiplied by acceleration. Without looking at specific numbers, just consider that formula and the danger of high impact streams becomes readily apparent. In fact, the closer to the wall the stream is, the more intense the impact of the stream will be. High velocity streams can knock slate shingles loose, collapse chimneys and parapets, and send coping stones flying. They also have the power to break loose a freestanding masonry façade. In addition, be cognizant of the depth of the building. A stream can be directed right through the interior and strike the rear wall, collapsing it and the roof into the rear yard.

The operator of the stream will probably not be in the best position to judge the stream. For this reason, it is advantageous to use a spotter to direct the stream. Hand signals between the spotter and the operator should be worked

The nozzle operator may not be in the best position to judge where to direct the stream. For this reason, a spotter should be utilized to guide the operator. This applies not only to aerial devices, but also to deck guns and large diameter exterior handlines.
(Bob Scollan, NJMFPA)

out in advance as the noise of the fireground will likely prevent verbal communication between the two. Also, be aware of personnel in the area of the stream. There may be personnel operating on the roof of an exposure. A misdirected master steam has the power to knock a firefighter right off the roof.

Let the building settle after master stream use. Once master streams are shut down, do not rush right into the building, especially if they have been in operation for a period of time. Give the building a chance to stabilize. This time will also allow some of the water, (hopefully most of it) to run out of the building. Then, allow only a safety or chief officer or both (always at least two members) to enter and assess the structural stability of the building. They should look for signs of structural compromise and indicators that the building is losing the battle against gravity. If it is deemed safe to enter, before allowing anyone inside, an effort should be made to identify and mark unsafe areas. Barrier tape, announcements over the radio, and directly showing the companies who are to go inside where the hazards are, are a few of the ways to make personnel aware of the dangers. In addition, light the area as necessary. Only the minimum number of personnel should enter to finish the task of extinguishment and overhaul. Strict control and supervision should be exercised over this operation. The more hazards, the stricter the control should be.

If building integrity is doubtful, do not enter. There will come a time when entering for overhaul and final extinguishment will not be worth the effort due to the damage inflicted upon the structure by the fire. Remember that no building, especially a burned-out building, is worth the life of a firefighter. In this case, hydraulic overhauling using master streams should be ordered if necessary, a fire watch should be set up and heavy equipment should be called in to pull the building apart for final extinguishment. It will usually be acceptable to leave only an engine company to perform this task. Keeping a large amount of companies and manpower at the scene of a fire watch operation unnecessarily tires the men who have been there a long time already and

Buildings that have suffered serious fires may be deemed unsafe to enter. It may be best to overhaul utilizing a master stream. No building is worth the life of a firefighter.

continues to strip the district and the other districts who are relocated of proper fire protection. Know when it is time to send the troops home and de-escalate the operation.

There may be times where the building is considered safe to enter, but other considerations cause the incident commander to delay the entry. One may be where night is falling and it is getting dark. I have found that it is safer to maintain a fire watch on the overnight with rotating crews and wait for the morning to enter the building. Remember to do another safety assessment in the morning before companies enter the building. Conditions may have changed overnight. Daylight is safer than even the best lighting at night. Another reason to wait to enter the building may be when fresh crews are not available. Sending tired crews into a questionable structure is dangerous and improper. Fatigue-related injuries are preventable injuries. It is better to wait until new crews arrive than to risk injury to exhausted crews, even if it means waiting until the next morning.

Emergency fireground communication

It is imperative that all fire departments have an emergency fireground communications system in place. This includes the procedure for transmitting a Mayday message. It is best to develop joint procedures of this nature between those departments who work together on a routine basis. When

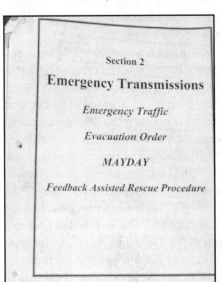

buildings are falling and/or personnel are missing or in distress, SOP-induced control of the fireground will often make the difference between chaos and order.

Emergency and Mayday transmissions are used when a message of an extreme nature is required as in the case of withdrawal and strategy change (top-down) or a downed or trapped firefighter (bottom-up).

There are two general types of emergency transmissions on the fireground:

1. "Top-down" Emergency Transmission (Incident Command to Operating Personnel)
2. "Bottom-up" Emergency Transmission (Operating Personnel to Incident Command)

Top-down emergency transmission. There will be times during the operation where the incident commander needs to transmit a message of urgency to all operating personnel. This may be a warning about a dangerous building condition, a withdrawal from the building, or any other critical message that is essential to the fireground operation. This is called a "top-down" emergency transmission. All firefighters should be aware of when this type of message is about to be transmitted across the air. An emergency transmission should be initiated with a special tone. This tone should be tested each day at as specific time so that firefighters become conditioned to immediately stop what they are doing, listen up, and maintain radio silence when they hear it. In North Hudson, the tones are tested every day in the morning when the house vocal system tones are tested. In Jersey City, it is tested several times a day. The aim is ensure that firefighters are conditioned to take notice of the tone, so no one is caught off guard at the fire scene. An emergency transmission has absolute priority over all other transmissions, no matter where it is originating. However, the use of the "Mayday" transmission, whether it be "top-down" or "bottom-up," must be restricted to those situations where life or personal injury are at stake, or in a critical situation requiring immediate assistance.

In North Hudson, we have been using the same system for about ten years, and it has worked well. After the emergency transmission is initiated by someone on the fireground, usually the incident commander, Fire Control (North Hudson Regional Communications Center) transmits a series of four distinct tones followed by a boilerplate statement announcing the emergency transmission. After the boilerplate announcement, the emergency transmission is repeated several times. An example of an emergency transmission and procedure would be: (series of four tones), "Fire Control to all companies operating on the fireground, stand by for an emergency transmission. By order of Park Avenue Command, all companies operating in the fire building, evacuate the fire building immediately, companies operating in exposures B and D, hold your position. Repeat, (tones again) by order of Park Avenue Command, all companies operating in the fire building, evacuate the fire building immediately, companies operating in exposures B and D, hold your position. 15:33 hours." This message may be repeated several times.

Bottom-up emergency transmission. Just as incident command has a method for transmitting a "top down" emergency transmission, personnel on the fireground and in the fire building must have a way of initiating what is called a "bottom-up" emergency transmission. This is basically a "firefighter down" transmission. A firefighter lost, trapped, in need of some type of assistance, or has the need to get some information to incident command or dispatch should have a system that is recognized by all personnel as an emergency transmission, and get the same priority as a "top-down" emergency transmission.

Again, in North Hudson, a system has been adopted that is tested weekly over the air. The member who needs to send the message initiates the transmission in the following manner.

STEP 1: The member manually activates his PASS unit, causing the PASS alarm to sound.

STEP 2: The portable radio is keyed for transmission.

STEP 3: The keyed radio mike is placed next to the PASS unit for a period of ten (10) seconds. The PASS tone is heard all over the fireground, signaling the initiation of a "bottom-up emergency transmission.

STEP 4: The PASS device is turned back to the "Arm" position.

STEP 5: The emergency message is broadcast using the following format:

Mayday, Repeat, Mayday, this is Firefighter _____. Repeat Mayday, this is Firefighter _____. (Message is then broadcast).

If the emergency transmission is initiated to signal that an entire company is in distress, then the company name is substituted for the individual firefighter.

Neither of these methods is foolproof, but it is a start; an established procedure to allow the incident commander (top-down) and an individual firefighter or officer (bottom-up) to let someone know that something extraordinary or unusual is occurring. It is hoped that these procedures are never used. However, to be of any use, they must be taken seriously, be no secret to anyone, and must be practiced.

Vacant Building Dangers

Vacant and abandoned buildings are just plain dangerous. Not only are they often open and exposed to the elements, but they may also be the site for previous fires. They should be torn down as soon as possible. Fire officials should make it a point to ensure all personnel are aware of which buildings are vacant and what some of the inherent dangers are. Using the aforementioned "Exceptional Response Report" is one method. Another is via dispatch when companies are responding to these addresses. The third way is a symbol system used to alert fire personnel to dangers in and around a building. This will be discussed later in this chapter. Some of the other hazards inherent in vacant structures follow.

Are they really "vacant"?

Many times, vacant buildings are not really vacant. They become shelters for the homeless and havens for drug dealing and prostitution. Homeless people often have few possessions, of which they guard jealously, often to the point of inflicting harm to intruders. Firefighters searching these buildings are often seen as intruders and have had debris thrown at them, been stabbed, shot at, and have become victims of booby traps. The same is true where drugs and/ or money is stashed. Often hypodermic needles and used condoms are strewn about the building. A firefighter crawling in smoke may be exposed to a deadly disease without knowing it. It may be necessary for the nozzle man to sweep the floor with the stream prior to hose advancement in an attempt to clear the area for the attack. Firefighters should wear all protective clothing and not take unnecessary risks in regard to operations in these buildings.

As conditions permit, a primary search must still be conducted in a "vacant" building. Clothes, cooking materials, and makeshift bedding are all indicators that the "vacant" is occupied and a rescue problem may be present.

Illegal manufacturing

Illegal manufacturing used to mean drugs and illegal drug labs. Firefighting personnel must be aware that other illegal manufacturing may be going on in these buildings as well. Vacant buildings may be used by terrorists to produce bombs, and other weapons such as biological and chemical agents. The bomb that was detonated at the World Trade center in New York City in 1993 was reported to have been manufactured in a vacant building in Jersey City. Even in the raw form, the substances used to design these weapons are dangerous.

Buildings used to produce and store terrorist items may be booby-trapped to kill or maim law enforcement personnel. Sadly, firefighters are often the recipients of these traps. These may include holes in the floor and missing stairs covered over with carpets to hide their presence, pressure-release spikes and firearms such as those used in war zones, razor blades tainted with poison or deadly chemicals imbedded into the banisters, electrified fire escapes, and countless other diabolical measures aimed at keeping authorities out of the area.

If entry must be made, it should be with extreme caution. Any evidence that illegal activity is being conducted at the building must cause an immediate shift in strategy to exterior, shielded, defensive positions.

Dogs

Not only are dogs used to protect homeless people, but are also routinely used by drug dealers to keep unwanted intruders out. Generally, these dogs are of the large and nasty variety. Not only is rabies a danger, but you may not hear them attacking at all. Often, dealers take out their voice boxes to prevent growling and barking. In addition, the claws are also removed to eliminate the sound of the paws on the wood floors. The result is a large, vicious animal in your face without warning. By then, it is too late to avoid serious injury. Be aware that dogs may not be the only deadly threat. Poisonous snakes, frogs, and spiders as well as other ferocious and exotic animals may replace dogs and be even more deadly to responders.

Illegal utility connections

When sizing up a vacant structure, look around to see if any wires are strung from one building to another. Vacant buildings are often powered by stealing electricity and gas from an adjacent building. Be sure to check enclosed light and airshafts. Because this area is not visible from the street, it is a favorite place to run illegal zip cords from one building to another. Other times, the cords are run from one cellar window to another, where the wiring is then run through the floor to the apartments above.

Fire escapes are another preferred place to run the wires because it is easier to disguise the wires if they are wrapped around the fire escape railings. Improperly insulated wire that has been exposed to the elements over a period of time may cause the metal fire escape to become charged. If you contact the fire escape ladder while still maintaining your grip on a ground or aerial ladder, you may complete the charge to ground and be electrocuted or shocked to the point where you are thrown off the ladder to the ground. Heed any reports of tingling sensations when climbing fire escapes. Utility stealing is not just exclusively done in vacant buildings. Sometimes in an occupied building, electricity is run from one apartment to the apartment above via the fire escape. The tenants then share the electrical costs. Report any such condition to the incident commander immediately.

Flammable liquids and gases/excessive combustibles

This is a lethal combination and is especially a problem in the cold weather. Often the only way to keep warm and/or cook meals is to use fire. Sterno cans, propane torches, and more often than not, open flame fires are used for this purpose. This has been the cause of many fires in these buildings. With the excessive amount of debris and other combustibles in the building as well as the usually open construction, the place is ripe for ignition and rapid fire spread.

Compromised structural elements

Building elements that have been exposed to the elements are prone to deterioration. Many of these vacant buildings were ill-maintained while they were occupied. Due to missing interior stairs, holes in floors and walls, dangerous fire escapes, and rotted roofs, parapets, and coping stones, firefighters must be on the alert for building failure any time a fire strikes a vacant building. These missing elements also make it easier for fire to spread throughout the building.

Vacant buildings are also prone to failure due to theft of building materials. Often, bricks are removed from walls in unseen areas and sold illegally, compromising structural integrity. Copper wiring and piping are also ripped out, creating voids that allow fire to spread vertically. The department must ensure that all personnel are informed of dangerous conditions as soon as they are discovered.

Previous fires

Buildings that have suffered previous fires should be treated with extra caution. Not only will the fire compromise the building, but often firefighting activities will also have a detrimental effect on the structure. Master streams will often knock structural elements out of plumb, making them unstable. Remember that a building on fire is basically a building under demolition. More than one fire in a vacant building will often cause firefighters to employ a "no-entry" strategy when another fire strikes. This means to use exterior streams from outside established collapse zones once a fire is discovered. No building is worth injuring a firefighter, especially one where the demolition by fire has already been started.

This fire in a vacant four-story was the second in this building in the same night. The first, was an offensive, manpower-intensive operation on a lower floor. Many voids had to be opened. The second, a fully involved top floor upon arrival, utilized a strictly defensive operation.

Dangerous fire escapes

Fire escapes will not only present a collapse hazard, but will be a serious life hazard if any "occupants" are on them attempting to escape. The drop ladder and stairs are not to be trusted. The same must be said for the gooseneck ladder at the rear of the building. Fire department ground ladders or aerial devices are the tools of choice for access when operations must be conducted in the vicinity of fire escapes. It should be absolutely forbidden to stretch hoselines up a fire escape of a vacant building. In addition, don't forget to check for zip cords supplying illegal power to the building via the fire escape. In freezing weather, ice from hose streams accumulating on a weakened fire escape may be the straw that breaks the camel's back.

This vacant building has had the fire escape drop ladder completely removed at the rear of the building. Incident command, upon being notified of the condition, must issue an emergency transmission to all operating personnel regarding the danger.

Boarded-up windows and doors

This will not only create an entry and egress problem, but can lead to the development of backdraft conditions. Personnel arriving on the scene must perform a thorough size-up regarding fire and smoke conditions before deciding on a strategy. Where backdraft conditions are present, weigh the consequences of placing men on the roof. It will usually be better to operate from an aerial device. Under no circumstances should firefighters attempt to remove the window boards under these conditions.

If backdraft conditions are not present or have been strategically alleviated, boarded-up structures will require total ventilation to not only give personnel the best chance of accessing the structure, but more importantly, to reveal hazards that may exist in the building. *(Bob Scollan, NJMFPA)*

Even where backdraft conditions are not present, buildings that are boarded-up create an extremely smoky fire due to the amount of combustible debris in the building coupled with inadequate ventilation openings. Also, buildings that are boarded-up prevent us from initially seeing the interior conditions. It will be necessary to completely vent the building by removing as many of the boards as possible. This will give the operating crews a chance to see what they are facing. This will also allow the fire to light up, which may cause you to lose control of the fire. This is a risk worth taking. The consequences of not completely opening up the building will create a situation where firefighters may have to operate in a dangerous building in zero visibility. This could lead to members falling through weakened or holes in floors, falling through missing steps, and the potential for lost firefighters. A vacant building is certainly not worth this risk.

So how do you best remove the boards. There are two types of boards used to seal up vacant buildings. The first, plywood nailed into the frame, is simple and will require some muscle and sweat to remove, but using a Halligan tool, it should not create much of a problem. It may even be possible to use a power saw to cut holes in them, but it is better to completely remove them in case they are needed for egress.

The second type are known as HUD windows and are placed there by the Housing and Urban Development Department until the building is either demolished or renovated, both of which can take a long time to accomplish. HUD windows are also made of plywood, but are reinforced by attaching the plywood to 2" x 4" planks. The planks are secured on the inside behind the window frame by threaded rod screwed down tight to hold the planking and plywood in place. These are extremely difficult to remove. The power saw is again the answer. Here, you can cut a window in a window. This is probably quick-

HUD windows are extremely difficult to open and will require both additional tools and manpower to accomplish. To most efficiently operate, firefighters must familiarize themselves with the problems associated with these windows.

est. Either cut deeply so you slice through the 2" x 4" or cut through the plywood where the plank is exposed and then cut through the planking. You can also use a metal blade and cut the threaded rod, which is more difficult as it is usually flush with the plywood. Another method, the brute force method, focuses on using a sledge or maul to hammer the bolts and rods right through the wood they are securing. Both methods will be difficult if the windows are in elevated positions. Extreme caution is necessary. If these windows are in your jurisdiction, make sure you familiarize with them as they may have a significant impact on your operational strategy.

Alerting Our Own

There are many dangerous buildings out there. Many times, these buildings are stumbled upon by accident, creating hazardous conditions for responders. It is the duty and responsibility of all firefighters who discover a building hazard to bring it to the attention of his superior so that all members may be made aware of it. There are several ways of doing this. One is by completing and posting an "Exceptional Response Report" or similar form. Another is to notify dispatch to include this information in the CADS information. This information should be included in the over-the-air response information. A third is by placing a symbol on the building. This works well, as long as all members are aware of the system. Using an established system alerts respon-

This multi-winged building is partially occupied and is a serious collapse hazard. The rear wing has been shored up by this bracing. Note also that the window frames are out of plumb as well as the cracked spandrel wall.

ders to unsafe building conditions as soon as they arrive on the scene. This assists in the prevention of unnecessary injuries and potentially deadly encounters with unsafe buildings and structural features.

Building hazard symbols can be used alone or in conjunction with one another. It should be the responsibility of the incident commander or the safety officer to make sure that these symbols are placed on dangerous buildings as soon as the hazard is discovered.

Symbols should be placed in conspicuous positions on the front of the building. It would be best to place them in the same place on all buildings. This way, at night, firefighters will know where to look for the symbol. If necessary, place the symbols on as many sides as is necessary to alert firefighters to a particular hazard. Highly visible spray paint such as dayglow orange or red should be used. All symbols should be placed in a box with the department's initials below the box. This way there will be no mistaking the symbol for routine graffiti. Some symbols may be placed on buildings during routine inspections while others are placed as the incident is wrapping up.

Before we look at the various symbols, let's look at a way the department can alert members to dangerous conditions from day one of a building's existence.

Truss construction

New Jersey has instituted a symbol system to denote a building where a truss construction system is used. It is used as follows:

The symbol is placed in proximity to the front door and is displayed in a highly visible color, usually international orange with black letters or luminescent white with red or black lettering.

Vacant buildings

The decision to enter the building should be based upon the fire conditions, building integrity, time of day, and most importantly, the previous knowledge members have of the structure. Officers must make it a point to investigate and familiarize their assigned members as well as pass on to the entire department any hazard information on a vacant building. Vacant buildings should carry the symbol of the letter "V" inside a box. Below the box, as in any building hazard notification symbol, should be the department's initials or other adopted distinguishing mark to alert members that the symbol is department approved.

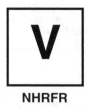

No entry

Due to previous fire damage or other destructive forces or circumstances, should be considered off limits to all members at all times. A building which has inherent hazards so severe that firefighters should not enter it under any circumstance should be denoted by a box with an "X" covering the whole box (like a "strike" in bowling).

NHRFR

The symbols on the door of this building prohibits entry as well as brings attention to hazards created by an open roof and floor. Symbol systems such as this allow us to alert our own of dangers before they enter the building.

Enter with extreme caution

Other buildings may be entered, but only under extreme caution. This is denoted by a box with one line cutting diagonally across it (like a "spare" in bowling). This building may have suffered damage from previous fires, vandalism, and rot due to neglect. Firefighters should not take unnecessary risks in this building. Any attack should be carried out with extreme caution and with a pessimistic prediction for success. This symbol and the "no entry" symbol may be accompanied by another symbol.

NHRFR

Unsafe building features

Some buildings may be safe to enter, but due to some safety problem, firefighters should use extreme caution in the particular area denoted inside the box. Here is an example of a building with a dangerous fire escape on the left and one with dangerous stairs and/or landings on the right. Departments are not limited to these and may create their own symbols to address a particular hazard. If the building owner objects to the department putting spray paint on his building, tell him the fire department objects to his unsafe building. The choice is his. Fix the item or have his building marked up.

Dangerous Fire Escape   **Dangerous Stairs/Landing**

NHRFR   NHRFR

Structural damage (post-fire)

Both primary (the damage done by the fire) and secondary damage (the damage done by firefighting) may leave the building in a dangerous condition. Firefighters from other shifts who may have another fire in the same building should be made aware of the hazards. Holes in the roof and/or the floors must be noted. The building should be marked to warn firefighters of these hazards. "Roof Open" is denoted by the symbol "RO". Floor open is denoted by the symbol "FO". Floor hole notations should include the floor that is in question. A building with holes in both the floors and the roof may use two separate symbols or they may be placed in the same box. If there are multiple floors with holes, the symbol "FO-M" may be used. These symbols may be used in conjunction with limited or no entry symbols.

Roof Open   **Holes in the 4th Floor**

```
┌──────┐         ┌──────┐
│  RO  │         │ FO-4 │
│      │         │      │
└──────┘         └──────┘
```

NHRFR   NHRFR

If both the roof and the floor(s) are opened, the symbol may look like the one below. In this case, there are multiple floors with holes as well as an open roof.

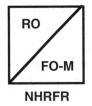

Multiple hazards exist in this fire building. There are holes in the fourth floor as well as an open roof. Their presence may cause incident command to consider a more cautious strategy.

Establishing a building hazard awareness program is an effective way of making responding personnel aware of unsafe building conditions. Any discovery of this type by any company should prompt the response of the safety officer or battalion commander. Other notifications may include the Building Department, Fire Prevention, or the Health Department.

Demobilization

Once the fire is placed under control, it is time to start thinking about how you are going to get the companies back in service in a safe, logical, and sane manner. Lines need to be picked up, ladders taken down, bedded and stored, and water supplies terminated. If necessary, a plan for fire watch must also be devised. To best accomplish this, the incident commander should assign a demobilization officer. This officer should survey what equipment needs to be picked up and how to best accomplish it. What may work well is to call officers to the command post and find out what has to be picked up by each individual company. Then companies, based on their "stuff-off-the-rig" status, can be doubled or tripled up to place the most stripped apparatus back in service.

It is important to remember that firefighters are likely to be fatigued and the last thing they will want to do is pick up miles of hose. Fatigued firefighters rapidly become injured firefighters if not properly relieved. It may be wise for incident command to request additional companies to the scene to help in the demobilization process and relieve fatigued crews. Companies not called to work the fire as well as relocated companies are candidates for this. The fresh companies may not like the pick-up detail, but they will appreciate it when it is their turn in the barrel and fresh companies assist them in packing up.

It is also important, as the incident de-escalates, to get the most fatigued companies released from the scene as soon as possible. Leaving on their own apparatus may not be possible due to their position at the fire. Many times, the apparatus committed to the fire area is also used by the companies assigned to fire watch, so it may be easier to leave initial responding apparatus where they are for the time being. It may be necessary to allow those companies to switch to an apparatus that responded on a later alarm and is parked on the perimeter of operations. They can also take the apparatus of the companies responding to assist with the demobilization. In any case, have a plan to release battle-weary companies in a timely manner and leave the demobilization to less fatigued personnel.

A safe and orderly demobilization is the result of effective planning and cooperation among the participants. A demobilization officer should be assigned and all officers made aware of the plan. *(Bob Scollan, NJMFPA)*

Incident Critique Considerations

Something of value can be learned at every fire, thus all fires should receive a critique. What is revealed and learned in the critique can be of value for the rest of a firefighter's career. The department that does not hold incident critiques because they have "seen it all" is usually the ones with the largest parking lots and casualty tolls. Each fire should be a learning experience that should go beyond the kitchen table or bar. The problem with critiques is that by the time the critique is held, not everyone is there due to vacation, sick leave, or other commitments. Other members who are present may not remember exactly what happened. Sometimes the best lessons are forgotten forever.

How can the fire department effectively hold a critique where all the pertinent information is offered? Remembering that the palest ink is better than the sharpest memory, notes should be jotted down as soon as possible after the incident. This will make the critique more meaningful. An adopted form will work well and put everyone on the same page during the critique. It should be simple and user-friendly. Asking for input from the officers who will be using it is the best way to ensure that personnel will "buy into the program". A form adopted by the North Hudson Regional Fire & Rescue is the "After Action Report". The form is filled out anytime that a hoseline is charged at a structure fire. If lines are dropped and they are not charged, the officer is not required to fill out the form. The forms are filled out as soon as the company returns to quarters and is forwarded to the battalion commander. That way, if the officer is off the next tour, his input will still be included in the critique. What is likely to happen in this instance is that the acting officer takes the place of his company officer and speaks on his behalf at the critique. This form has worked out well as a way to run more effective critiques.

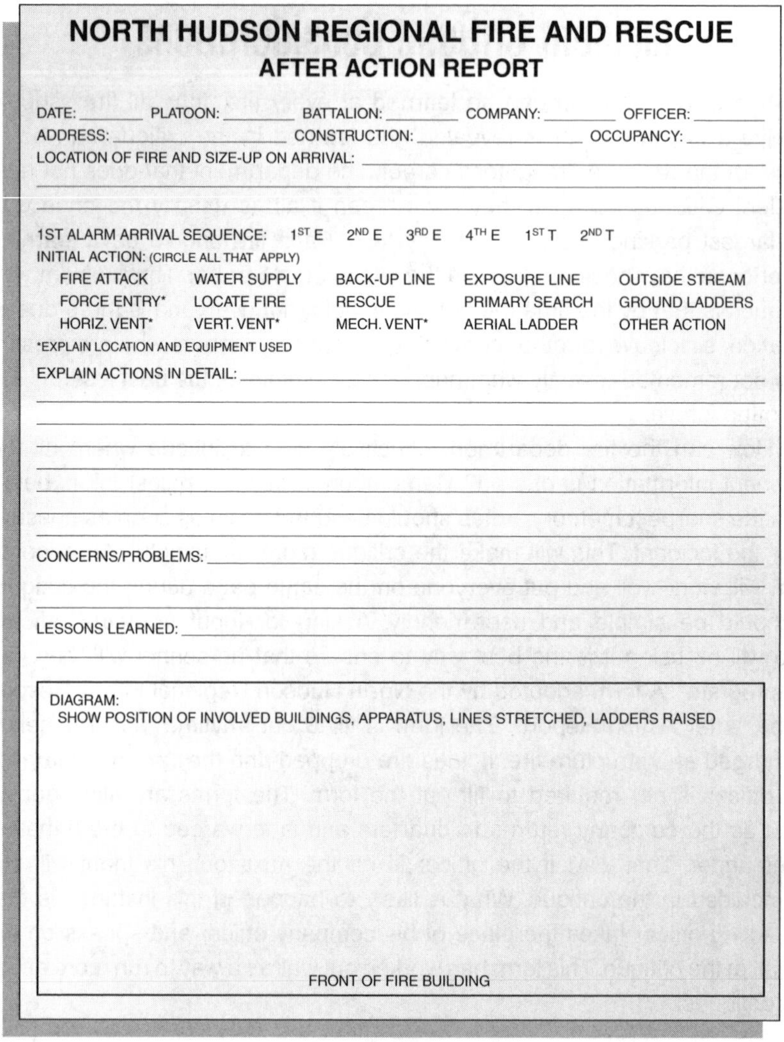

NORTH HUDSON REGIONAL FIRE AND RESCUE
AFTER ACTION REPORT

DATE: _____ PLATOON: _____ BATTALION: _____ COMPANY: _____ OFFICER: _____
ADDRESS: _____ CONSTRUCTION: _____ OCCUPANCY: _____
LOCATION OF FIRE AND SIZE-UP ON ARRIVAL: _____

1ST ALARM ARRIVAL SEQUENCE: 1ST E   2ND E   3RD E   4TH E   1ST T   2ND T
INITIAL ACTION: (CIRCLE ALL THAT APPLY)

FIRE ATTACK	WATER SUPPLY	BACK-UP LINE	EXPOSURE LINE	OUTSIDE STREAM
FORCE ENTRY*	LOCATE FIRE	RESCUE	PRIMARY SEARCH	GROUND LADDERS
HORIZ. VENT*	VERT. VENT*	MECH. VENT*	AERIAL LADDER	OTHER ACTION

*EXPLAIN LOCATION AND EQUIPMENT USED

EXPLAIN ACTIONS IN DETAIL: _____

CONCERNS/PROBLEMS: _____

LESSONS LEARNED: _____

DIAGRAM:
SHOW POSITION OF INVOLVED BUILDINGS, APPARATUS, LINES STRETCHED, LADDERS RAISED

FRONT OF FIRE BUILDING

Exhibit G

An often overlooked portion of the critique, but no less important is the critique by the safety officer. Most line officers that fought the fire concern themselves with their assigned tactics during the critique. The real value of the safety officer's input into the critique is that it should be from a point of view of someone that is standing back and taking in the big picture at the fire scene. His point of view should be from a different angle than the operating companies. Some of the concerns of the safety officer that should be included in the critique are:

- Was everyone properly wearing protective equipment?
- Were department SOPs followed?
- Was the fire attack coordinated with proper and timely support?
- Did companies stay together or was freelancing evident?
- Did the operation provide for emergency firefighter egress from the interior? From the roof?
- Were communications conducive to safe operations?
- Were tools and equipment used safely and properly?
- Was the FAST team properly equipped and ready to be deployed?

Remember that the safety officer's critique must not be a scolding. It should be a constructive session, aimed at improving the department's performance from a safety standpoint. A safer operation will always translate into a better, more effective operation. Together with the safety officer, the shift officers should set goals that are both measurable and realistic in regard to a safer fireground operation. Unsafe practices as well as sub-par scene performance should be looped back into the training process and approached in a positive manner. Remember that the further you are from the last "Big One", the closer you are to the next.

Conclusion

Scene safety is everyone's business on the fireground. Fire departments can take many steps prior to, during, and after the incident to make the emergency experience as safe as possible. Firefighting will never be a safe profession. It is attention to detail in regard to safety that prevents injuries and death. We must be each other's keepers. Everyone must do their part to make safety the overriding concern in all fire department operations.

Questions for Discussion

1. What are some of the ways in which a department can effectively transfer and distribute information to all members?
2. Discuss some of the guidelines for effective apparatus positioning on the fireground
3. Discuss problems of dead-end street response and some solutions to those problems.
4. What is span of control and how can it be most effectively used on the fireground?
5. Name some of the safety-oriented duties the FAST team can accomplish while waiting at the Command Post.
6. What are some of the advantages of the Tactical Worksheet?
7. What are some of the ways the battalion chief can be most effectively utilized by the incident commander?
8. Discuss the importance of perimeter control and how to best make operating personnel aware of established control zones.
9. Discuss guidelines and methods in which building search can be more safely conducted?
10. Describe some of the considerations to be taken to safely operate master streams.
11. Discuss the importance and use of "top-down" and "bottom-up" emergency transmissions.
12. Discuss some of the problems inherent in vacant buildings.
13. Discuss how a building hazards awareness program can increase safety on the fireground.
14. What are some of the ways that a post-incident critique can be made more effective?
15. Discuss the concerns the safety officer must address during the post-incident critique.

CHAPTER THIRTEEN
Conclusion

"Set the main sail
Steady the course
We are headed full speed ahead
The helm will someday be yours"

When now-retired Weehawken Chief of Operations Gerald Huelbig promoted me to captain, these words were written on the inside cover of a book he gave to me. They were an inspiration and are more than appropriate for the fire service as we head into the next millennium. We will be called upon to perform many tasks, both routine and unusual, not only due to economic restraints, but because we are the only ones who can handle many of these jobs. We are and will continue to be respected leaders in the field of emergency service.

Remember that the first job of the fire service is the prevention of fire, not fire suppression. Only after fire prevention efforts have failed do the use of fire suppression techniques become necessary. Realize that when a fire breaks out that it is a failure of the system.

Be a student of the job. Understand that to learn the profession of fire-fighting, there are no easy roads or magic formulas. Hard work will always pay off. Narrow-minded and uninformed fire service leaders have no place in today's fire service.

Realize too that invincibility is not part of the formula. We will all lose buildings and make mistakes. We are not Captain James T. Kirk, James Bond, or Bugs Bunny. We all lose sometimes.

Mistakes are acceptable. Mistakes become intolerable when repeated because a lesson was not learned or paid attention to. The definition of insanity is repeating the same act over and over again and expecting different results. Fire officers who lose buildings and unfortunately people time and time again due to improper strategies and tactics from unlearned lessons are truly psychotic and a danger on the fireground.

I would like to end this text on the light side as gratuity to the reader that has persevered to this point. There was a story of a fire chief who had been in that position for over 25 years. He was revered in his field and considered *the* expert on all matters related to fire. He finally retired and because life had nothing more to offer, died soon after.

Upon arriving at the Pearly Gates, he found himself at the back of a long line. Since he was a chief and not used to this treatment, he pushed his way to the front of the line. There at the front was St. Peter. He explained to St. Peter that he was a fire chief for over 25 years and that he was not used to waiting for anything. He stated that he did not take orders, he gave orders, and people obeyed them without question. St. Peter's reply was that in this place, there was no rank, and chief or no chief, he would have to return to the back of the line. Grumbling to himself, the chief obeyed and returned unhappily to the end of the line. As he did, he became aware of an approaching siren in the distance. Sure enough, a shiny red chief car zoomed past him to the front of the line. Out of the vehicle stepped a fire chief, with shiny white helmet and all. He said a few words to St. Peter, who immediately opened the gate and waived this chief and vehicle inside. The chief, now completely miffed, again pushed his way to the front of the line and asked who that was and why that chief got special treatment. To this St. Peter replied, "Oh that, that was God, he just thinks he's a fire chief."

Maintain a healthy sense of humor. Egos have no place in the fire service. None of us are bigger or more important than the job. The fire service was here before any of us were born, hired, or appointed. It will be around long after we've retired and died. Our job, if we do it properly, is to make it a little better while we're here, leaving it better than we found it for future generations. Someday, believe it or not, someone at the kitchen table will say, "Avillo who?" You can replace my name with yours.

Remember the three R's of leadership:

- Respect for yourself
- Respect for others
- Responsibility for your actions

Also, remember that your personnel are your greatest resource. Treat them and the job with respect. Go to any lengths to protect them both. Good luck, God bless, and stay safe.

The ultimate goal of every fire officer must be the safe return of the firefighting personnel to quarters when the incident is over.

Index

A

Access difficulty, 85, 274–276, 326–332:
stairs, 85

Access stairs, 85

Accountability (personnel), 228, 256–262, 285–287, 348–351, 359–360

Actions, 34–35, 45, 47

Aerial device positioning, 390, 427–428

Aiming stream at fire, 430–431

Air conditioning/refrigeration shops, 311

Air shafts, 186–190

Alarm apparatus commitment, 390–391

Alarm response, 36–40

Alerting firefighters , 442–447:
truss construction, 444;
vacant building, 444;
no entry, 444–445;
enter with extreme caution, 445;
unsafe building features, 445–446;
structural damage (post-fire), 446–447

Aluminum balconies, 244–245

Anchoring ground-mounted streams, 403

Answering questions, xxxiii–xxxiv

Apartment buildings/tenements.
SEE Multiple-dwellings fire.

Apartment numbers, 204–207

Apartments (garden), 265–297.
SEE ALSO Contiguous-structures fire.

Apparatus and manpower factors, 6–8

Apparatus positioning (safety), 279, 386–391:
traffic flow, 388;
bumping responsibilities/positions, 389;
blocking intersection/road, 389;
engine behind ladder company, 389–390;
ladder truck and aerial device, 390;
alarm apparatus commitment, 390–391;
radio silence, 392

Area and height factors, 20–22

Area source (heat), 65

Assignment (personnel), 44

Atmosphere testing, 75

Attached buildings, 190–192

Attack, 87, 132–145, 173–176, 211–213, 262–263, 294–295, 362, 379:
direction change, 143–144
strategies, 132–141;
water supply, 174–176

F

G

H

I–J

M

N

O

P

S